Study Guide and Casebook for
Applied Microeconomics

Study Guide and Casebook for

Applied Microeconomics

EDWIN MANSFIELD
University of Pennsylvania

W • W • NORTON & COMPANY
NEW YORK • LONDON

Printed in the United States of America

FIRST EDITION

Composition by Gerry Wilson

Manufacturing by Capital City Press

ISBN 0-393-96482-5

W. W. Norton & Company, Inc., 500 Fifth Avenue, New York, N.Y. 10110
W. W. Norton & Company Ltd., 10 Coptic Street, London WCIA 1PU

1 2 3 4 5 6 7 8 9 0

Contents

Preface

This study guide and casebook contains a wide variety of materials to help the student learn applied microeconomics. In particular, it provides a comprehensive set of questions which cover the entire range of topics taught in intermediate-level microeconomics courses. The inclusion of empirical material as well as problems based on real-world situations demonstrates to students the power of microeconomic theory as an aid to decisionmakers in both the private and public sectors of the economy. The result is a combination of theory on the one hand with measurement and application on the other. In all, about 1,000 questions are included in this book.

In addition, the following five full-length cases are provided. Each of these cases supplements a chapter of my text, *Applied Microeconomics.* Case One by J. David Hunger of Iowa State University ("The Standard Oil Company: British Petroleum Loses Patience") supplements and extends Chapter 6 of my text. Case Two by Richard F. Fallert, Don P. Blayney, and James J. Miller of the U.S. Department of Agriculture ("The Dairy Industry") supplements and extends Chapter 9 of my text. Case Three by Joseph Wolfe of the University of Tulsa ("Cineplex Odeon Corporation") supplements and extends Chapter 14 of my text. Case Four by Shaker Zahra, Daniel Hurley, Jr., and John Pearce II of George Mason University ("Airbus Industrie: A Wave of the Future") supplements and extends Chapter 18 of my text. Case Five by Dorothy Robyn of Harvard University ("California Water Pricing") supplements and extends Chapter 20 of my text. These cases should provide a valuable extra dimension to classes in applied microeconomics.

Philadelphia, 1994 E.M.

Part 1 Introduction

Chapter 1 The Nature of Microeconomics

Case Study: Michael S. Dell

In 1993, Michael S. Dell, then 28 years old, was the chief executive officer and founder of Dell Computer Corporation, the fourth largest maker of personal computers. According to *Business Week*, he ". . . has become one of the most mythic heroes in an industry full of Homeric figures."[1] Nine years after starting his business in a college apartment in Austin, Texas, he was head of a major firm with billions of dollars in sales in 1993. To accomplish this, he ". . . has cut out the dealers and distributors, designs and assembles most of his PCs from off-the-shelf components, and runs a no-frills operation whose main focus is customer service."[2]

a. What types of resources does Dell Computer use? Are the services of its 4,654 employees (in 1993) a resource? When it raised $31 million by selling stock to the public in 1988, did Dell Computer obtain additional resources?

b. Since most personal computers are rather similar, manufacturers have come to compete for sales on the basis of price, brand name, and extra features such as 24-hour service. In early 1993, Compaq Computer Company, a rival of Dell's, cut its prices by as much as 15 percent. How did this influence the allocation of resources?

c. Compaq's operating margin (that is, the difference between price and cost) was about 10 percent in late 1992, whereas Dell's was about 7 percent. If a price war were to break out, are these facts of significance? If so, how?

d. When Dell entered the Japanese market, it undercut the prices of other companies, including Compaq, with the result that its small Japanese staff had great difficulties in keeping up with the demand. Did this have an effect on the distribution of income in Japan? In the United States? If so, what were its effects?

e. Because of their pricing policies and other factors, Compaq's share of the PC market increased from about 4 to 9 percent during 1991–93, and Dell's market share increased from about 1 to 5 percent.[3] Does it seem likely that the Department of Justice or other federal agencies will regard these market shares as big enough to argue that they are in violation of the antitrust laws?

Completion Questions

1. National defense is an example of a ______________ good. Such goods (will, will not) ________________________ be provided in the right amounts by private industry, so ______________ tends to intervene.

---

1. *Business Week*, March 22, 1993, p. 83.
2. Ibid.
3. *New York Times*, March 18, 1993, p. D 4.

2. Economics is concerned with the way in which ______________ are allocated among alternative uses to satisfy ______________.
3. Microeconomics deals with the economic behavior of ______________.
4. Wants ______________ from individual to individual and over time for the same person.
5. Resources are used to produce ______________.
6. There are two types of resources: economic and ______________.
7. Resources are often classified as land, labor, and ______________.
8. Technology is directed toward ______________; science is directed toward ______________.
9. In a free-enterprise economy, ______________ choose the amount of each good that they want, and ______________ act in accord with these decisions.
10. In our system, the income of each individual depends largely on how much he or she owns of each resource and ______________.
11. A nation's rate of growth of per capita income depends on the rate of growth of its resources and the ______________.

True or False

____ 1. When economists refer to goods as being scarce, they mean that these items are monopolized by a few people. If goods were distributed equitably, there would be no problem of scarcity.

____ 2. Models generally are not useful in practice; if they were, they would not be called models.

____ 3. Microeconomics is helpful in promoting an understanding of the powerful modern tools of managerial decision-making.

____ 4. The lawyer who argues an antitrust case, and the judge who decides one, must both rely on and use the principles of microeconomics.

____ 5. Microeconomics is concerned only with solving practical problems.

____ 6. Economic resources always have a zero price.

____ 7. There usually is only one way of producing a commodity so it is easy to figure out which way is best.

____ 8. The price system plays some role, but only a minor one, in allocating resources in a free-enterprise economy.

____ 9. In the acquisition of new weapons, society relies exclusively on the price system.

____ 10. A model cannot be useful if it simplifies and abstracts from reality.

____ 11. Models are used by economists, but not physicists.

____ 12. One reason for using a model is that it may be the cheapest way of getting needed information.

Multiple Choice

1. In choosing among alternative models, economists generally have the strongest preference for models that
   a. have assumptions that are close to exact replicas of reality.
   b. predict better than any other that is available.
   c. have few assumptions and are as simple as possible, even if they cannot predict very well.
   d. are detailed and complex, with every available fact and figure included.
   e. all of the above.
2. Microeconomics is concerned with
   a. optimal production decisions.
   b. pricing policy.
   c. optimal resource allocation.
   d. antitrust policy.
   e. all of the above.
3. In a free-enterprise economy, profits and losses are
   a. the stick used to eliminate less efficient firms.
   b. the carrot used to reward the proper decisions.
   c. both *a* and *b*.
   d. neither *a* nor *b*.
   e. always equal to zero.
4. If a model is to be any good, it must
   a. make assumptions that are exact replicas of reality.
   b. refrain from referring to things that are not directly measurable.
   c. predict phenomena in the real world reasonably well.
   d. all of the above.
   e. none of the above.

Review Questions

1. In December 1992, about 7 percent of the labor force was unemployed. How can anyone say that labor is scarce?
2. On most questions of policy, one can find disagreements among economists. Does this prove that economics is not a science?
3. If a certain proposition holds true for a part of a system, must it hold true for the whole system? For example, suppose that a farmer will benefit from producing a larger crop. Does it follow that all farmers will benefit from producing a larger crop? Explain.
4. In evaluating the accuracy of their statements, should one distinguish between (1) economists' descriptive statements, propositions, and predictions about the world and (2) their statements about what policies should be adopted? Explain.
5. According to Adam Smith, "Monopoly . . . is a great enemy to good management." What do you think he meant? Do you agree or disagree?

6. In aeronautical engineering, models of an airplane are used to investigate its aerodynamic properties in a wind tunnel. Must such models have seats for passengers? Must they be hollow? What functions must they serve?
7. According to Alfred P. Sloan, who was president of General Motors from 1923 to 1937 and chairman of its board of directors from 1937 to 1956, "The great difference between the industry of today as compared to that of yesterday is what might be referred to as the necessity of the scientific approach, the elimination of operation by hunches."[4] Indicate how microeconomics plays a role in this more scientific approach to management.
8. What concerns does economics deal with? What is the difference between microeconomics and macroeconomics?
9. Give several examples of the particular types of problems that microeconomics helps to solve.
10. Is microeconomics concerned solely with the solution of practical problems? Give examples of questions that are dealt with in microeconomics that do not take the form of practical problems. In what sense is microeconomics like mathematics?
11. Define human wants. What role do human wants play in microeconomics?
12. Define economic resources. What role do economic resources play in microeconomics?
13. Describe the various types of economic resources.
14. Define technology. Is there any difference between science and technology? If so, what is the difference? What role does technology play in microeconomics?
15. Describe the four basic tasks that must be performed by any economic system.
16. How does our system determine the level and composition of output in the society?
17. How does our economic system allocate its resources among competing uses and process these resources to obtain the desired level and composition of output?
18. How does our economic system determine how much in the way of goods and services each member of the society is to receive?
19. How does our economic system determine the rate of growth of per capita income?
20. What is a model? Can the usefulness of a model be deduced from the realism of its assumptions? Why do economists use models?
21. What considerations must be taken into account in judging or evaluating a model?
22. Are the models contained in this book sufficiently accurate to solve all of the problems faced by governments and firms? Have all of them been tested completely? Are they the best available at this time?

---

4. A. Sloan, *Adventures of a White Collar Worker* (Garden City, N.Y.: Doubleday, 1941).

Problems

1. In 1990, the world witnessed an interesting and highly unusual development: Poland, as well as its East European neighbors, began to shift from a centrally planned economy to a capitalistic economy. But unlike its neighbors, Poland seemed to make a rapid transition. In early 1990, it adopted a bold and controversial plan calling for the transfer of industry from government to private hands, for an end to government subsidies, and for reliance on market prices. Bankruptcy and unemployment would be tolerated. However, by December 1992, only 54 out of 3,500 government-owned companies had been sold to private investors.[5]
   a. According to the Council of Economic Advisers, in centrally planned economies "prices of labor, goods, and services do not adjust to reflect supplies and demands, and production decisions are not motivated by profitability." What determines the level of a particular good's price in a centrally planned economy?
   b. The Council of Economic Advisers also says that "the lack of private ownership implies that individuals have little stake in improving resource allocation." Why?
   c. Will a rapid transition to capitalism raise Poland's per capita output (now about one-fourth of that in the U.S.) to the U.S. level in a year or two?
   d. Of what use might microeconomic theory be to Poland's leaders in making strategic decisions of this sort?
   e. Why did Poland move more slowly toward capitalism?

2. After World War II, Japan's Ministry of International Trade and Industry (MITI) played an important role in guiding the nation's industrial development. According to many accounts, it provided government aid (such as subsidies and tax reductions) for industries that were expected to grow rapidly (so-called sunrise industries) and adjustment assistance for those that were declining (sunset industries). In recent years, electronic computers, lasers, robotics, and biotechnology are some of the industries that MITI and others regard as sunrise industries. In the computer field, MITI has allocated over $500 million to try to beat the United States to the development of a new generation of computers. Research is being carried out on advanced computer architecture and artificial intelligence, among other things.

   During the 1980s, a substantial number of American politicians and business executives, impressed by Japan's rapid economic growth since World War II, proposed that the U.S. government should intervene in a similar way.
   a. If you were asked to argue in favor of such intervention, what points would you make? What arguments do you think are made by the proponents of such intervention?
   b. If you were asked to argue against such intervention, what points would you make? What arguments to you think are made by the opponents of such intervention?
   c. Can microeconomics shed light on a question of this sort?

5. *New York Times*, March 19, 1993, p. A3.

3. A baguette is a long, thin, crispy loaf of French bread. From the French Revolution to 1978, the price of a baguette was controlled by the French government. Then it was decontrolled, and it rose to about 40 cents in 1980. Subsequently, Albert Rodriguez, a baker in southern France, cut the price to about 22 cents. Afterwards bakers everywhere were following his example.
   a. According to a Parisian baker, "It is going to kill the small-business man. It will mean going to a system of commercial baking in a huge central factory. You have got to pay the baker and his costs." What is this baker's implicit assumption about the relationship between the size of a bakery and its cost per baguette?
   b. After he cut his price to 22 cents, some of Mr. Rodriguez's competitors tried to sell baguettes outside his shop for 11 cents each. Why do you think they did this?
   c. Most of Mr. Rodriguez's customers ignored Mr. Rodriguez's competitors, when they tried to sell baguettes outside his shop for 11 cents each. Why do you think they did this?
   d. What are some of the principal inputs required to produce baguettes?
   e. If you were assigned the task of constructing a model to predict whether a particular baker would cut the price he or she charges for a baguette, what factors would you stress?
4. As part of the Revenue Act of 1978, Congress passed a tax credit for families that invested in solar energy equipment. For example, if a family spent $10,000 on solar energy equipment, it could reduce its income tax by $2,200. (For expenditures up to $2,000, the tax credit was equal to 30 percent of the expenditure.) The Office of Tax Analysis of the U.S. Treasury Department, in analyzing this proposal, estimated that the cost of heating a 1,850-square-foot house over a 20-year period with solar and other types of heating systems was as follows:

Type of heating system	*Cost (dollars)*
Solar	12,907
Electric furnace	5,440
Electric resistance	4,968
Oil	3,659
Gas	2,582

   a. Some observers feel that solar energy will play an important role in solving our energy problems. How does the price system determine how rapidly the use of solar energy will spread?
   b. Based on the information provided above, how rapidly do you think that the use of solar energy will spread?
   c. What reasons can you give for the government's intervening in this way to encourage the use of solar energy?
   d. Did this tax credit cut the income tax liabilities of the poor more than those of the rich?

e. Construct a model to predict what type of heating system a consumer will install in a new house.

Key Concepts for Review

Economics	Capital
Microeconomics	Technology
Macroeconomics	Tasks performed by an economic system
Human wants	The price system
Labor	Models
Land	Model-building

ANSWERS

Case Study: Michael S. Dell

a. Labor, capital, and land. Yes. Yes.
b. Because of this price cut, Compaq expanded its sales and increased its output. It tended to grow and gain resources. As indicated in part (*e*) of this question, Compaq's share of the market was growing during this period, due in part to price cuts of this sort.
c. Compaq can cut its price by a larger percentage than Dell before incurring losses.
d. To the extent that Dell took profits away from Japanese firms, it affected the distribution of income in Japan. To the extent that it took profits away from other U.S. firms, it affected the distribution of income in the U.S. Also, it may have had some effect on employment and wages in both countries.
e. No.

Completion Questions

1. public; will not; government
2. resources; human wants
3. individual consumers, firms, and resource owners
4. vary
5. goods and services
6. free
7. capital
8. use; understanding
9. consumers; producers
10. the price of each resource
11. rate of increase of the efficiency with which they are used

True or False

1. False	2. False	3. True	4. True	5. False	6. False
7. False	8. False	9. False	10. False	11. False	12. True

Multiple Choice

1. *b* 2. *e* 3. *c* 4. *c*

Review Questions

1. Despite the fact that some labor is unemployed, labor is scarce. Certainly, the quantity of labor is not unlimited, and labor's price is nonzero.
2. No, because most of the disagreements stem from differences in ethical and political views.
3. No. This is the so-called fallacy of composition.
4. Yes, because the latter statements reflect the value judgments of the economist.
5. He meant that a monopolist, having succeeded in freeing himself from competition, is likely to take things easy, and worry less about efficiency.
6. No.

   No. They must provide information that is useful in predicting the characteristics of the final airplane.
7. It helps to indicate how firms should analyze their production, marketing, and financial problems in order to increase their profitability.
8. Economics is concerned with the way in which resources are allocated among alternative users to satisfy human wants. Microeconomics deals with the economic behavior of individual units like consumers, firms, and resource owners; macroeconomics deals with the behavior of economic aggregates such as gross domestic product and the level of employment.
9. Business firms are constantly faced with the problem of choosing among alternative ways of producing their product. One type of problem that microeconomics can help to solve is: which technique will maximize the firm's profits?

   Firms are also faced with the problem of pricing their product. Another type of problem that microeconomics can help to solve is: which price will maximize the firm's profits?

   Society as a whole must decide how it wants to organize the production and distribution of goods and services. Microeconomics can sometimes be useful in helping to indicate what changes society would be justified in making in this system.

   Public policy must also be concerned with the structure of individual markets and industries. Microeconomics plays an important role in helping to illuminate antitrust cases and to solve problems in this area.
10. No.

    Why is steak more expensive than hamburger? Why are physicians paid more than carpenters? How will an increase in the price of margarine affect the amount of butter purchased by Mrs. Smith? Why are there so many producers of wheat and so few producers of automobiles?

    Pure mathematics is not concerned with the solution of particular problems, but it has turned out that various branches of mathematics are of great value in solving practical problems. This is true as well of much of microeconomic theory. Also, microeconomics, like mathematics, is extremely important as a basis for understanding the world around us and for further professional training.

11. Human wants are the things, services, goods, and circumstances that people desire. Human wants—or, more precisely, their fulfillment—are the objective at which economic activity is directed.

12. Resources are the things or services used to produce goods which can be used to satisfy wants. Economic resources are scarce, and thus have a nonzero price. The essence of "the economic problem" is that some resources are scarce and must be allocated among alternative uses. If all resources were free, there would be no economic problem.

13. Land is a shorthand expression for natural resources.
    Labor is human effort, both physical and mental.
    Capital includes equipment, buildings, inventories, raw materials, and other nonhuman producible resources.

14. Technology is society's pool of knowledge regarding the industrial arts.
    Yes. Pure science is directed toward understanding, whereas technology is directed toward use. Technology sets limits on the amount and type of goods that can be derived from a given amount of resources.

15. First, an economic system must allocate its resources among competing uses and combine these resources to produce the desired output efficiently.
    Second, an economic system must determine the level and composition of output.
    Third, an economic system must determine how the goods and services that are produced are distributed among the members of society.
    Fourth, an economic system must determine the rate of growth of per capita income.

16. Consumers choose the amount of each good that they want, and producers act in accord with these decisions. Also, the production of some goods is a matter of political decision.

17. The price system is used to indicate the desires of workers and the relative value of various types of materials and equipment as well as the desires of consumers. To firms, profits are the carrot and losses are the stick. In addition, the government intervenes directly in some areas like weapons acquisition.

18. The income of an individual depends largely on the quantities of resources of various kinds that he or she owns and the prices he or she gets for them. In addition, the government modifies the resulting distribution of income by imposing income taxes and by welfare programs like aid to dependent children.

19. The rate at which labor and capital resources are increased is motivated, at least in part, through the price system. Increases in efficiency, due in considerable measure to the advance of technology, are also stimulated by the price system. But the government plays an extremely significant role in supporting research and development.

20. A model is composed of a number of assumptions from which conclusions—or predictions are drawn.
    No.
    The real world is so complex that it is necessary to simplify and abstract if any progress is to be made. A model may be the cheapest way of obtaining needed information.

21. The most important test of a model is how well it predicts. Another is whether its assumptions are logically consistent. Another important consideration is the range of phenomena to which the model applies.

22. No.
    No.
    Yes, according to a consensus of the economics profession.

Problems

1. *a.* Government agencies set prices.
   b. Since individuals gain little from improving resource allocation, there is little incentive to do so.
   c. It seems very unlikely, since Poland lacks modern equipment and sophisticated managers, among other things. But in the longer run, the price system may enable Poland to raise its standard of living substantially.
   d. Microeconomics can be of use in indicating how markets work and how much input of various kinds is required to produce a certain amount of output. Also, microeconomics is useful in describing the factors that promote increases in productivity, as well as those that influence the distribution of income.
   e. There is fear that workers will lose their jobs, that foreigners will buy the firms, and that existing plants will be worth little in the marketplace.

2. *a.* Some people believe that the price system does not work in a sufficiently dependable and timely fashion. Also, they argue that the government is already intervening in a variety of ways, and that what is needed is to coordinate its policies and make them more efficient.
   b. Many people are skeptical of the ability of government agencies to forecast which industries will grow rapidly and which will decline. MITI itself has made a considerable number of mistakes in this regard. More basically, these people believe that the price system will allocate resources more effectively than government intervention of this sort.
   c. Microeconomics shows the way in which the price system allocates resources and the conditions under which government intervention of various kinds can be justified. Clearly, a knowledge of these matters is essential in thinking about questions of this sort.

3. *a.* This baker is assuming that the cost per baguette falls as the size of the bakery increases.
   b. They did it to cut Mr. Rodriguez's sales and to punish him for cutting his price.
   c. They did it because they suspected that, once Mr. Rodriguez's competitors had hurt his business and brought him into line (or put him out of business),

they would raise their price to its original level. As one of Mr. Rodriguez's customers said, "We know that at his place it will still be [22 cents] tomorrow."

d. Flour, services of ovens, fuel, services of bakers, water, shortening, yeast.

e. The probability that a particular baker would cut its price would be expected to be directly related to how profitable he or she believes such an action to be.

4. *a.* The price system allows consumers and producers to decide whether or not they want to use solar energy. If potential users are willing to pay the amount required to use solar energy, they will adopt it; otherwise they won't. In general, one would expect that if other types of heating systems are cheaper, most consumers and firms will not adopt it.

b. It probably will not spread very fast because it seems to be much more expensive than other types of heating systems.

c. The government seemed to feel that there were social benefits from switching to solar energy (and thus reducing our dependence on foreign oil) that are not reflected in a comparison of the costs to the consumer of solar energy with those of other types of heating equipment.

d. The direct impact on the taxes of the rich probably tended to be relatively greater, since few poor people could afford to install solar energy.

e. It seems likely that the consumer would choose the least costly type of heating system that will be appropriate for his or her house. Thus the choice will depend on the relative costs of various types of heating systems, among other things.

Chapter 2 Demand and Supply

Case Study: In Vino Veritas

In 1979, the weather conditions in many parts of Europe were ideal for cultivating grapes. In France, there was a record output of more than 2.2 billion gallons of wine. In Bordeaux, wine-makers took about 10 million gallons of wine off the market in an attempt to buttress the price of their wine.[1] However, another way that industries can attempt to protect themselves against reductions in the price of their product is to prevail on the government to establish a price floor. Suppose that the French government had decreed that Bordeaux wine could not be sold in 1979 at less than its 1978 price.

a. If the market demand curve for Bordeaux wine was the same in 1979 as in 1978, would such a price floor have raised the price above the equilibrium price that otherwise would have prevailed?

b. If such a price floor raised the price, would the price increase result in an increase in the total amount of money received by the Bordeaux wine-makers?

Completion Questions

1. If the price elasticity of demand for refrigerators is 2, and the price of a refrigerator increases by 1 percent, there will be an (increase, decrease) ________________ of about ________________ percent in the amount spent on refrigerators.

2. The market demand curve for grade-A widgets is a vertical line at a quantity of 1,000 units per year. The price elasticity of demand for grade-A widgets equals ____________.

3. The equation for the market supply curve of grade-A widgets is $P = 0.1Q$, where P is the price (in dollars) of a grade-A widget and Q is the output (in units per year) of grade-A widgets. The price elasticity of supply of grade-A widgets equals ____________________.

4. Based on the information in Questions 2 and 3, the equilibrium price of a grade-A widget equals ________________.

5. Based on the information in Questions 2 and 3, the equilibrium output of grade-A widgets equals ________________.

6. If the government sets a price floor of $150 for a grade-A widget, the excess supply of grade-A widgets will be ________________, based on Questions 2 and 3.

7. If the government sets a price ceiling of $80 for a grade-A widget the excess demand of grade-A widgets will be ________________, based on Questions 2 and 3.

---

1. *New York Times,* May 7, 1980.

8. If a tax of $1 is imposed on each grade-A widget, the effect is to increase the equilibrium price of a grade-A widget by ________________, based on Questions 2 and 3.

9. If a tax of $10 is imposed on each grade-A widget, the effect is to reduce the equilibrium output of grade-A widgets by ____________________ units per year, based on Questions 2 and 3.

10. If the price elasticity of demand for gasoline is 0.50, a ______________ percent increase in the price of gasoline will be required to reduce the quantity demanded of gasoline by 1 percent.

11. If the actual price of gasoline equals the equilibrium price, the difference between the quantity of gasoline supplied and the quantity of gasoline demanded equals __________.

True or False

____ 1. No equilibrium price nor equilibrium quantity exists if a good's demand curve is a vertical line and its supply curve is a horizontal line.

____ 2. The demand curve for a free good (a good with a zero price) must be a horizontal line.

____ 3. If actual price exceeds equilibrium price, there is a tendency for actual price to rise.

____ 4. A shift to the right of the market supply curve tends to increase the equilibrium price.

____ 5. A product's market demand curve generally slopes upward and to the right, if the product's price elasticity of demand is very large.

____ 6. The slope of the market demand curve equals the price elasticity of demand.

____ 7. In any market, the seller alone determines the price of the product that is bought and sold. Since the seller has the product, while the buyer does not have it, the buyer must pay what the seller asks.

____ 8. If the price elasticity of demand for a good is infinite, a tax on this good will reduce the equilibrium price of the good.

____ 9. If the price elasticity of demand for a good is infinite, a tax on this good will increase the equilibrium price of the good.

____ 10. If the price elasticity of demand for a good is zero, a tax on this good will increase the equilibrium output of this good.

____ 11. If the price elasticity of demand for a good is zero, a tax on this good will reduce the equilibrium output of this good.

____ 12. If the price of coffee goes up it is likely that the market demand curve for tea will shift to the left.

Multiple Choice

1. If the President of the United States were to announce that the government would no longer allow private parties to buy gold (so that the government would be the sole buyer) and if he were to announce that the government would buy any and all gold at $500 per ounce, which of the diagrams below would represent the demand curve for gold in the United States?

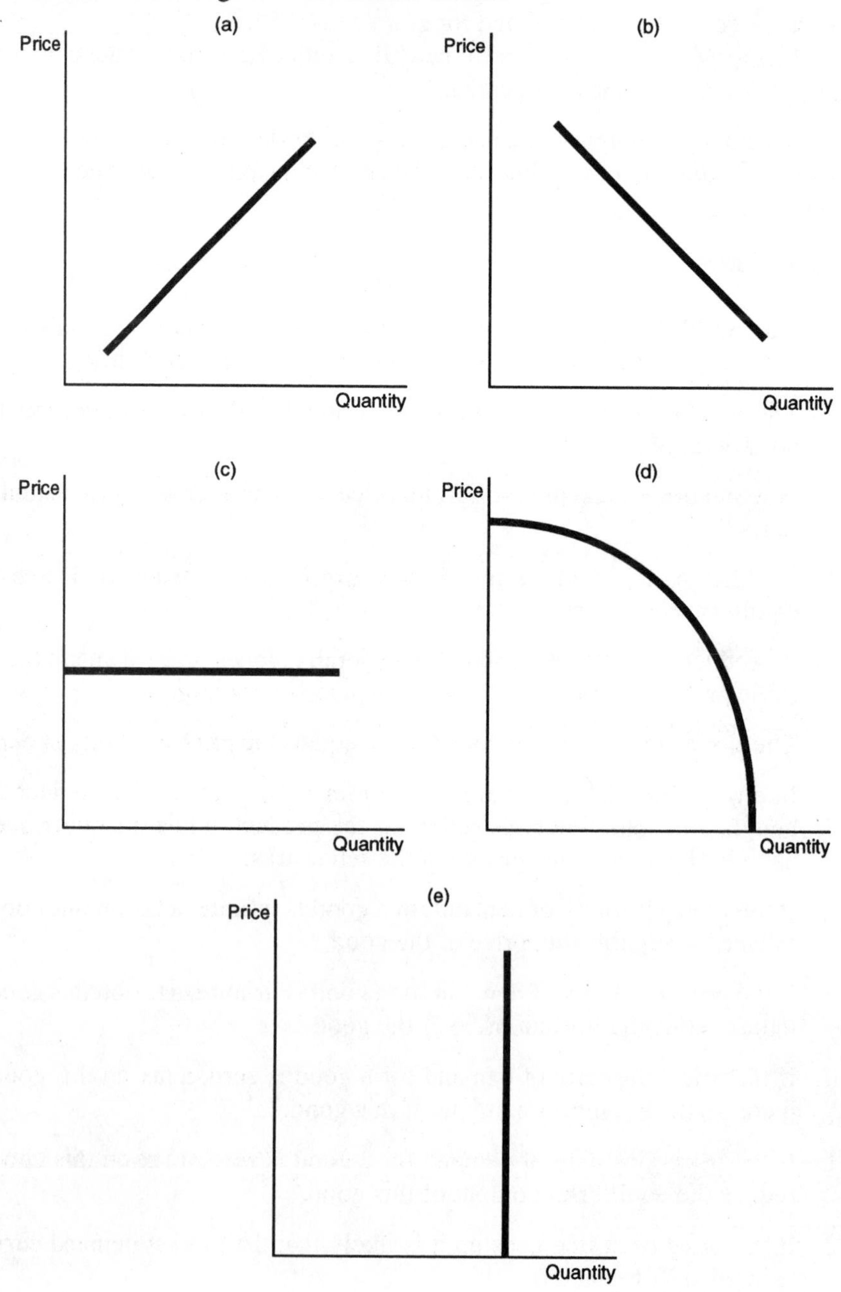

2. Suppose that the market demand curve and the market supply curve for asparagus are as shown in the graph below.

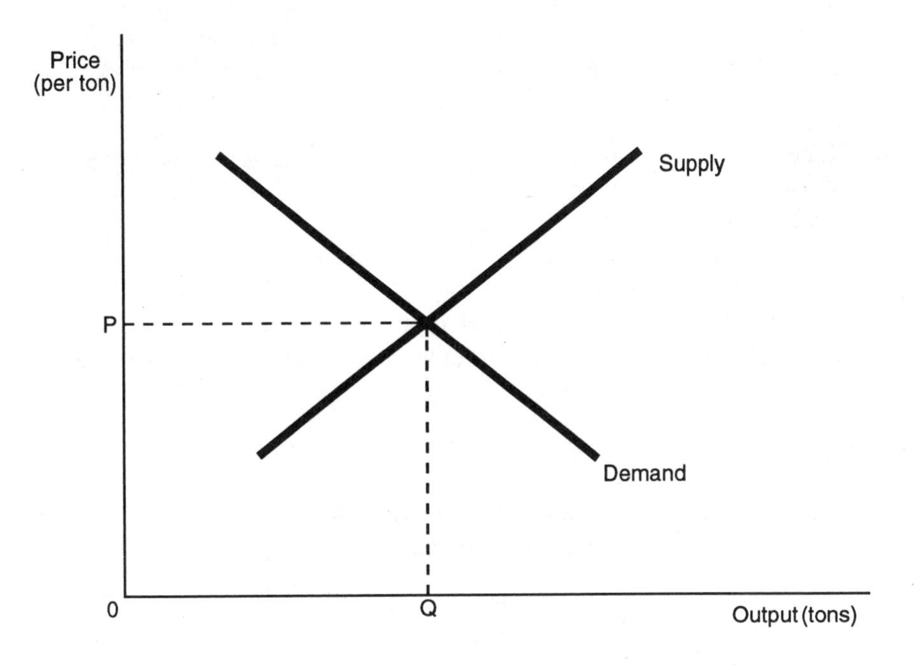

If the government sets a price floor equal to $2P$ per ton, the result will be

a. a reduction in the quantity of asparagus supplied.
b. an excess supply of asparagus.
c. a shift to the right of the market supply curve for asparagus.
d. all of the above.
e. none of the above.

3. In the previous question, if a government report indicates that asparagus contains important, hitherto unrecognized ingredients that can safeguard health and prolong life, the result is likely to be

a. a shift to the right in the market supply curve for asparagus.
b. a shift to the left in the market supply curve for asparagus.
c. a shift to the right in the market demand curve for asparagus.
d. all of the above.
e. none of the above.

4. If the government imposes a tax of $1 per ton on asparagus, it is clear from the diagram in Question 2 that the equilibrium price of asparagus will

a. fall.
b. increase by $1.
c. increase by less than $1.
d. increase by more than $1.
e. be unaffected.

5. If the price of broccoli falls considerably, the result is likely to be
   a. a shift to the right in the market demand curve for asparagus.
   b. a shift to the left in the market supply curve for asparagus.
   c. a shift to the left in the market demand curve for asparagus.
   d. an increase in the equilibrium price of asparagus.
   e. none of the above.

6. If the price of broccoli falls considerably, and if the government freezes the price of asparagus at the level prevailing before the fall in the price of broccoli, the result is likely to be
   a. a surplus of asparagus.
   b. a shortage of asparagus.
   c. a shift to the left in the market supply curve for asparagus.
   d. a shift to the right in the market supply curve for asparagus.
   e. none of the above.

Review Questions

1. Do you think that the market demand curve for rubber is more price elastic in a period of 10 years than in a period of 10 days? Explain.

2. Do you think that the market supply curve for rubber is more price elastic in a period of 10 years than in a period of 10 days? Explain.

3. Suppose that the market demand curve for fish shifts to the right. Does this mean that consumers are willing to buy more fish than previously at any given price? Does it mean that a given quantity of fish put on the market will now bring a higher price?

4. If the market demand curve for cocoa shifts to the left, does this mean that less money will be spent on cocoa?

5. Is it likely that the market demand curve for cocoa will shift to the right because of a technological innovation that reduces the cost of producing cocoa? Explain.

6. Suppose that a plague attacks the nation's beef cattle, but has no effect on its pork production. What effect will this have on:
   a. the market supply curve for pork?
   b. the market demand curve for pork?
   c. the equilibrium price of pork?

7. Suppose that a major increase occurs in the cost of producing hogs (but not in the cost of producing beef). What effect will this have on:
   a. the market supply curve for pork?
   b. the market demand curve for pork?
   c. the equilibrium price of pork?

8. Suppose that the transit authority in a major city permits a large increase in fares on bus, subway, and trolley lines. Do you think that this will shift the market demand curve for taxis in this city to the left or to the right? Explain.

9. Define a market. Must the market demand curve for a commodity always slope downward to the right?
10. Describe the difference between the point elasticity of demand and the arc elasticity of demand.
11. "Since the actual price does not equal the equilibrium price, it is useless to figure out the equilibrium price." Comment on this statement, and indicate whether or not you agree with it.
12. "In order to protect the poor, it is essential that the price of natural gas be kept at its existing level. Any increase in price would only go into the pockets of the rich." Comment on this statement, and indicate whether or not you agree with it.
13. "The laws of supply and demand are immutable. No one, including the government, can affect a commodity's demand curve or supply curve." Comment on this statement, and indicate whether or not you agree with it.

Problems

1. According to unofficial estimates by economists at the U.S. Department of Agriculture, the market demand curve for wheat in the American market in the early 1960s was (roughly) as follows

Farm price of wheat (dollars per bushel)	*Quantity of wheat demanded (millions of bushels)*
1.00	1,500
1.20	1,300
1.40	1,100
1.60	900
1.80	800
2.00	700

In the middle 1970s, the market demand and supply curves for wheat in the American market were (roughly) as follows:

Farm price of wheat (dollars per bushel)	*Quantity of wheat demanded (millions of bushels)*	*Quantity of wheat supplied (millions of bushels)*
3.00	1,850	1,600
3.50	1,750	1,750
4.00	1,650	1,900
5.00	1,500	2,200

a. If the market demand curve in the middle 1970s had remained as it was in the early 1960s, can you tell whether the farm price of wheat in the middle 1970s would have been greater or less than $2.00? Explain.

b. Did a shift occur between the early 1960s and the middle 1970s in the market demand curve for wheat? If so, was it a shift to the right or the left, and what factors were responsible for this shift?

c. What was the price elasticity of demand for wheat during the middle 1970s if the price was between \$3.00 and \$3.50?

d. Suppose that the government had supported the price of wheat at \$4.00 in the middle 1970s. How big would have been the excess supply? What would have been some of the objections to such a policy?

2. The following two demand curves, D_1 and D_2, are parallel straight lines:

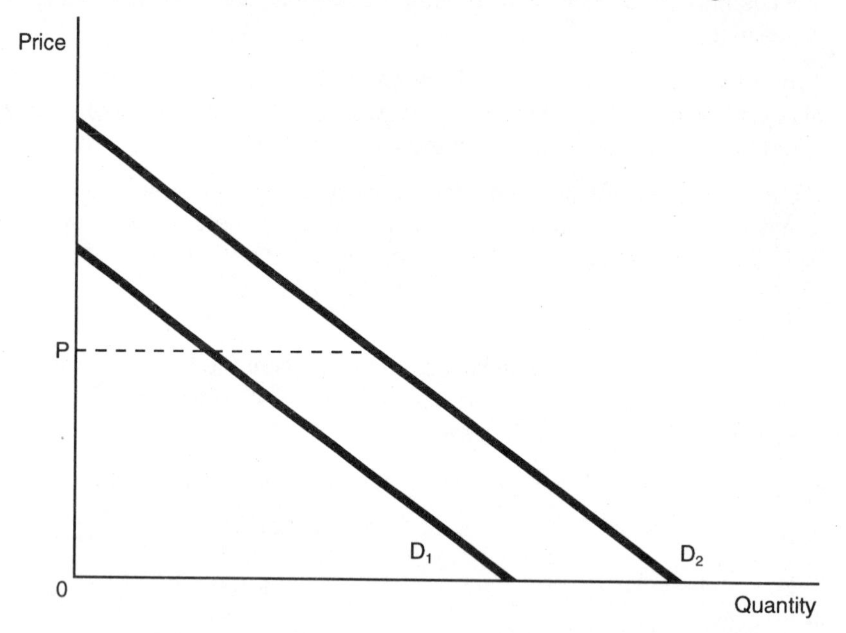

Prove that, if the price is held constant (say, at OP), the price elasticity of demand is not the same if the demand curve is D_1, as it is if the demand curve is D_2.

3. If both the demand and supply curves in a competitive market shift to the right, one can predict the direction of change of output but not of price. If the supply curve shifts to the right, but the demand curve shifts to the left, one cannot, without further knowledge, be certain about the direction of either the price or the quantity change. Do you agree with these statements? Explain.

4. *a.* The demand curve for beer is

$$Q_D = 500 - 5P$$

where Q_D is the quantity demanded of beer (in millions of barrels per year) and P is its price (in dollars per barrel). If the supply curve for beer is a vertical line at $Q_S = 400$ million barrels of beer per year, what is the equilibrium price of a barrel of beer?

b. Under the conditions described in part (a), what would be the effect on the price of a barrel of beer if a tax of \$5 per barrel is imposed by the government on beer?

5. *a.* Suppose that the supply curve for butter is

$$Q_S = 100 + 3P$$

where Q_S is the quantity supplied of butter (in millions of pounds per year) and P is the price of butter (in dollars per pound). If the demand curve for butter is a vertical line at $Q_D = 106$ millions of pounds per year, and if the government imposes a price floor of $1 per pound on butter, will there be an excess supply or excess demand of butter, and how big will it be?

b. If the government's price floor is set at $3 per pound, will there be an excess supply or excess demand of butter, and how big will it be?

c. Under the conditions described in part (*a*), what is the price elasticity of demand for butter? Do you regard this as a realistic value for this price elasticity? Explain.

6. Suppose that the number of cameras demanded in the United States in 1994 at various prices is as follows:

Price of a camera (dollars)	*Quantity demanded per year (millions of cameras)*
80	20
100	18
120	16

a. Draw three points on the demand curve for cameras in the graph below.

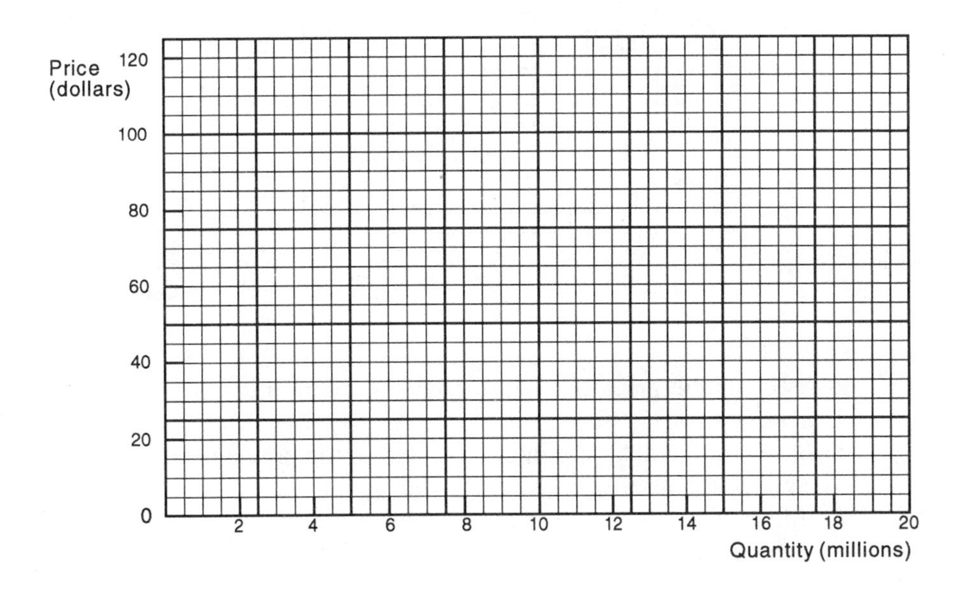

b. Calculate the arc price elasticity of demand when (1) the price is between $80 and $100 and (2) the price is between $100 and $120.

7. Suppose that the relationship between the quantity of cameras supplied in 1994 in the United States and the price per camera is as follows:

Price of a camera (dollars)	*Quantity supplied per year (millions of cameras)*
60	14
80	16
100	18
120	19

a. Draw four points on the supply curve for cameras in the graph below.

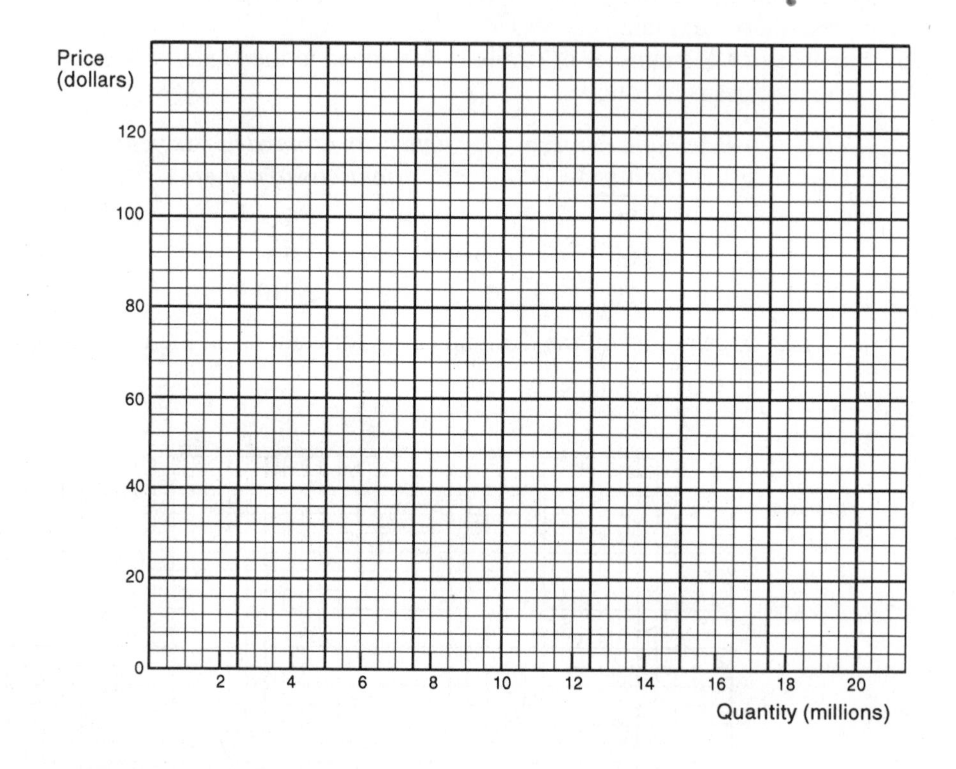

b. Estimate the price elasticity of supply when the price is between \$80 and \$100.

c. Based on the data presented here and in the previous question, what is the equilibrium price of a camera in the United States?

d. If the price is \$80, will there be an excess demand? An excess supply?

e. If the price is \$120, will there be an excess demand? An excess supply?

8. The government imposes an excise tax of \$40 on each camera (in the previous problem). Then it decides to set a price ceiling of \$100 on the price of a camera. Will there be a surplus or shortage of cameras? If so, how big a surplus or shortage? (Hint use the data in Question 6.)

9. Suppose that the demand curve for cantaloupes is

$$P = 120 - 3Q_D$$

where P is the price per pound (in cents) of a cantaloupe and Q_D is the quantity demanded per year (in millions of pounds). Suppose that the supply curve for cantaloupes is

$$P = 5Q_S$$

where Q_S is the quantity supplied per year (in millions of pounds). What is the equilibrium price per pound of a cantaloupe? What is the equilibrium quantity of cantaloupes produced?

10. C. Nisbet and F. Vakil estimated the demand curve for marijuana among students at the University of California at Los Angeles. The estimated demand curve for a representative student (who was in the market for the drug) is shown below.

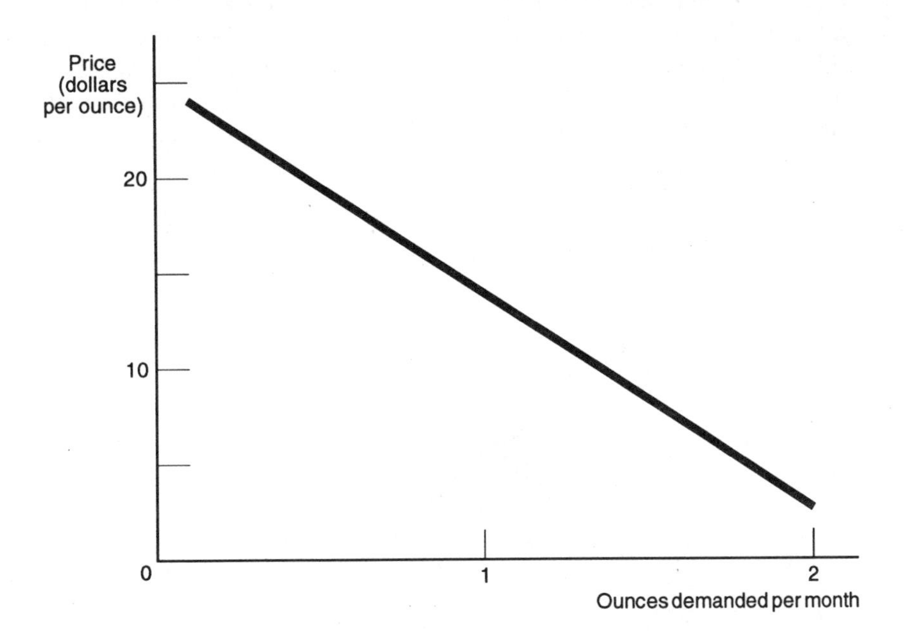

a. How many ounces per month of marijuana were demanded by the student when the price was \$10 per ounce? \$15 per ounce?
b. What was the arc elasticity of demand when the price was between \$10 and \$15 per ounce?
c. In recent years, the United States government has intensified its efforts to restrict the inflow of marijuana in order to cut back the supply, thus increasing the price. If the price were driven sky-high, would this affect the market demand curve for marijuana? If so, in what way?
d. If marijuana were legalized, what effect would it have on the market supply curve for marijuana?
e. If research were to indicate that marijuana causes birth defects, what effect would it have on the market demand curve for marijuana?

Key Concepts for Review

Market demand curve	Price elasticity of demand
Market supply curve	Price elasticity of supply
Equilibrium	Price ceiling
Equilibrium price	Price floor
Surplus	Shortage

ANSWERS

Case Study: In Vino Veritas

a. If the market demand curve for Bordeaux wine was the same in 1979 as in 1978, and if the 1979 market supply curve for Bordeaux wine (S_{79}) is to the right of the 1978 supply curve (S_{78}), the situation is as shown in the graph below. The equilibrium price in 1979 would be OP_0, and the price floor would be OP_1 (since this was the 1978 price). Thus the price floor would be above the equilibrium price.

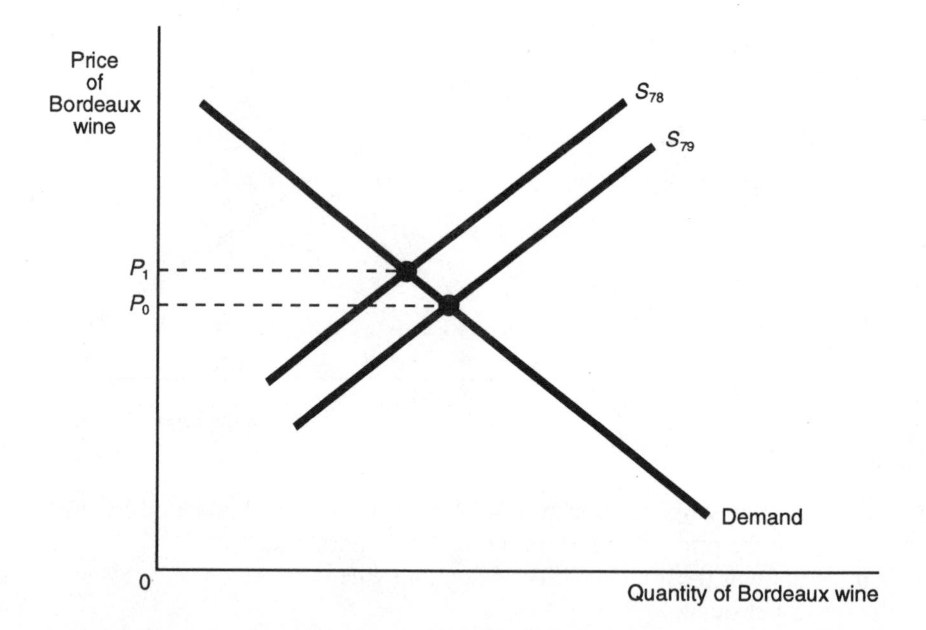

b. Whether such a price increase would result in an increase in the total amount of money received by the Bordeaux wine-makers depends on the price elasticity of demand for their wine. For example, if this price elasticity equals 1, such a price increase would have no effect on the amount of money they receive. To see this, note that the total amount of money they receive is the quantity of wine sold multiplied by its price. If the price elasticity of demand equals 1, a 1 percent increase in price results in a 1 percent decrease in quantity sold. Thus the total amount they receive is not affected by the price increase. (More will be said on this score in Chapters 4 and 5.)

Completion Questions

1. decrease; 1
2. zero
3. 1
4. $100
5. 1,000 units per year
6. 500 units per year
7. 200 units per year
8. $1 per unit
9. zero
10. 2
11. zero

True or False

1. False	2. False	3. False	4. False	5. False	6. False
7. False	8. False	9. False	10. False	11. False	12. False

Multiple Choice

1. *c*	2. *b*	3. *c*	4. *c*	5. *c*	6. *a*

Review Questions

1. It is likely to be more price elastic in a period of 10 years because consumers will have more time to adapt to changes in the price of rubber.
2. It is likely to be more price elastic in a period of 10 years because producers will have more time to adapt to changes in the price of rubber.
3. Yes. Yes.
4. Yes, if the supply curve for cocoa slopes upward to the right.
5. No. Such an innovation would be expected to influence the supply curve, not the demand curve, for cocoa.
6. *a.* None.
   b. It will shift to the right.
   c. It will increase.
7. *a.* It will shift to the left.
   b. None.
   c. It will increase.
8. It may shift them to the right because bus, subway, and trolley transportation is a substitute for taxis.
9. A market is a group of firms and individuals that are in touch with each other in order to buy or sell some good. No.
10. The point elasticity of demand refers to a case where the change in price is small; the arc elasticity of demand refers to a case where the change in price is not small.

11. Since the actual price tends to move toward the equilibrium price, it is useful to figure out the equilibrium price, because this helps to indicate whether the actual price will rise or fall.

12. This statement is obviously a vast oversimplification of the effects of such a price increase. For example, an increase in price might increase natural gas supplies which might be socially beneficial.

13. The government can influence a commodity's demand or supply curve by subsidies, taxation, and other means.

Problems

1. *a.* It is impossible to tell. At $2, the quantity demanded would have been 700 million bushels but we are not given the quantity supplied at $2. If it were greater than 700 million bushels, the equilibrium price would have been less than $2; if it were less than 700 million bushels, the equilibrium price would have been greater than $2.

   b. Yes. At $3, the quantity of wheat demanded during the 1960s would have been less than 700 million bushels, if the demand curve was downward sloping to the right. During the 1970s, 1,850 million bushels were demanded at $3. Thus the demand curve seemed to shift to the right. The reason was increased foreign demand due partly to poor harvests in the Soviet Union, Australia, and Argentina, as well as devaluation of the dollar.

   c.
   $$\eta = -\frac{1{,}850 - 1{,}750}{(1{,}850 + 1{,}750)/2} \div \frac{3.00 - 3.50}{(3.00 + 3.50)/2} = 0.36$$

   d. 250 million bushels. The problems involved in such farm supports are discussed in more detail in later chapters. Such surpluses have been an embarrassment, both economically and politically. They suggest that society's scarce resources are being utilized to produce products that consumers do not want at existing prices. Also, the cost of storing these surpluses can be large.

2. As pointed out in footnote 6 of the text, the slope of the demand curve does not equal the price elasticity of demand. While the slope is the same if the demand curve is D_1 as it is if the demand curve is D_2 the price elasticity is not. Specifically, whereas dQ_D/dP and P are the same under these circumstances, Q_D is not the same, so the price elasticity is not the same. (Q_D is quantity demanded, and P is price.)

3. The first sentence is correct. The second sentence is incorrect; one can be certain that price will fall.

4. *a.* Since the quantity supplied must equal the quantity demanded if equilibrium is achieved, it follows that
   $$500 - 5P = 400$$
   or
   $$100 = 5P$$
   which means the $P = 20$. Thus, the equilibrium price is $20 per barrel.

   b. No effect, because the supply curve would be unaffected by the tax.

5. *a.* In equilibrium, $Q_S = Q_D$, so

$$100 + 3P = 106$$
$$3P = 6$$
$$P = 2$$

Thus the equilibrium price is \$2 per pound. Since this exceeds the government's price floor, there will be no excess supply or excess demand of butter.

b. If $P = 3$, the quantity supplied equals $100 + 3(3) = 109$, and the quantity demanded equals 106, so the excess supply equals $109 - 106$, or 3; that is, it equals 3 million pounds per year.

c. Zero.

No, because decreases in the price of butter are almost certain to increase the quantity demanded.

6. *a.*

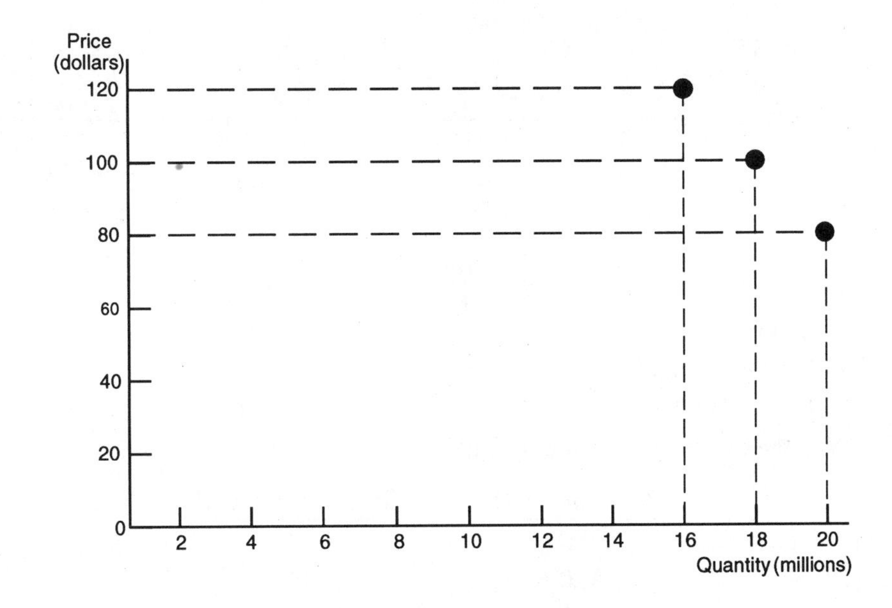

b.

$$\eta = -\frac{20 - 18}{(20 + 18)/2} \div \frac{80 - 100}{(80 + 100)/2} = \frac{2}{19} \div \frac{20}{90} = 0.47$$

$$\eta = -\frac{18 - 16}{(18 + 16)/2} \div \frac{100 - 120}{(100 + 120)/2} = \frac{2}{17} \div \frac{20}{110} = 0.65$$

7. *a.*

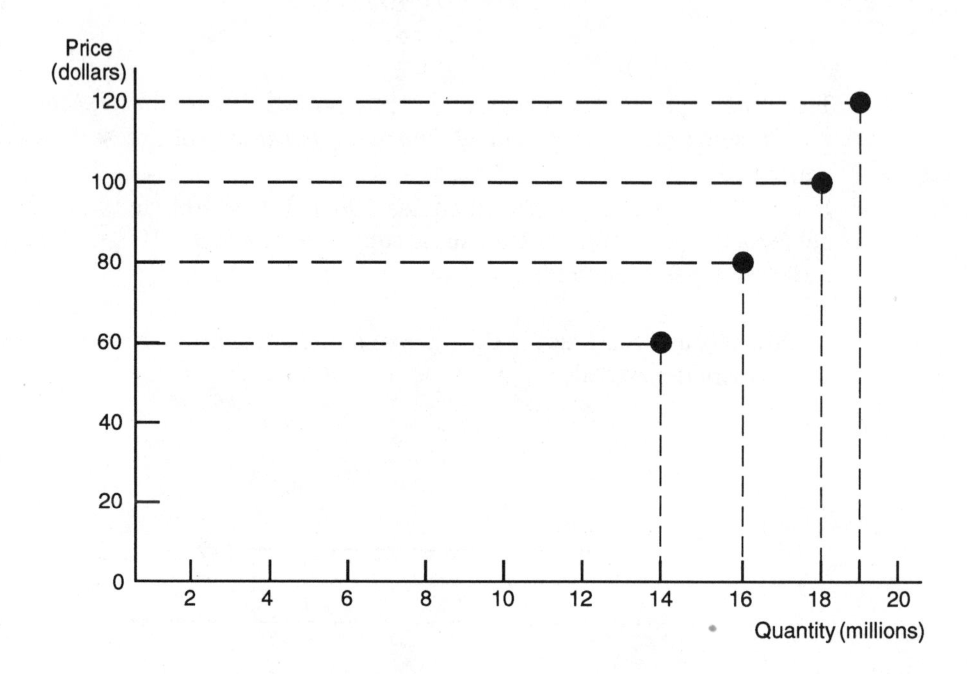

b.

$$\eta_S = \frac{18-16}{(18+16)/2} \div \frac{100-80}{(100+80)/2} = \frac{2}{17} \div \frac{20}{90} = 0.53$$

c. \$100

d. Excess demand.

e. Excess supply.

8. With the excise tax, the supply curve is:

Price (dollars)	*Quantity supplied*
100	14
120	16
140	18
160	19

Thus the equilibrium price of a camera is \$120, since at this price the quantity supplied equals the quantity demanded. If the government sets a price ceiling of \$100, there will be a shortage of 4 million cameras per year, since the quantity demanded will equal 18 million while the quantity supplied will equal 14 million. (See the data in Question 6.)

9. Since the quantity supplied must equal the quantity demanded, we have two equations to be solved simultaneously

$$120 - 3Q_D = 5Q_S$$
$$Q_D = Q_S$$

Letting quantity equal Q, it follows that

$$\begin{aligned} 120 - 3Q &= 5Q \\ 120 &= 8Q \\ 15 &= Q \end{aligned}$$

Since $P = 5Q$, it follows that $P = 75$. Thus the equilibrium price is 75 cents and the equilibrium quantity is 15 million pounds per year.

10. *a.* About 1.3 ounces per month at $10. About 0.9 ounces per month at $15.

b. Let $P_1 = \$10$, $Q_1 = 1.3$ ounces, $P_2 = \$15$, and $Q_2 = 0.9$ ounces. Then the arc elasticity equals

$$-\frac{(1.3 - 0.9)}{(1.3 + 0.9)/2} \div \frac{(10 - 15)}{(10 + 15)/2}$$

or 0.91.

c. Such an increase in price would reduce the *quantity demanded*, but it would *not* shift the demand curve. In other words, there would be a *movement along* the demand curve from a point corresponding to the current price to a point corresponding to a sky-high price, but no *shift* in the demand curve.

d. Legalization would probably shift the market supply curve to the right. Suppliers could supply more at a given price because their costs would be lower. They would no longer have to pay for secrecy and protection, and the risks would be less.

e. It would shift to the left for obvious reasons.

Part 2 Consumer Behavior and Market Demand

Chapter 3 The Tastes and Preferences of the Consumer

Case Study: Nuclear-Powered vs. Conventionally Powered Aircraft Carriers

During recent decades, a basic decision confronting the United States Navy has been whether to build nuclear-powered or conventionally powered aircraft carriers. Suppose that the effectiveness of the United States Navy would be the same if the navy had each of the following combinations of numbers of aircraft carriers of each type.

	Number of aircraft carriers	
Combination	*Nuclear*	*Conventional*
A	4	9
B	5	6
C	6	4
D	7	2

Suppose too that the price of each nuclear-powered aircraft carrier is more than double the price of a conventionally powered aircraft carrier.

a. Can you suggest how the theory of consumer behavior might be applied to this case? How would utility be defined? Who would be the consumer? What determines the consumer's budget line?

b. Given the (hypothetical) data presented above, can you prove that some of the combinations of numbers of aircraft carriers of each type are nonoptimal?

c. Suppose that the people who are closely involved in this decision prefer nuclear-powered carriers because they are large and impressive and more consistent with their idea of what the "warship of the future" should look like. If these people are influenced by such considerations, as well as by the data presented above, is this a violation of the assumptions underlying the theory?

d. In the table above, what would be the meaning of the marginal rate of substitution? To maximize the effectiveness of the navy, what should be the value of the marginal rate of substitution?

Completion Questions

1. All other things equal, the rational consumer is assumed to prefer __________ of a good to ________________ of a good.

2. All market baskets possessing the same utility are said to be on the same ________________ curve.

3. The negative of the slope of the tangent to an indifference curve is termed the

_______________.

4. The consumer attempts to _______________ his or her utility subject to a _______________ constraint.
5. The condition for consumer equilibrium is that the budget line be _______________ an indifference curve (if some of both commodities is consumed).
6. If the rational consumer always prefers more of a good to less, it follows that all indifference curves have a _______________ slope.
7. If the marginal rate of substitution of good X for good Y at constant utility decreases with increases in the quantity of X, then the indifference curve is _______________.
8. One of the most important determinants of a consumer's behavior is his or her _______________.
9. The three basic assumptions an economist makes about the nature of consumer tastes are: If the consumer is presented with two market baskets, he or she can decide _______________. If the consumer prefers oranges to bananas and bananas to apples, then he or she _______________. The consumer always prefers _______________ of a commodity to _______________ of it.
10. Different market baskets on the same indifference curve should be given the _______________ values of utility.
11. Market baskets on higher indifference curves should receive _______________ utilities than market baskets on lower indifference curves.

True or False

____ 1.[1] If the total utility from consuming caviar is 5 times the number of ounces of caviar consumed and the total utility from consuming hot dogs is 2 times the number of hot dogs consumed, the consumer should buy caviar, no hot dogs.

____ 2.[2] A consumer who is rational equates the marginal utility of all goods consumed.

____ 3. Two indifference curves can intersect only when one of the goods being studied is a high-priced item.

____ 4. Economists generally assume that indifference curves always lie above their tangents.

____ 5. Indifference curves are always concave to the origin.

____ 6. The marginal rate of substitution of good X for good Y is the number of units of good X that a customer will accept instead of good Y to increase his or her satisfaction.

1. This question pertains to the chapter appendix.
2. This question pertains to the chapter appendix.

____ 7. Utility theory assumes that market baskets on higher indifference curves have higher utilities.

____ 8. Any numbers can be attached to a set of market baskets to represent utility so long as market baskets higher up on the same indifference curve have higher values.

____ 9. If a consumer's income rises, he or she will probably buy the same amount of a good.

____ 10. The shape of a consumer's indifference curve is generally assumed to be unaffected by price changes.

____ 11. A person's tastes are like his or her fingerprints: They don't change.

Multiple Choice

1.[3] David Howe has 8 hours to spend during which he can either play backgammon or read. The marginal utility he obtains from an hour of reading is 8 utils. The total utility he obtains from 1, 2, 3, 4, and 5 hours of backgammon is as follows:

Hours	*Total utility*
1	20
2	33
3	40
4	40
5	35

If he maximizes utility (and if he can allocate only an integer number of hours to each activity), he will spend
a. 4 hours playing backgammon and 4 hours reading.
b. 3 hours playing backgammon and 5 hours reading.
c. 2 hours playing backgammon and 6 hours reading.
d. 1 hour playing backgammon and 7 hours reading.
e. none of the above.

2. If point *B* lies above and to the right of point *A* on a two-commodity indifference map, and the indifference curve passing through point *A* is characterized by a utility level of 1, then the utility level of the indifference curve passing through point *B* has utility
a. greater than 1.
b. equal to 1.
c. less than 1.
d. equal to zero.
e. equal to infinity.

3. This question pertains to the chapter appendix.

3.[4] Modern microeconomic theory generally regards utility as
   a. cardinal.
   b. ordinal.
   c. independent.
   d. Republican.
   e. Democrat.

4. The consumer is likely to find the market basket that maximizes his or her utility
   a. immediately.
   b. if time is allowed for him or her to adapt and learn.
   c. never.
   d. if he or she has studied economics.
   e. none of the above.

5. A basic assumption of the theory of consumer choice is that
   a. the consumer tries to get on the highest indifference curve.
   b. the consumer tries to get the most of good *Y*.
   c. the budget line is concave.
   d. none of the above.
   e. all of the above.

Review Questions

1. Suppose that in 1994 consumers in San Francisco pay twice as much for apples as for pears, whereas consumers in Los Angeles pay 50 percent more for apples than for pears. If consumers in both cities maximize utility, will the marginal rate of substitution of pears for apples be the same in San Francisco as in Los Angeles? If not, in which city will it be higher?
2. During recent decades, American wine producers "were encouraged by the whole changing social role of their product." Specifically, the American consumer was becoming much more attuned to wine. How did the growing acceptance of wine by American consumers affect their indifference curves between wine and other kinds of alcoholic beverages?
3. In recent year, great numbers of Americans have traveled to Europe. What effect do you think that this has had on American tastes for wine? How could you test your hypothesis?
4. What are the basic assumptions that economists make about the nature of consumers' tastes?

4. This question pertains to the chapter appendix.

5. Draw the indifference curve that includes the following market baskets. (Use the grid below.) Each of these market baskets gives the consumer equal satisfaction.

Market basket	*Meat (pounds)*	*Potatoes (pounds)*
1	2	8
2	3	7
3	4	6
4	5	5
5	6	4
6	7	3
7	8	2
8	9	1

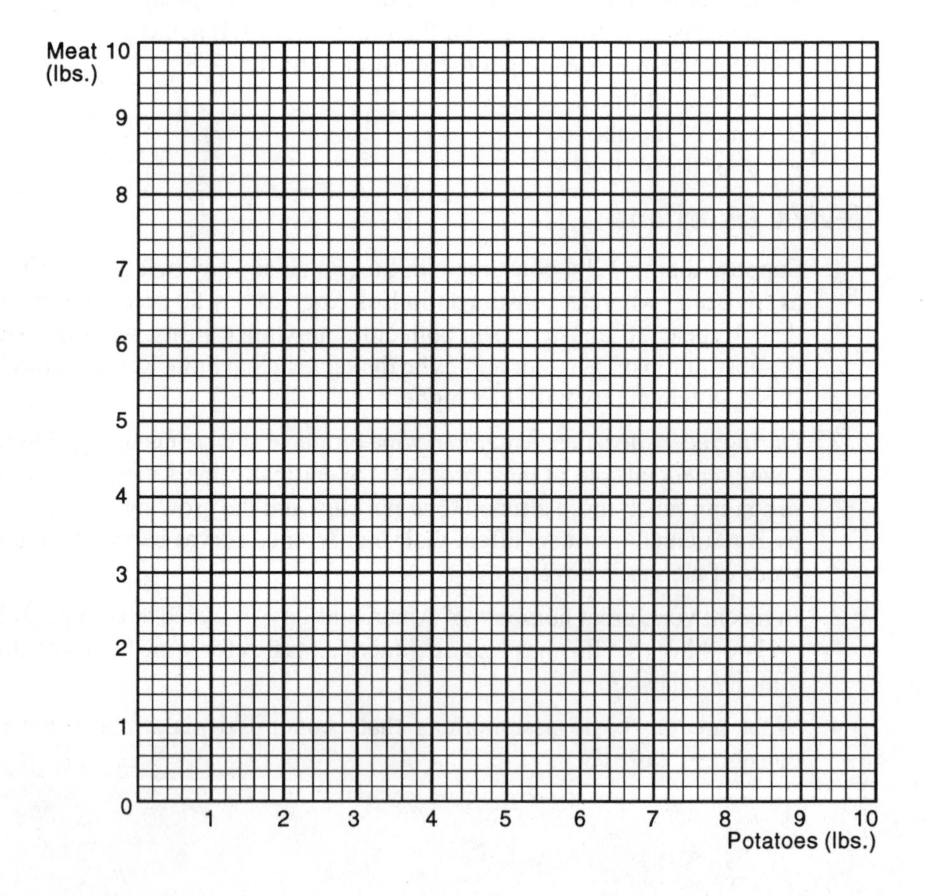

6. In the previous question, what is the marginal rate of substitution of potatoes for meat? How does the marginal rate of substitution vary as the consumer consumes more meat and less potatoes? Is this realistic?

7.[5] Define utility. How does cardinal utility differ from ordinal utility? Which concept is generally used by economists today?

8. Suppose that the consumer has an income of $10 per period and that he or she must spend it all on meat or potatoes. If meat is $1 a pound and potatoes are 10 cents a pound, draw the consumers's budget line.

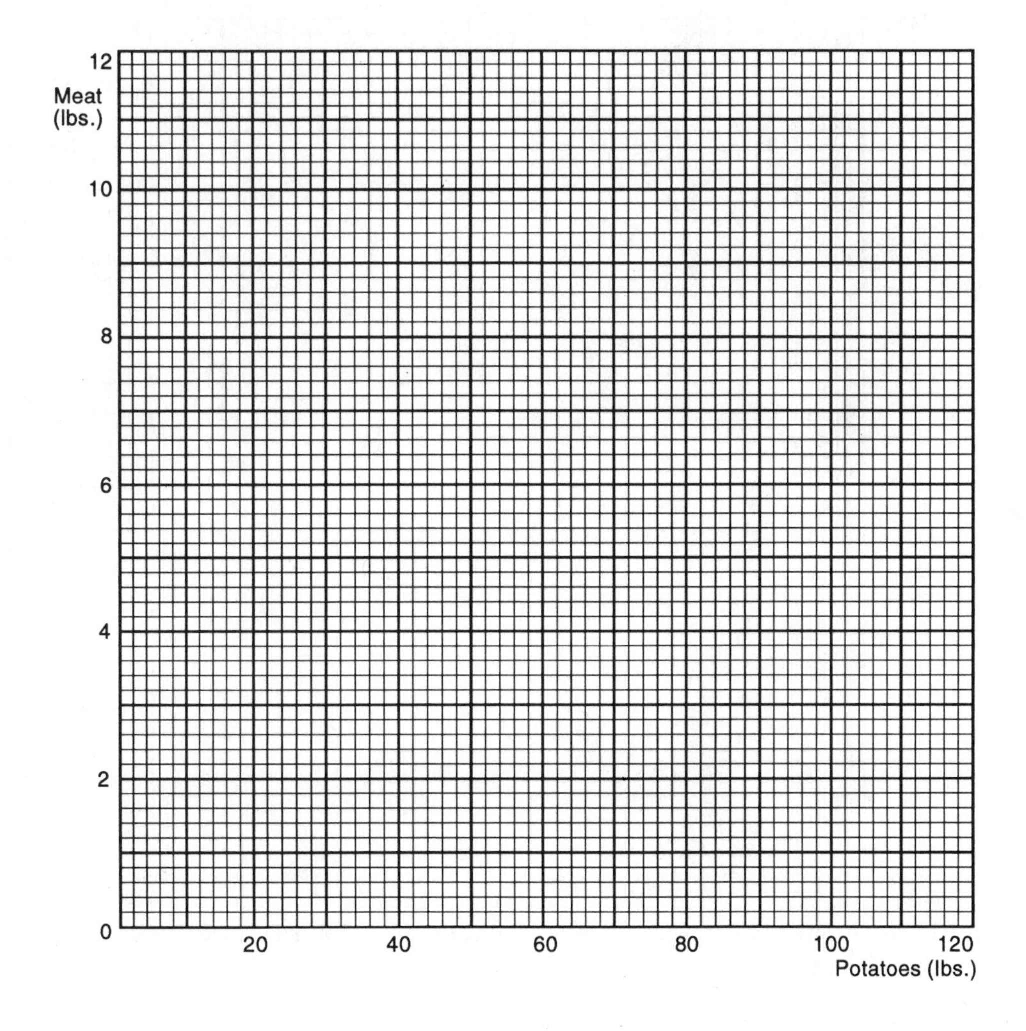

5. This question pertains to the chapter appendix.

9. In the previous case, what will be the budget line if the consumer's income increases to $12? What will be the budget line if the consumer's income is $10, but the price of meat increases to $2 per lb.? What will be the budget line if the consumer's income is $10, but the price of potatoes increases to 20 cents per lb.?

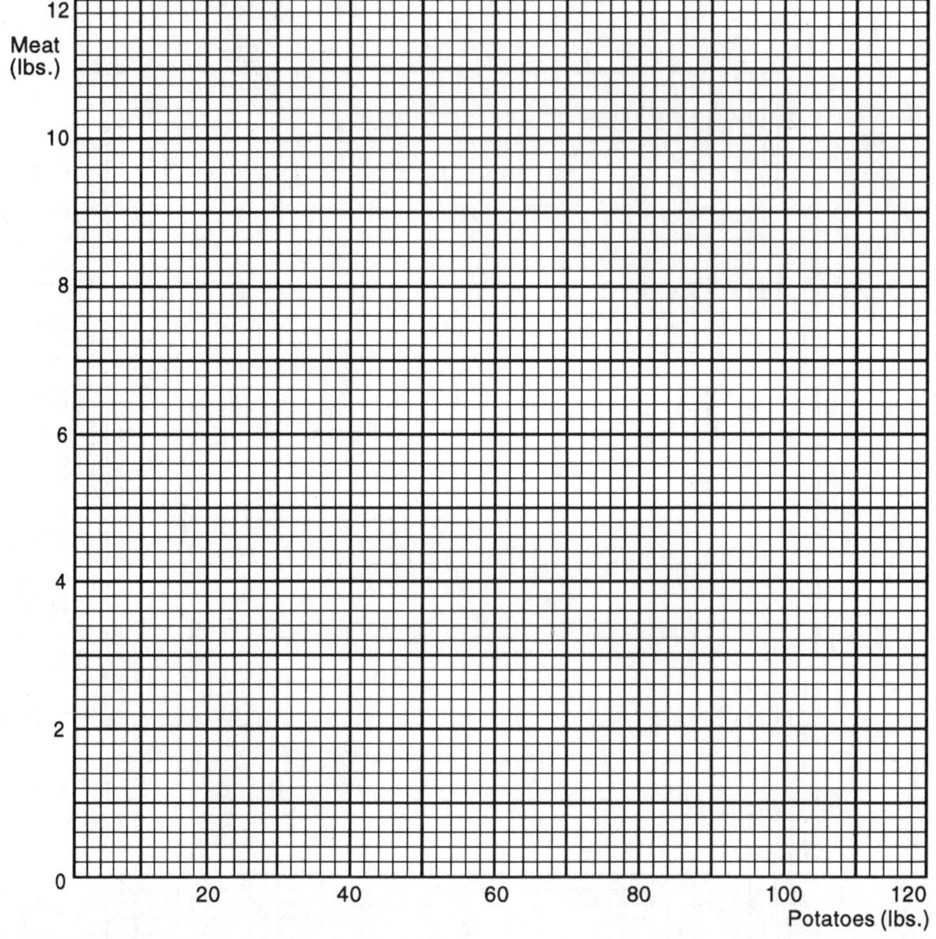

10. What determines consumer tastes and preferences? What is meant by conspicuous consumption?

Problems

1. A consumer has budgeted a total of $100 to spend on two goods, X and Y. She likes to consume a unit of good X in combination with a unit of good Y. Any unit of good X that she cannot consume in combination with a unit of good Y is useless. Similarly, any unit of good Y that she cannot consume in combination with good X is useless. If the price of a unit of good X is $5 and the price of a unit of good Y is $10, how many units of each good will the consumer purchase?

2. John Jones has 2 fifths of scotch and 2 fifths of gin. His friend Bill Bailey (who, as is well known, sometimes must be beseeched to come home) has 1 fifth of

scotch and 2 fifths of gin. The marginal utilities of each good are shown in the table below. For example, to Jones, the marginal utility of the first fifth of scotch is 20 utils, the marginal utility of the second fifth of scotch is 18 utils, the marginal utility of the third fifth of scotch is 15 utils, and so on.

John Jones				*Bill Bailey*			
Scotch		*Gin*		*Scotch*		*Gin*	
Number of fifths	*Marginal utility*	*Number of fifths*	*Marginal utility*	*Number of fifths*	*Marginal utility*	*Number of fifths*	*Marginal utility*
1	20	1	30	1	15	1	7
2	18	2	25	2	14	2	5
3	15	3	20	3	13	3	3
4	10	4	15	4	11	4	1

a. Will Jones be better off if he trades a fifth of his scotch for a fifth of Bailey's gin?

b. Suppose that Jones decides to ask for two of Bailey's fifths of gin in exchange for a fifth of his scotch. Would Bailey be better off to accept this deal than none at all?

3. Suppose that, if a family is eligible for food stamps, it pays $100 per month for $200 worth of food. The price of a pound of food is $5; the price of a pound of nonfood items is $3.

   a. Draw a family's budget line on a graph where the quantity of food consumed per month is measured along the horizontal axis and the quantity of nonfood items consumed per month is measured along the vertical axis, if the family has a cash income of $300 per month and it is not eligible for food stamps. Use the grid below.

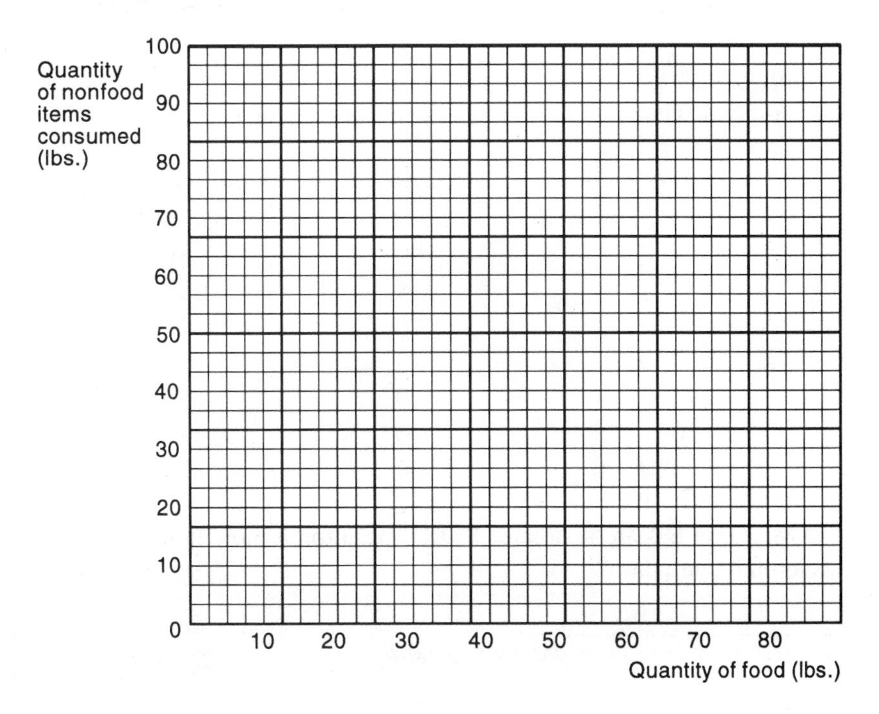

b. Draw (in the grid below) the family's budget line under these conditions if it takes part in the food stamp program described above.

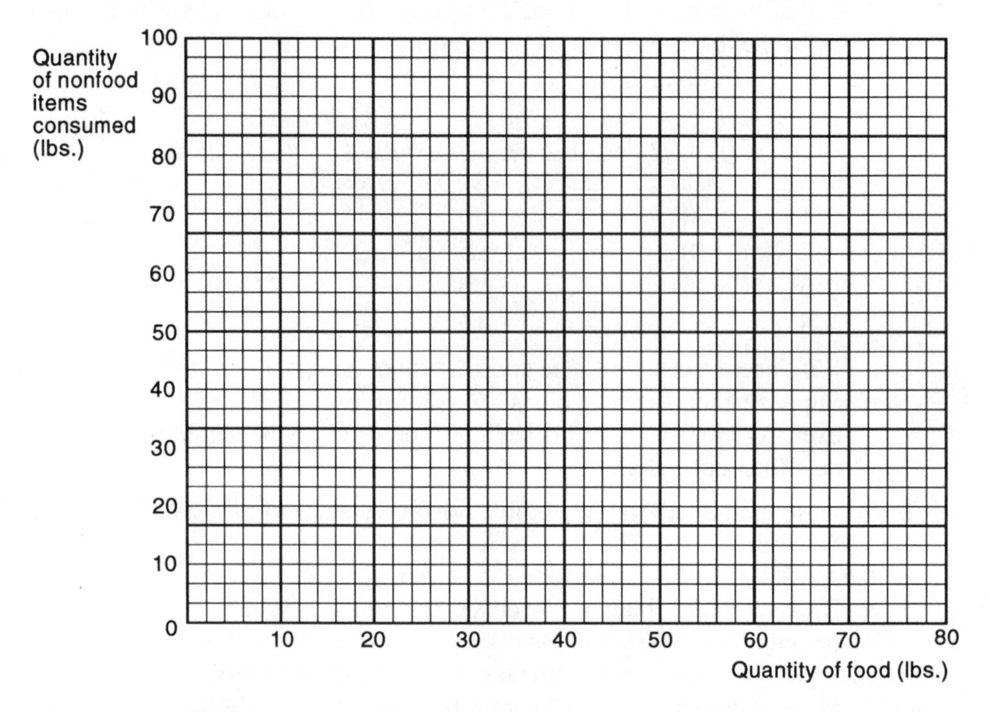

c. Under what circumstances would the family be better off if it were given $100 per month in cash, rather than in food via the above food stamp program?

4.[6] Suppose that James Gray spends his entire income on goods *X* and *Y*. The marginal utility of each good (shown below) is independent of the amount consumed of the other good. The price of *X* is $100 and the price of *Y* is $500.

Number of units	*Mr. Gray's Marginal utility (utils)*	
of good consumed	*Good X*	*Good Y*
1	20	50
2	18	45
3	16	40
4	13	35
5	10	30
6	6	25
7	4	20
8	2	15

If Mr. Gray has an income of $1,000 per month, how many units of each good should he purchase?

6. This question pertains to the chapter appendix.

5. In the diagram below, we show one of Ellen White's indifference curves and her budget line.
   a. If the price of good *A* is $50, what is Ms. White's income?
   b. What is the equation for her budget line?
   c. What is the slope of the budget line?
   d. What is the price of good *B*?
   e. What is her marginal rate of substitution in equilibrium?

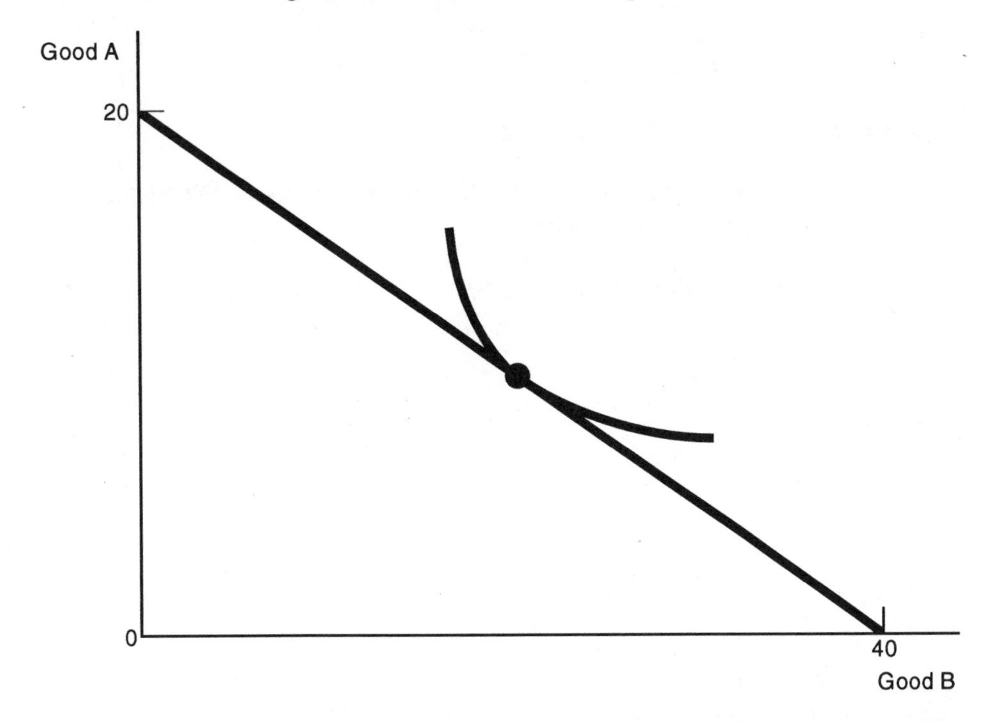

6. (*Advanced*) Suppose that Mrs. Brown has $50 to be divided between corn and beans and that the price of beans is 50 cents per lb. What will be the relationship between the price of corn and the amount of corn she will buy, if her utility function is $U = \log Q_c + 4 \log Q_b$, where U is her utility, Q_c is the quantity of corn she consumes (in lbs.) and Q_b is the quantity of beans she consumes (in lbs.)?

7.[7] The fact that each person consumes many different goods supports the theory of diminishing marginal utility. Explain.

7. This question pertains to the chapter appendix.

Key Concepts for Review

Indifference curves	Market income
Marginal rate of substitution	Conspicuous consumption
Convexity	Determinants of tastes
Utility	Rationality
Budget line	Advertising and selling expenses
Money income	* Cardinal utility
Equilibrium	* Ordinal utility
Indifference map	* Marginal utility

ANSWERS

Case Study: Nuclear-powered vs. Conventionally Powered Aircraft Carriers

a. The various combinations can be viewed as points on an indifference curve, as shown below.

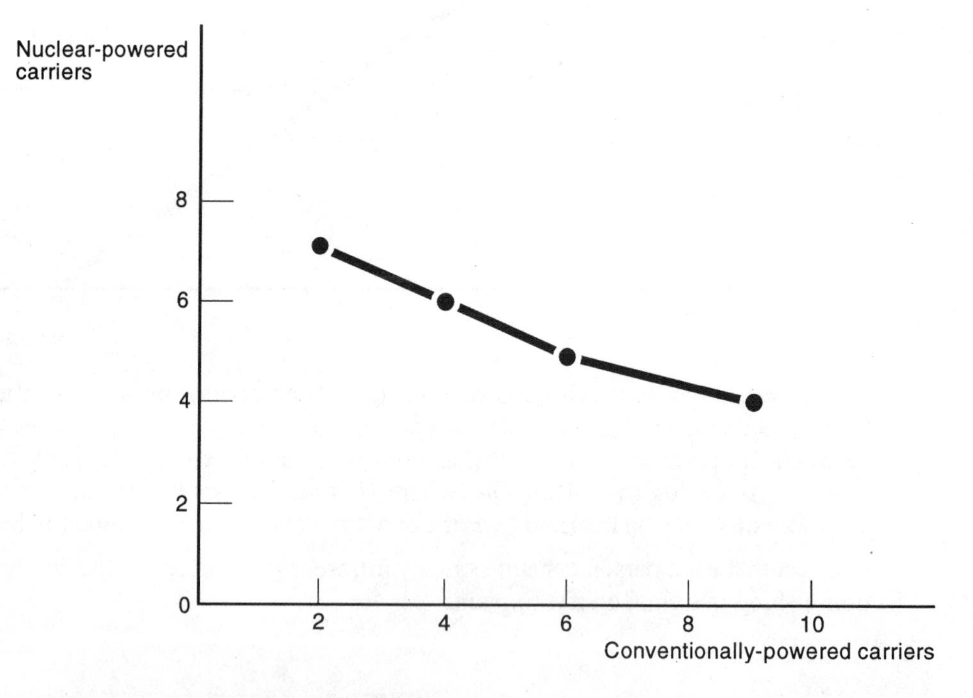

Utility would be defined as effectiveness of the navy. The consumer would be the navy (or the Defense Department or the American people). The budget line would be determined by the total amount of money allocated for aircraft carriers (of either type) and the relative price of each type of aircraft carrier.

b. If the price of a nuclear-powered aircraft carrier is more than double the price of

* This concept pertains to the chapter appendix.

a conventionally powered aircraft carrier, the budget line must have a slope that is between $-\frac{1}{2}$ and zero. Such a budget line could be tangent to the indifference curve shown above at points *B* or *A*, but not at points *C* or *D*. Thus, combinations *C* or *D* seem to be poorer than combinations *B* or *A*. Another way of proving this is to note that, when we move from combination *B* to combination *C* or *D*, we must increase cost (since a nuclear-powered aircraft carrier is assumed to cost more than twice as much as a conventionally powered aircraft carrier), but the effectiveness is the same.

c. No, but the indifference curves representing these people's tastes (not the indifference curves representing navy effectiveness) may be the relevant ones in shaping the decision.

d. The marginal rate of substitution is the number of aircraft carriers of one type that can be substituted for one aircraft carrier of the other type with navy effectiveness (or some other measure of utility) held constant. It should be set equal to the ratio of the prices of the two types of aircraft carriers.

Completion Questions

1. more; less
2. indifference
3. marginal rate of substitution
4. maximize; budget
5. tangent to
6. negative
7. convex
8. tastes (his or her income)
9. which he or she prefers; prefers oranges to apples; more; less
10. same
11. higher

True or False

1. False	2. False	3. False	4. True	5. False	6. False
7. True	8. False	9. False	10. True	11. False	

Multiple Choice

1. *c*	2. *a*	3. *b*	4. *b*	5. *a*

Review Questions

1. No.

   Los Angeles, because the price of a pear divided by the price of an apple is higher in Los Angeles.

2. Consumers valued wine more highly. They tended to be willing to exchange more of other types of alcoholic beverages for a pint of wine.

3. By coming into contact with wine in France, Italy, and other European countries, American travelers have become exposed to wine, and many have found that they liked it. Possibly one could test this hypothesis by analyzing the buying habits of

people before and after trips abroad to determine if there were any evidence of a change in tastes.

4. First, we assume that the consumer can decide whether he or she prefers one market basket to another or whether he or she is indifferent between them.
   Second, we assume that the consumer's preferences are transitive.
   Third, we assume that the consumer always prefers more of a good to less.
   Fourth, we assume that, by adding a certain amount of one of the goods to the market basket that is not preferred, we can make it equally desirable in the eyes of the consumer.
5. The indifference curve is drawn below.

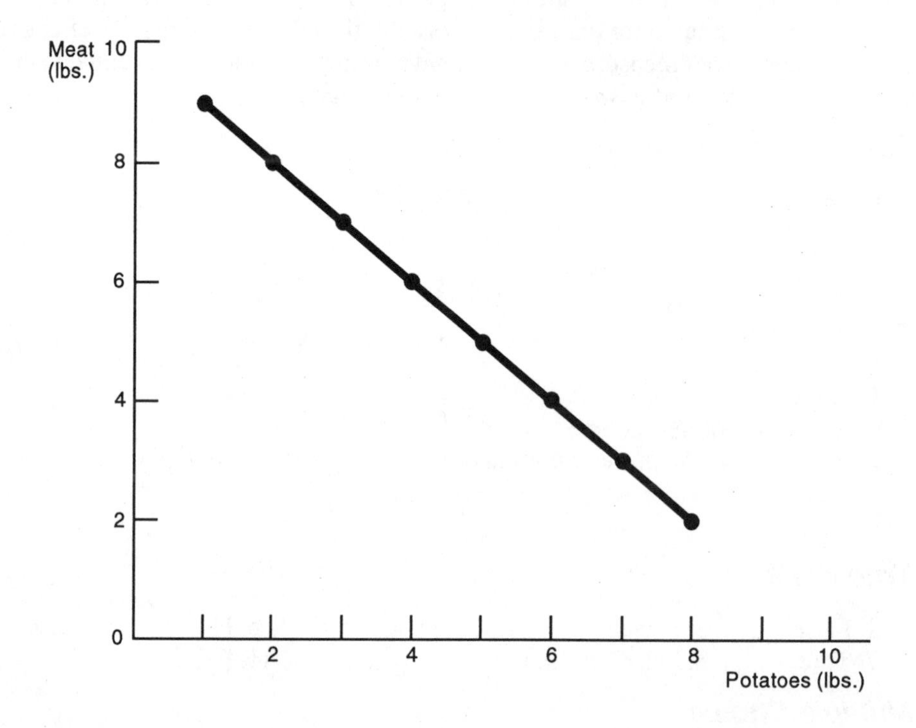

6. One.
   It does not vary at all, at least in this range.
   No.
7. Utility is a number that indicates the level of enjoyment or preference attached to a market basket.
   In the case of ordinal utility, no particular meaning or significance attaches to the scale which is used to measure utility or to the size of the difference between the utilities attached to two market baskets. In the case of cardinal utility, it is assumed that utility is measurable in the same sense as a man's height or weight is measurable.
   Ordinal utility.

8. The consumer's budget line is drawn on the graph below.

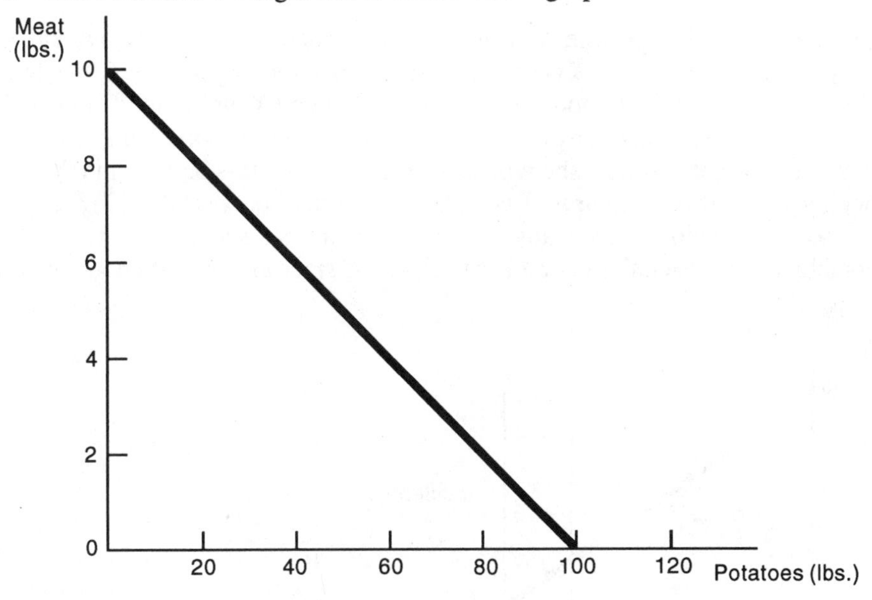

9. Under the first set of circumstances, the budget line is *A*. Under the second set of circumstances, it is *B*. And under the third set of circumstances, it is *C*.

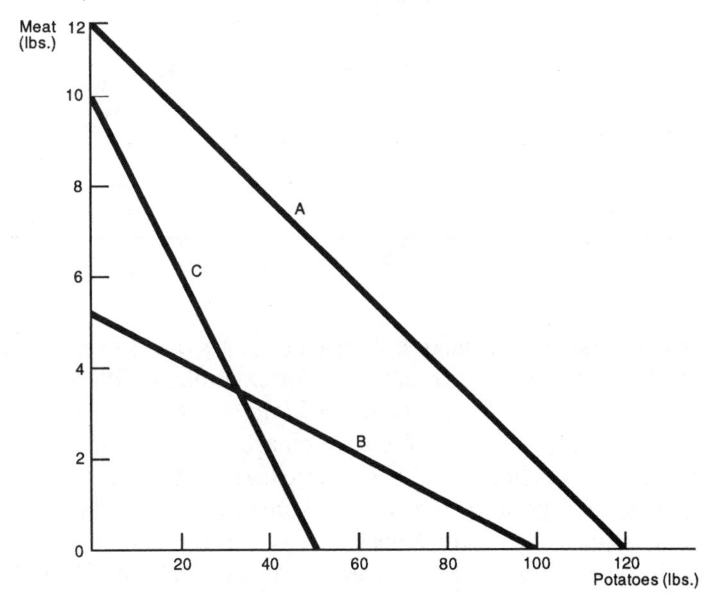

10. A consumer's tastes depend on his or her age, education, experience, and observation of others, as well as on advertising and other selling expenses, and sometimes on price.

    Conspicuous consumption refers to cases where goods are consumed because they are expensive and quality is judged by price.

Problems

1. The consumer's budget line is as shown below, since she has \$100 to spend, and the price of a unit of good *X* is \$5 and the price of a unit of good *Y* is \$10. Clearly, she can buy 20 units of good *X* if she buys only good *X*, or 10 units of good *Y* if she buys only good *Y*, or any combination of quantities of good *X* and good *Y* on this line. If she is rational, she will choose the point on this budget line that is on her highest indifference curve. Two of her indifference curves (1 and 2) are shown in the graph below. Since any unit of good *X* that she cannot consume in combination with a unit of good *Y* is useless and since any unit of good *Y* that she

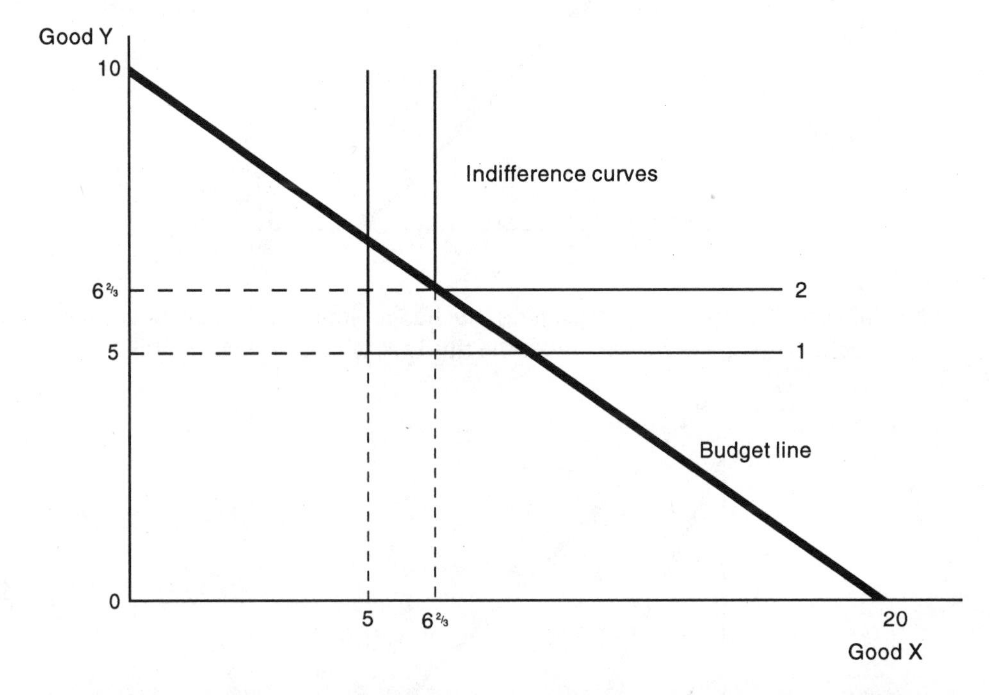

cannot consume in combination with a unit of good *X* is useless, her indifference curves have the shape shown above. (For example, indifference curve 1 shows that, if she purchases 5 units of good *X* and more than 5 units of good *Y*, her satisfaction is the same as if she purchases 5 units of each. Similarly, this indifference curve shows that, if she purchases 5 units of good *Y* and more than 5 units of good *X*, her satisfaction is the same as if she purchases 5 units of each.) Given the shape of her indifference curves, it is obvious from the graph that she will achieve her highest indifference curve if she purchases $6\frac{2}{3}$ units of each good.

2. *a.* Since the marginal utility of the second fifth of scotch to Jones is 18 utils and the marginal utility of a third fifth of gin to him is 20 utils, the change in his total utility if he trades a fifth of scotch for a fifth of gin is 20 – 18, or 2 utils. Thus, Jones would be better off if he makes this trade.

   b. If Bailey gives up 2 fifths of gin, his total utility will decrease by 7 + 5, or 12 utils (since the marginal utility of the first fifth of gin is 7 utils, and the marginal utility of the second fifth of gin is 5 utils). If he gains 1 fifth of scotch, his total utility will increase by 14 utils (since this is the marginal utility to him of a second fifth of scotch). Thus, the exchange of 2 fifths of gin for 1 fifth of scotch will change his total utility by 14 – 12, or 2 utils. Thus, Bailey would be better off to accept this deal than none at all.

3. *a.*

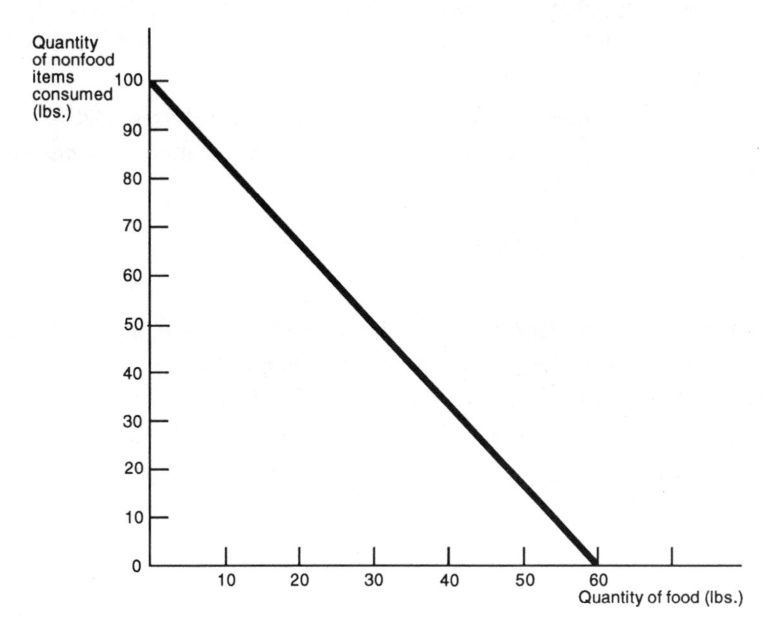

b.

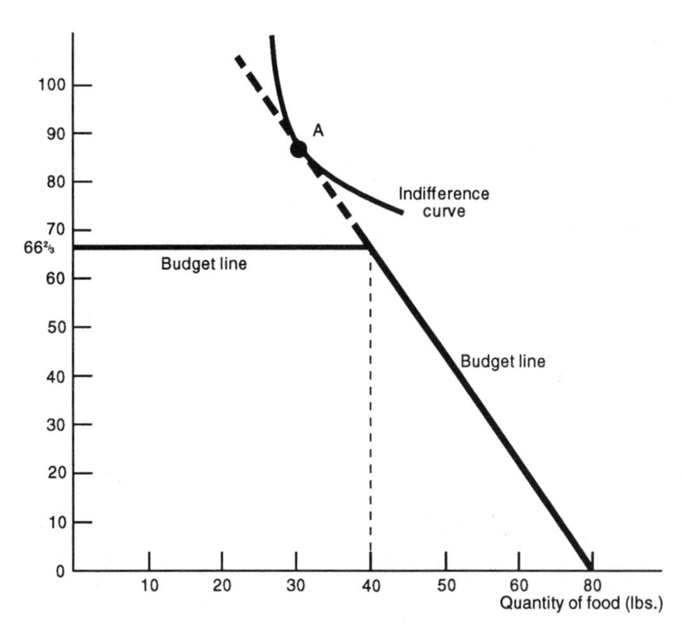

c. The family would be better off with \$100 in cash if its indifference curves were such that it could reach its highest indifference curve by choosing point *A* in part (b) above. Point *A* would be on the family's budget line if it received \$100 in cash; it is not on the family's budget line under the food stamp program.

4. He should set the ratio of the marginal utility of good *X* to its price equal to the ratio of the marginal utility of good *Y* to its price. If he buys 5 units of good *X* and 1 unit of good *Y*, the total amount spent is \$1,000, and this condition is met, since for each good this ratio equals 1 util ÷ \$10.

5. *a.* \$1,000, since the budget line intersects the vertical axis at 20.
   b. $Q_A = 20 - 0.5Q_B$, where Q_A is the quantity consumed of good *A* and Q_B is the quantity consumed of good *B*.
   c. –0.5.
   d. It must be \$1,000 ÷ 40, or \$25.
   e. 0.5.

6. If P_b is the price of beans and P_c is the price of corn, $P_bQ_b + P_cQ_c = 50$. Since $P_b = 0.50$, $0.5Q_b = 50 - Q_cP_c$. Also $\partial U/\partial Q_b \div P_b = \partial U/\partial Q_c \div P_c$. This means that $4/Q_b \div 0.5 = 1/Q_c \div P_c$. Thus, $8/Q_b = 1/Q_cP_c$, or $Q_cP_c = Q_b/8$. Since $Q_b = 100 - 2Q_cP_c$, $8Q_cP_c = 100 - 2Q_cP_c$, or $Q_cP_c = 10$. Thus the relationship is $Q_c = 10/P_c$.

7. If a consumer received increasing marginal utility from a good, he or she might spend all of his or her income on this good.

Chapter 4 Consumer Behavior and Individual Demand

Case Study: The Benefits from a New Highway

A state is considering building a new highway. John Marshall lives and works near the site of the new highway. He makes a number of trips each month between towns *A* and *B*. If the new highway were built, it would reduce the cost of each such trip from 30 cents to 20 cents. The graph below shows Mr. Marshall's individual demand curve for trips between towns A and B.

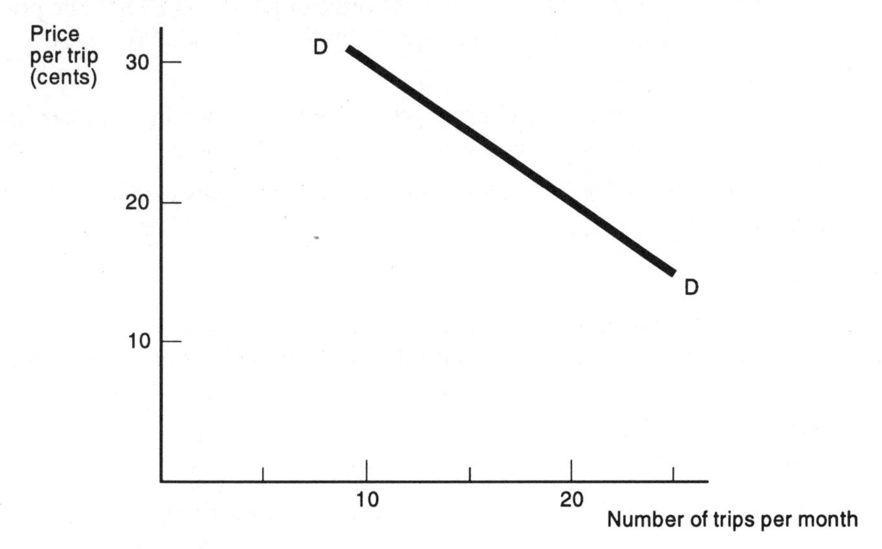

As indicated, Mr. Marshall would make 10 trips per month when the price of a trip is 30 cents.

A state legislator asks Mr. Marshall whether he favors the construction of the new highway. Mr. Marshall replies that it would save him $1 per month since he makes 10 trips per month between towns *A* and *B*, and the saving per trip is 30 cents minus 20 cents, or 10 cents. Thus, he says that the most that he would be willing to pay is $1 per month in taxes to support the new highway. If he is not asked to pay more than this amount, he favors the construction of the new highway. Otherwise he opposes it.

Do you agree with Mr. Marshall's reasoning? If not, what changes would you make in his argument?

Completion Questions

1. If the income effect of a price change is zero, but the substitution effect is nonzero, the quantity demanded of the good must be (directly, inversely, neither directly nor inversely) ______________ related to the good's price.
2. The substitution effect is always ______________.
3. Cost-of-living indexes have often been closely associated with ______________.

4. An Engel curve is the relationship between the ____________ and ____________.
5. The demand for a commodity is said to be price ______________ if the elasticity of demand exceeds 1.
6. The demand for a commodity is said to be price ______________ if the elasticity of demand is less than 1.
7. The demand for a commodity is said to be of ______________ if the price elasticity of demand is equal to 1.
8. The total effect of a change in price is the sum of the ______________ effect and the ______________ effect.
9. If the consumer is in equilibrium and if the price of good X is $2 and the price of good Y is $1, then the marginal rate of substitution of good X for good Y must equal ______________.
10. If increases in real income (with no price changes) result in increases in the consumption of a good, this good is a ______________ good.
11. If you can buy a good for $5 but you would be willing to pay as much as $50, the consumer surplus is $______________.

True or False

____ 1. If John Jones's demand curve for water is $P = 4 - Q$, where P is the price of water (in cents per gallon) and Q is the number of gallons of water demanded per day, the maximum amount that he will pay for a fourth gallon of water per day is 2 cents.

____ 2. If a good is a Giffen good, it must be inferior.

____ 3. The income-consumption curve is of no use in determining the Engel curve.

____ 4. The price-consumption curve is of use in determining the individual demand curve.

____ 5. A consumer's demand curve for a commodity generally will shift if his or her income changes.

____ 6. A consumer's demand curve for a commodity generally will shift if the prices of other commodities change.

____ 7. A consumer's demand curve for a commodity generally will shift if the consumer's tastes change a great deal.

____ 8. The price elasticity of demand is measured by the slope of the demand curve.

____ 9. If a good is price elastic, a decrease in its price will result in a decrease in the amount of money spent on it.

____ 10. If a good is price elastic, an increase in its price will result in a decrease in the amount of money spent on it.

____ 11. If the demand for a good is of unitary elasticity, the same amount of money is spent on it regardless of its price.

____ 12. Giffen's paradox is a frequent occurrence.

___ 13. Consumer surplus can never be positive.

Multiple Choice

1. If the quantity demanded equals 10 divided by the price of the commodity,
   a. the demand curve for the commodity is downward sloping to the right.
   b. the price elasticity of demand for the commodity is 1.
   c. the amount spent on the commodity is the same, regardless of its price.
   d. all of the above.
   e. none of the above.

2. The substitution effect must always be
   a. positive.
   b. negative.
   c. zero.
   d. bigger than the income effect.
   e. none of the above.

3. The income effect
   a. must always be negative.
   b. must always be positive.
   c. can be positive or negative.
   d. must be smaller than the substitution effect.
   e. none of the above.

4. Normal goods experience an increase in consumption when
   a. real income increases.
   b. real income falls.
   c. price rises.
   d. tastes change.
   e. none of the above.

5. The Laspeyres index
   a. measures the change in the cost of the market basket purchased in the original year.
   b. measures the change in the cost of the market basket purchased in the later year.
   c. always exceeds 1.
   d. always is less than 1.
   e. none of the above.

6. The Paasche index
   a. measures the change in the cost of the market basket purchased in the original year.
   b. measures the change in the cost of the market basket purchased in the later year.
   c. always exceeds 1.
   d. always is less than 1.
   e. none of the above.

Review Questions

1. According to Karl Fox, "An increase of 10 percent in the farm price of the 'average' food product would be associated with something like a 4 percent increase in the retail price and perhaps a 2 percent decrease in per capita consumption." Is the price elasticity of demand different at the farm level than at the retail level? Why?
2. James B. Hendry has pointed out, in connection with the demand for fuel, that "Fuel-burning equipment tends to be specialized and costly, and change overs are generally not made frequently." What are the implications of this fact for a household's demand for fuel?
3. Show why, if the consumer is to be in equilibrium, the marginal rate of substitution of good X for good Y must equal the ratio of the price of good X to the price of good Y.
4. Suppose the following relationship exists between a consumer's income and the amount of eggs he or she consumes:

Income (dollars per week)	*Eggs (number per week)*
100	12
150	24
200	36
250	42
300	48

Graph the consumer's Engel curve for eggs below.

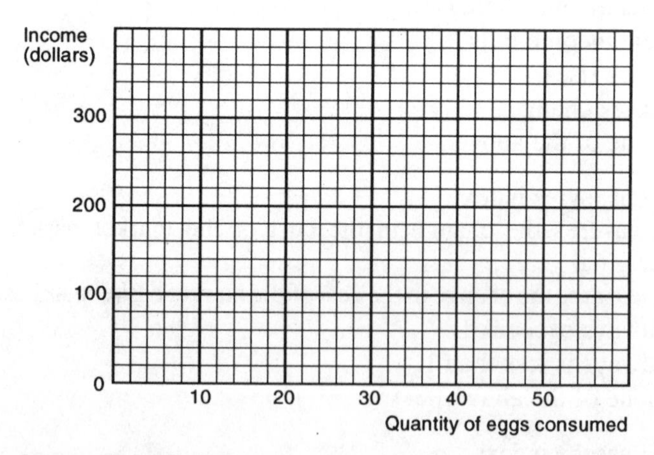

5. Describe the factors that will influence the shape of a consumer's Engel curve for a particular good.
6. Describe what is meant by a price-consumption curve. How can it be used to help determine the individual demand curve?

7. In the case of John Jones, the relationship between the price of eggs and the amount that he will purchase is shown below.

Price of eggs (cents per dozen)	*Quantity of eggs consumed per week*
50	15
60	14
70	13
80	12
90	11
100	10

Plot John Jones's individual demand curve for eggs in the graph below.

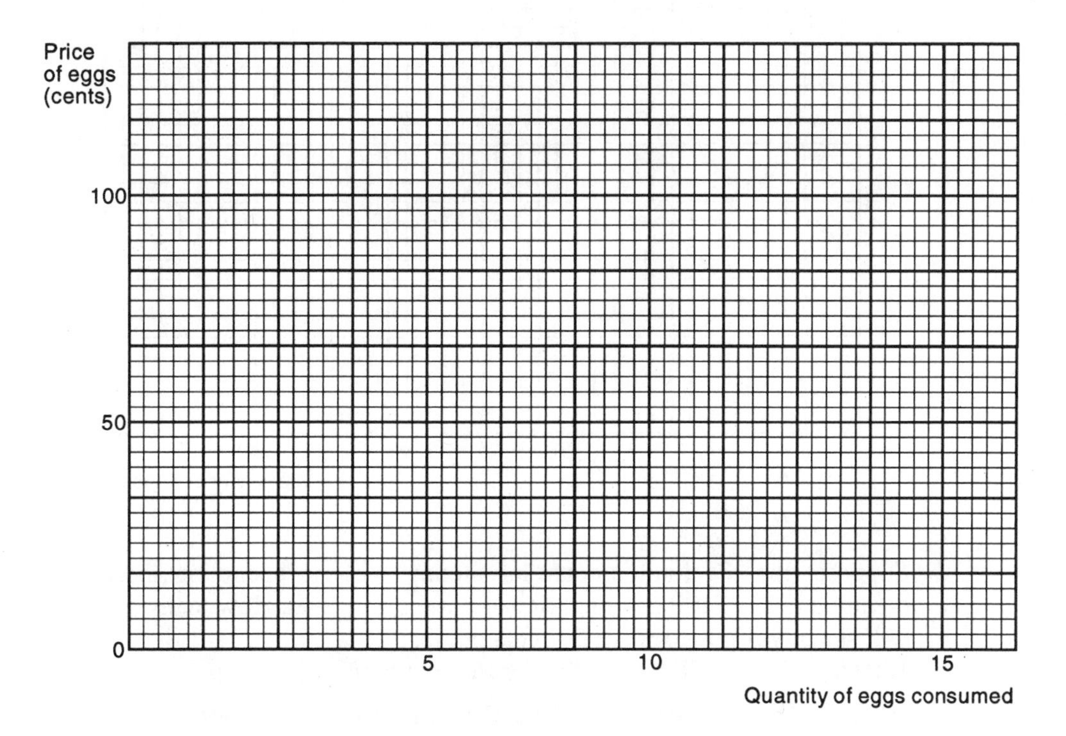

8. Explain what is meant by the "substitution effect" and the "income effect." Can the substitution effect be positive? Can the income effect be positive?
9. Explain the difference between normal and inferior goods. What is Giffen's paradox?
10. Explain the meaning of consumer surplus.

Problems

1. Suppose that a 1 percent increase in the price of pork chops results in Mrs. Smith's buying 3 percent fewer pork chops per week. What is the price elasticity of demand for pork chops on the part of Mrs. Smith? Is her demand for pork chops

price elastic or price inelastic? Will an increase in the price of pork chops result in an increase, or a decrease, in the total amount of money that she spends on pork chops?

2. *a.* Suppose Mrs. Smith's utility function can be described by $U = Q_C Q_P$. where U is her utility, Q_C is the amount of corn she consumes, and Q_P is the amount of potatoes she consumes. Draw her indifference curve when $U = 10$.

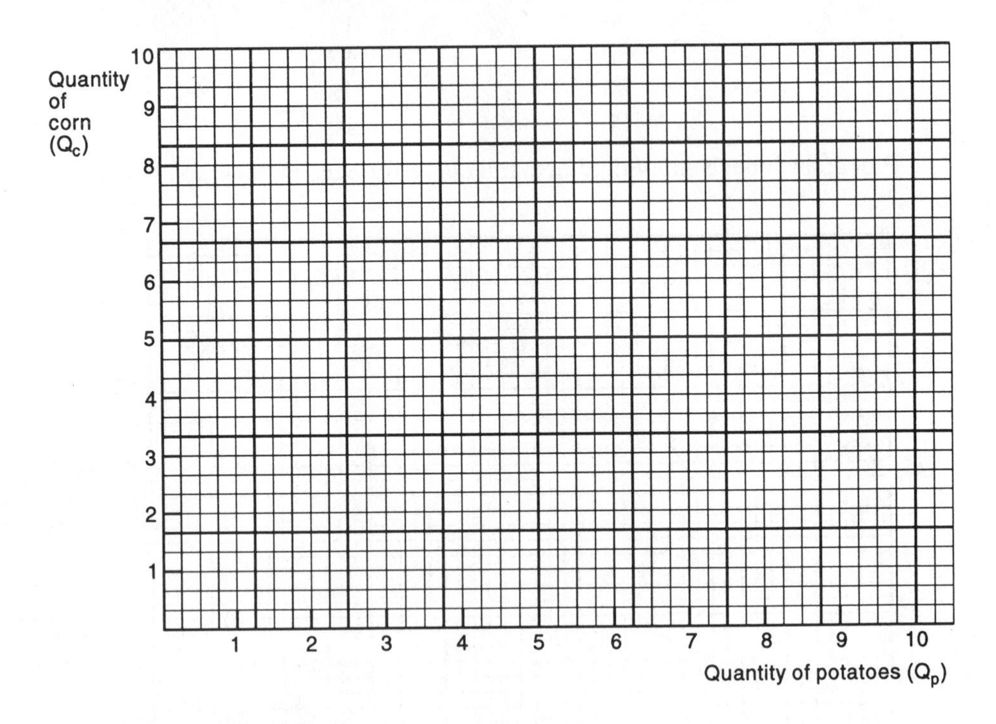

b. Suppose that the total amount of money she can spend on these two commodities is $100 and the price of corn is $1 per lb. How many potatoes will she buy if potatoes are 50 cents per lb.?

c. How much corn will she buy under these circumstances?

3. (*Advanced*) Derive a formula for Mrs. Smith's demand curve for potatoes. Let the price of potatoes be P_P and the price of corn be P_C. Let the total amount she spends on these two commodities be I. And assume that her utility function is $U = Q_C Q_P$.

4. The federal government is interested in purchasing two types of antipollution equipment. After extensive tests, government officials are convinced that two units of type A equipment are as effective as one unit of type B equipment. Assuming that the officials want to reduce pollution, draw their indifference curves for the two types of equipment on the graph at the top of the next page.

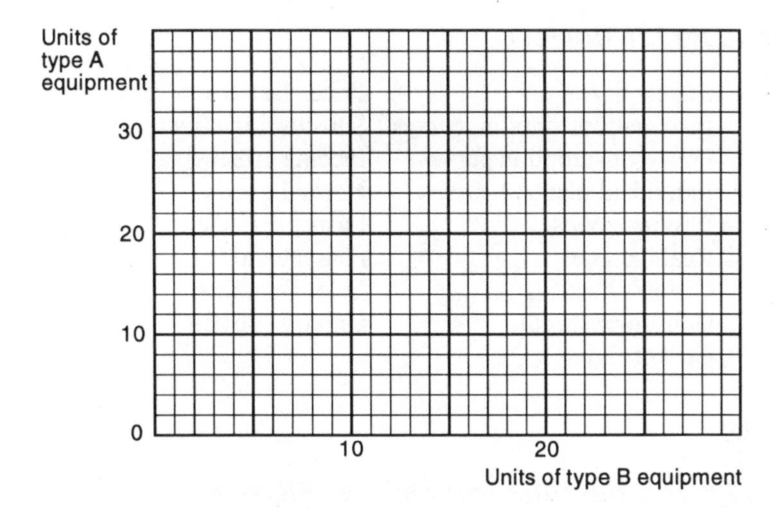

5. Assuming that the government has \$20 million to spend on either type A equipment, type B equipment, or a combination of both, draw the relevant budget line, and indicate the optimal choice of type of equipment if a unit of type A equipment costs \$1 million and a unit of type B equipment costs \$4 million.

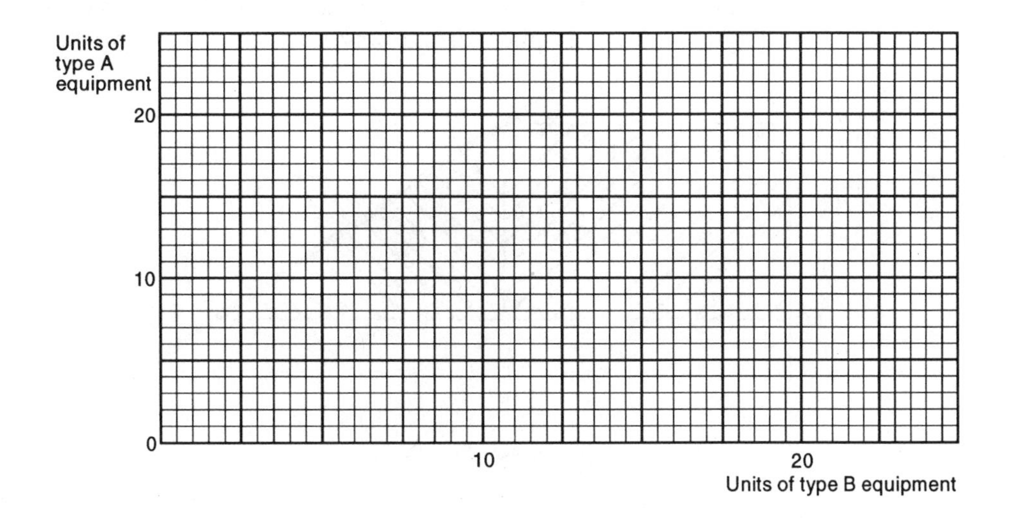

Key Concepts for Review

Engel curves	Price elasticity of demand
Income-consumption curves	Substitution effect
Price-consumption curves	Income effect
Individual demand curves	Normal goods
Inferior goods	Laspeyres index
Real income	Paasche index
Giffen's paradox	Index numbers
Consumer surplus	Unitary elasticity
Cost of living	

ANSWERS

Case Study: The Benefits from a New Highway

Mr. Marshall's reasoning is fallacious because he takes no account of the effect on the number of trips he makes of the reduction in the cost of a trip. If the cost of a trip is 30 cents, he makes 10 trips per month. The new highway reduces the cost of these trips, the saving being the area *ABCE*, which equals $1.00. In addition, however, Mr. Marshall will make 10 extra trips per month if the cost is 20 cents, and the consumer surplus from these trips is equal to the area *BCF*. In other words, he would be willing to pay an amount equal to the area *BCF* (in addition to the amount he does pay) for the extra 10 trips. The total benefit to Mr. Marshall is the sum of the two shaded areas in the following diagram (that is, it equals *ABFE*), not *ABCE* alone. Thus, Mr. Marshall

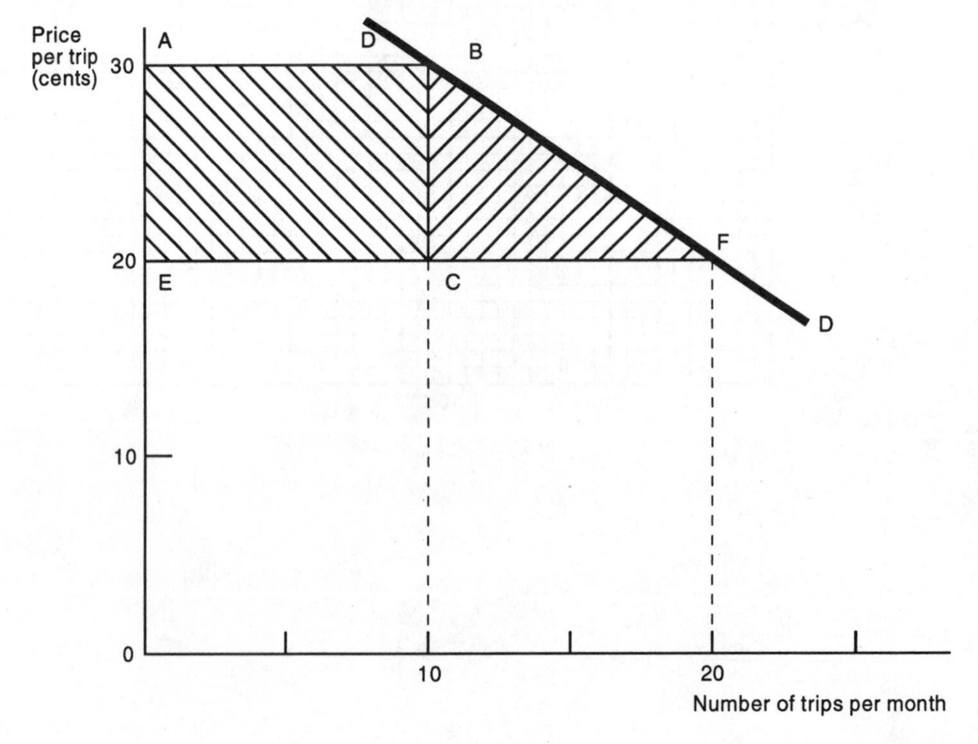

is underestimating the benefit to him from the new highway. Since the area *BCF* equals (1/2)(10) (10) cents, or 50 cents, the total area *ABFE* equals $1.50. This, not $1.00, is the maximum amount he should be willing to pay per month for the new highway.

Completion Questions

1. inversely
2. negative
3. inflation
4. consumer's income; the amount of a good that he or she demands
5. elastic
6. inelastic
7. unitary elasticity
8. substitution; income
9. 2
10. normal
11. $45

True or False

1. False	2. True	3. False	4. True	5. True	6. True
7. True	8. False	9. False	10. True	11. True	12. False
13. False					

Multiple Choice

1. *d*	2. *b*	3. *c*	4. *a*	5. *a*	6. *b*

Review Questions

1. Yes.

   Because the farm price and the retail price do not vary proportionately.

2. The price elasticity of demand for a particular fuel (like oil or natural gas) is probably greater in the long run than in the short run.

3. The marginal rate of substitution is the rate at which the consumer is willing to substitute good *X* for good *Y*, holding his or her total level of satisfaction constant. Thus, if the marginal rate of substitution is 3, the consumer is willing to give up 3 units of good *Y* in order to get 1 more unit of good *X*.

   If the consumer is in equilibrium, the rate at which the consumer is willing to substitute good *X* for good *Y* (holding satisfaction constant) must equal the rate at which he or she is able to substitute good *X* for good *Y*. Otherwise it is always possible to find another market basket that will increase the consumer's satisfaction.

4. The Engel curve is as shown at the top of the next page.

5. The shape of a consumer's Engel curve for a particular good will depend on the nature of the good, the nature of the consumer's tastes, and the level at which prices are held constant. For example, Engel curves for salt or shoelaces would generally show that the consumption of these commodities does not increase very much in response to increases in income. But goods like caviar or filet mignon

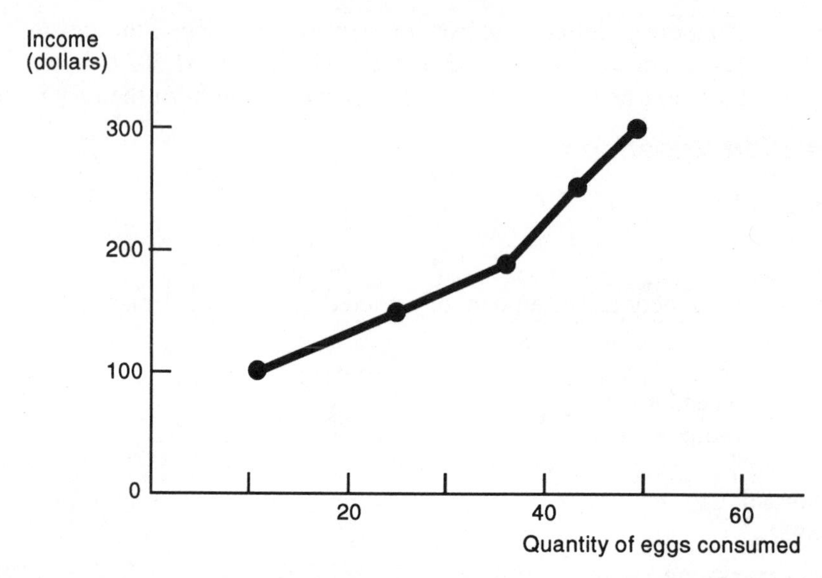

might be expected to have Engel curves showing that their consumption is much more sensitive to changes in income.

6. A price-consumption curve connects the various points at which the budget line is tangent to a consumer's indifference curves, given the level of his income and the price of good *Y*. (Only the price of good X varies.) Reading off the amount consumed of good *X* at each price of good *X*, one can get from the price-consumption curve the basic data needed to formulate the individual demand curve.

7. The demand curve is as follows:

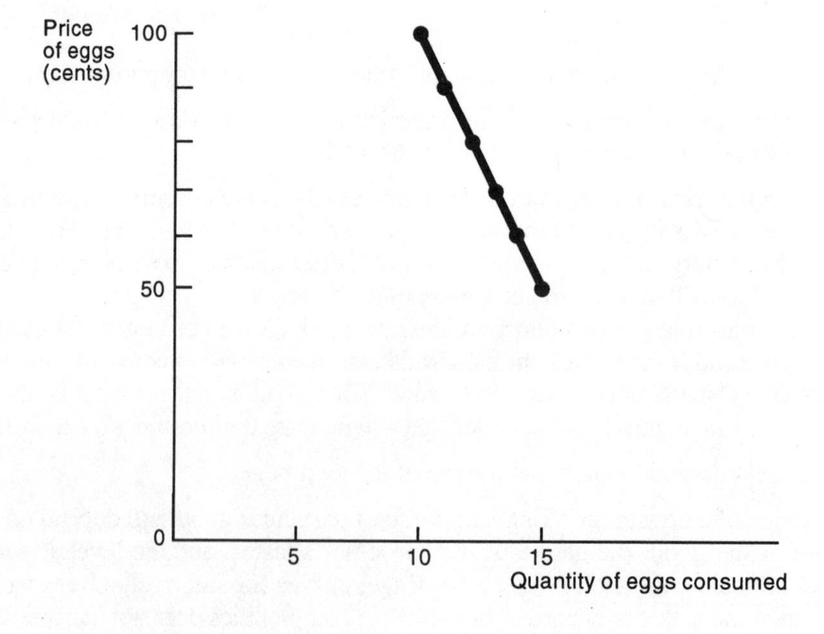

8. The substitution effect is the change in quantity demanded of a good resulting from a change in its price, when the level of satisfaction, or real income, is held constant. The income effect is the change in quantity demanded of a good due entirely to a change in real income, all prices being held constant.
   No, the substitution effect cannot be positive.
   Yes, the income effect can be positive.

9. Normal goods are goods where increases (decreases) in real income result in increases (decreases) in consumption of the good. Inferior goods are goods where the opposite is true.
   Giffen's paradox occurs when an inferior good's income effect is powerful enough to offset the substitution effect, the result being that quantity demanded is positively related to price, at least over some range of variation of price.

10. Consumer surplus is the difference between the maximum amount that a consumer would pay and the amount that he or she actually pays.

Problems

1. 3.
   Price elastic.
   Decrease.

2. *a.* This indifference curve is:

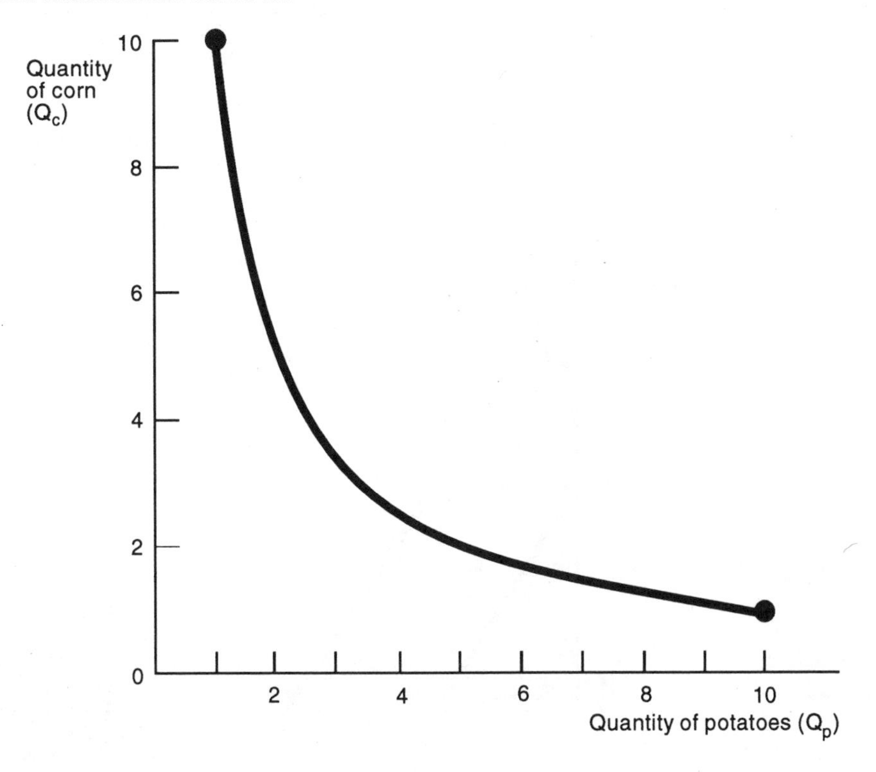

b. The budget line is $Q_C + 0.5Q_P = 100$. This line is tangent to an indifference curve when $Q_P = 100$.

c. The budget line in part (*b*) is tangent to an indifference curve when $Q_C = 50$ (and $Q_P = 100$).

3. Since $\partial U/\partial Q_C = Q_P$ and $\partial U/\partial Q_P = Q_C$, $Q_P \div P_C = Q_C \div P_P$. Moreover, $P_CQ_C + P_PQ_P = I$. Thus $2P_PQ_P = I$, and the demand curve is

$$P_P = I \div 2Q_P$$

4. The indifference curves are as follows:

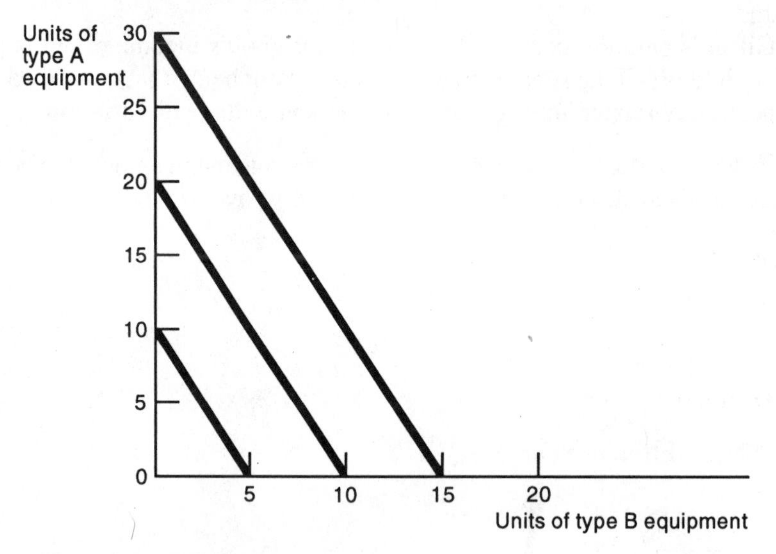

5. The budget line is the broken line.
Given the indifference map of solid lines, the optimal point is *A*, where 20 units of type A equipment are purchased.

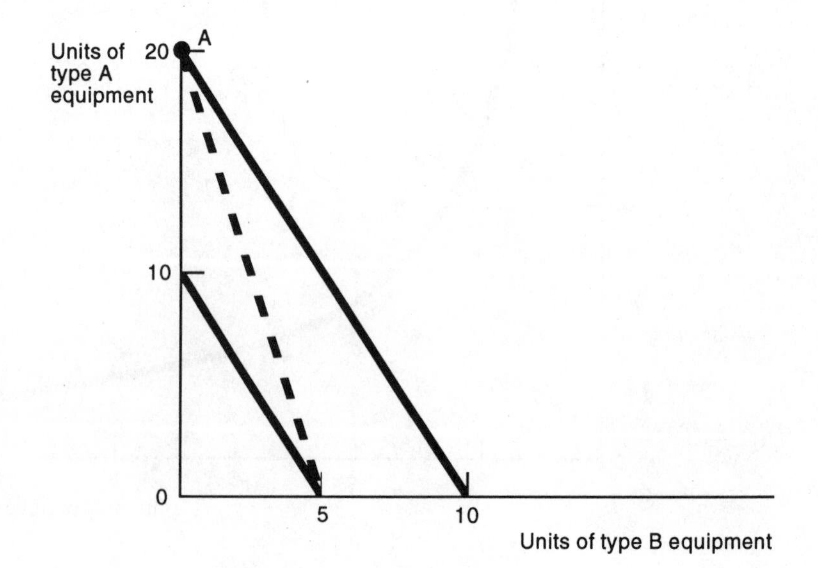

Chapter 5 Market Demand

Case Study: The Demand for Oranges

The Economic Research Service of the U.S. Department of Agriculture has reported the results of a study of the effects of the price of various types of oranges on the rate at which they were purchased.[1] In particular, three types of oranges were studied: (1) Florida Indian River, (2) Florida Interior, and (3) California. In nine test stores in Grand Rapids, Michigan, the researchers varied the price of each of these types of oranges for a month. The effect of a 1 percent increase in the price of each type of orange on the rate of purchase of this and each of the other types of oranges is shown below. For example, a 1 percent increase in the price of Florida Indian River oranges (holding other prices constant) seemed to result in a 3.1 percent decrease in the rate of purchase of Florida Indian River oranges, a 1.6 percent increase in the rate of purchase of Florida Interior oranges, and a 0.01 increase in the rate of purchase of California oranges.

	Results in the following percentage change in the rate of purchase of:		
A 1 percent increase in the price of:	*Florida Indian River*	*Florida Interior*	*California*
Florida Indian River	–3.1	+1.6	+0.01
Florida Interior	+1.2	–3.0	+0.1
California	+0.2	+0.1	–2.8

a. What seems to be the price elasticity of demand for each type of orange?
b. What seems to be the cross elasticity of demand for each pair of types of oranges?
c. Which types of oranges seem to be the closest substitutes?
d. Of what use might these results be to orange producers?
e. How accurate do you think that this study was? What improvements would you make in it?

Completion Questions

1. If the government imposes a tax on coal of 10 cents per ton, it will obtain the most revenue from the tax if coal's price elasticity of demand equals ________________. The largest burden of the tax is borne by consumers of coal if the price elasticity of demand equals________________.
2. If the cross elasticity of demand between goods *X* and *Y* is positive, these goods are classified as ________________.
3. The income elasticity of demand is the percentage change in quantity demanded resulting from a ________________ change in money income.

1. M.Godwin, W. Chapman, and W. Manley, *Competition Between Florida and California Valencia Oranges in the Fresh Market*, (Department of Agriculture). I have changed the numbers slightly.

4. Luxury goods are generally assumed to have a ________________ income elasticity of demand
5. If a commodity has many close substitutes its demand is likely to be ________________.
6. The price elasticity of demand equals ________________.
7. Engel's law states that ________________.
8. The demand curve for the individual firm under perfect competition is ________________.
9. The price elasticity of demand for nondurable goods often is ________________ in the long run than in the short run.
10. The total amount of money spent by consumers on a commodity equals the industry's ________________.
11. The ________________ curve shows marginal revenue at various quantities of output.
12. If the industry is not perfectly competitive, the firm's demand curve will not be ________________.
13. Direct experimentation can be a ____________ way to obtain data concerning a firm's or product's demand curve.

True or False

____ 1. The demand for open-heart surgery is likely to be less price elastic than the demand for aspirin.

____ 2. If a good's income elasticity exceeds 1, a decrease in the price of the good will increase the total amount spent on it.

____ 3. Summing horizontally the individual demand curves for all of the consumers in the market will produce the demand curve for the market.

____ 4. The demand curve for an individual firm under perfect competition is downward sloping, its slope being –1.

____ 5. The market demand curve for a product under perfect competition is horizontal.

____ 6. The demand for salt and pepper is likely to be price elastic.

____ 7. In general, demand is likely to be more inelastic in the long run than in the short run.

____ 8. The income elasticity of demand for food is very high.

____ 9. It is always true that $\eta_{xy} = \eta_{yx}$.

____ 10. The direct approach of simply asking people how much they would buy of a particular commodity is the best way to estimate the demand curve.

____ 11. The identification problem is the problem of identifying the person who knows what the demand curve looks like.

___ 12. The income elasticity of demand will always have the same sign regardless of the level of income at which it is measured.

___ 13. Marginal revenue is the ratio of the value of sales to the amount sold.

___ 14. When the demand curve is linear, the slope of the marginal revenue curve is twice (in absolute value) the slope of the demand curve.

Multiple Choice

1. The president of a leading producer of tantalum says that an increase in the price of tantalum would have no effect on the total amount spent on tantalum. If this is true, the price elasticity of demand for tantalum is
   a. less than zero.
   b. 1.
   c. 2.
   d. more than 1.
   e. none of the above.

2. The demand for a good is price inelastic if
   a. the price elasticity is 1.
   b. the price elasticity is less than 1.
   c. the price elasticity is greater than 1.
   d. all of the above.
   e. none of the above.

3. The relationship between marginal revenue and the price elasticity of demand is
   a. $MR = P(1 - 1/\eta)$.
   b. $P = MR(1 - 1/\eta)$.
   c. $P = MR(1 + \eta)$.
   d. $MR = P(1 + \eta)$.
   e. none of the above.

4. A demand curve with unitary elasticity at all points is
   a. a straight line.
   b. a parabola.
   c. a hyperbola.
   d. all of the above.
   e. none of the above.

5. Suppose we are concerned with the relationships between the quantity of food demanded and aggregate income. It seems most likely that this relationship will look like
   a. curve *A* on the next page.
   b. curve *B* on the next page.
   c. curve *C* on the next page.
   d. the vertical axis.
   e. the horizontal axis.

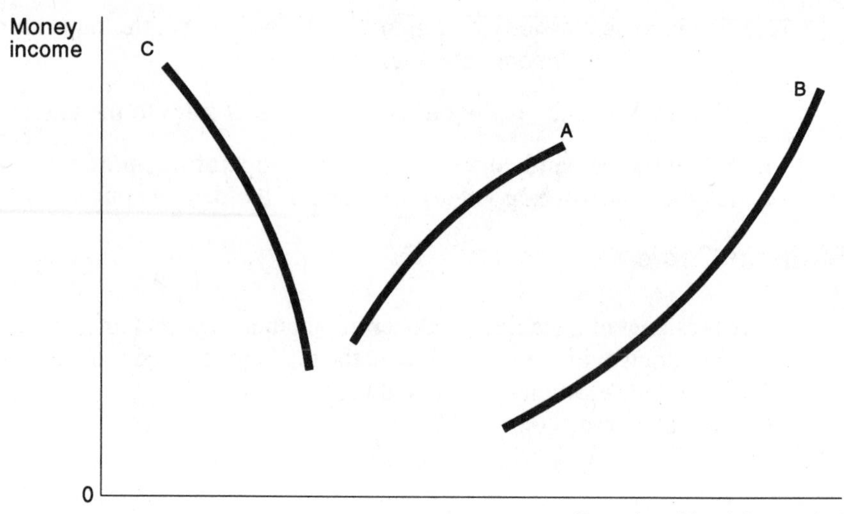

6. If goods X and Y are substitutes, the relationship between the quantity demanded of good X and the price of good Y should be like
   a. curve *A* below.
   b. curve *B* below.
   c. the vertical axis.
   d. the horizontal axis.
   e. none of the above.

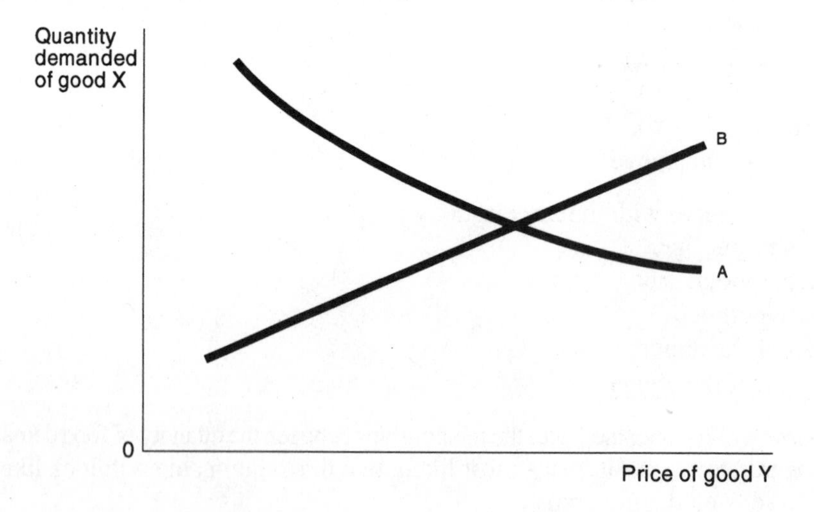

Review Questions

1. According to the president of Bethlehem Steel, the demand for steel is price inelastic because steel generally constitutes a very small percentage of the total cost of the product that includes it as a raw material.[2] If this is the case, will a price increase result in an increase or decrease in the amount of money spent on steel?

2. What is meant by a market? How can one derive the market demand curve from the demand curves of the individuals comprising the market?

3. What is the difference between the point elasticity of demand and the arc elasticity of demand?

4. Suppose that the relationship between the price of steel and the quantity of steel demanded is as follows:

Price (dollars)	*Quantity*
1	8
2	7
3	6
4	5
5	4

   What is the arc elasticity of demand when price is between $1 and $2? Between $2 and $3? Between $4 and $5?

5. Discuss in detail the determinants of the price elasticity of demand.

6. Define the income elasticity of demand. How does the income elasticity of demand differ between luxuries and necessities? What does Engel's law state?

7. Define the cross elasticity of demand. How does the cross elasticity of demand differ between substitutes and complements?

8. If the relationship between price and quantity is as given bellow, derive the marginal revenue at various quantities and plot it in the graph on the next page.

Price (dollars)	*Quantity*
10	1
9	2
8	3
7	4
6	5
5	6
4	7
3	8
2	9

9. What does the firm's demand curve look like under perfect competition? Why?

2. W. Adams, "The Steel Industry."

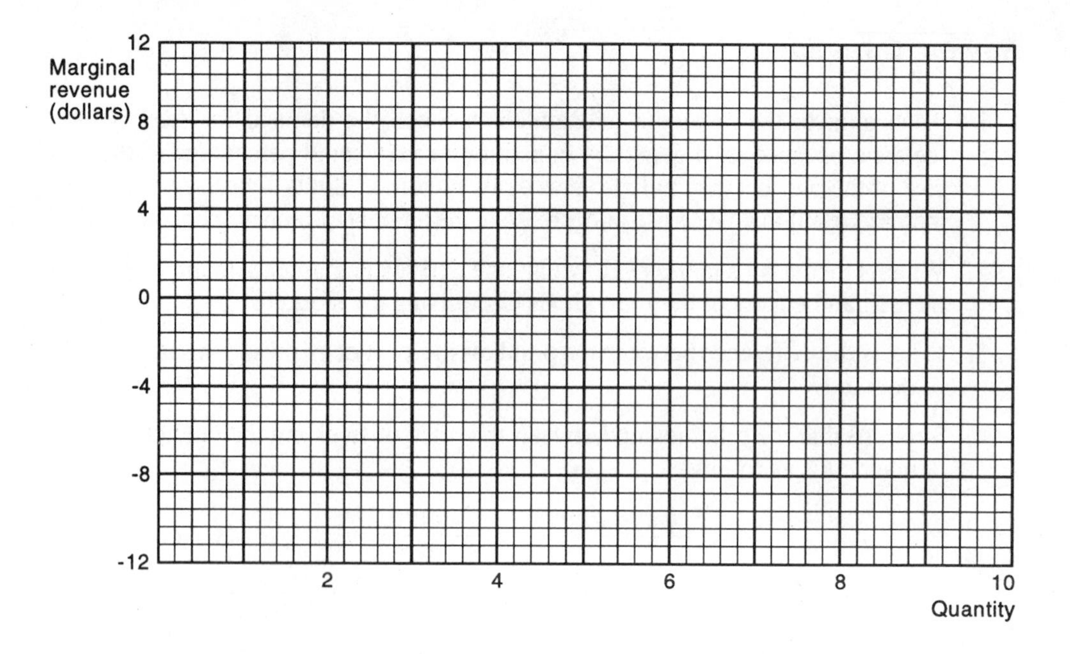

10. (*Advanced*) Use calculus to derive marginal revenue as a function of price (p) and the price elasticity of demand (η).
11. Discuss various ways that demand curves can be measured.
12. Suppose the Mayor of New York asked you to advise him concerning the proper fare that should be charged by the New York City subway. In what way might information concerning the price elasticity of demand be useful?
13. Suppose you are a business consultant and you become convinced that the U.S. steel industry underestimates the price elasticity of demand for steel. In what way might this information be useful to the steel companies? To the public?
14. Suppose you are a trustee of a major university. At a meeting of the board of trustees, one university official argues that the demand for places at this university is completely inelastic. As evidence, he cites the fact that, although the university has doubled its tuition in the last decade, there has been no appreciable decrease in the number of students enrolled. Do you agree? Comment on his argument.
15. Studies from cross-section data indicate that the income elasticity of demand for servants in the United States exceeds 1. Yet the number of servants has been decreasing during the last 50 years, while incomes have risen in the United States. How can these facts be reconciled?
16. *a.* According to Gregory Chow of Princeton University, the price elasticity of demand for automobiles in the United States is 1.2, and the income elasticity of demand for automobiles is 3. What would be the effect of a 3 percent decline in auto prices on the quantity of autos demanded, assuming Chow's estimates are right?

    b. What would be the effect of a 2 percent increase in income?

17. According to the Swedish economist Herman Wold's estimates, the income elasticity of demand for liquor is about 1. If you were an executive of a liquor firm, of what use might this fact be to you in forecasting sales?

Problems

1. *a.* According to J. Fred Bucy, former president of Texas Instruments, his firm continually makes detailed studies of the price elasticity of demand for each of its major products in order to determine how much its sales will increase if it changes its price by a particular amount. For example, Texas Instruments had to estimate the effect of a 10 percent reduction in the price of the TI-55, a hand calculator that Texas Instruments produces, and whether such a price reduction would increase sales by a large enough amount to be profitable. What sorts of techniques can Texas Instruments use to make such estimates?

   b. In the electronics industry, prices for many products have tended to drop dramatically. During the 1970s, simple four-function hand calculators declined in price from about $150 to less than $10. Other products that experienced similar price reductions were transistor radios and digital watches. Do the costs of producing such products tend to fall as larger and larger quantities of them are produced?

   c. In the 1980's, Texas Instruments cut the price of its 99/4A home computer from $299 to $199, and its rivals followed suit. If the price elasticity of demand was greater than 1, did the price cut increase the amount spent on such computers?

2. The price elasticity of good *Y* is 2, and marginal revenue for good *Y* is $2. What is the price of good *Y*?

3. (*Advanced*) The demand curve for screwdrivers shifts from D_1 in 1994 to D_2 in 1995. Use calculus to prove that the price elasticity of demand at any price less than $10 will be the same as it was before the shift in the demand curve.

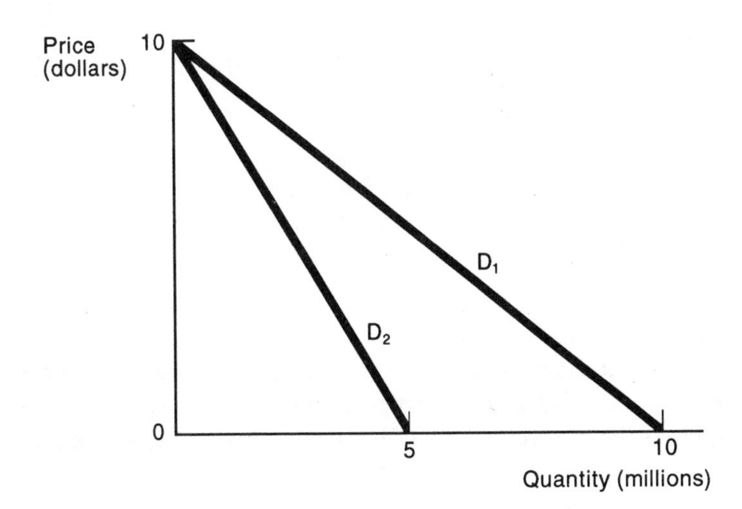

4. The Somalia Manufacturing Company believes that the demand curve for its product is

$$P = 5 - Q$$

where P is the price of its product (in dollars) and Q is the number of millions of units of its product sold per day. It is currently charging a price of $1 per unit for its product.

   a. The president of the firm asks you to comment on the wisdom of its pricing policy. Write a short paragraph on this score.

   b. A marketing specialist says that the price elasticity of demand for the firm's product is 1.0. Do you agree? Why or why not?

5. There are 15 people comprising the market for good X. The quantity of good X that each person demands at each price is shown below.

	Price of good X (dollars per pound)				
Person	*1*	*2*	*3*	*4*	*5*
	Pounds of good X per year				
Allen	20	18	16	14	12
Bach	15	13	11	9	7
Creamer	100	90	80	70	60
Dow	30	28	26	24	22
Easton	10	9	8	7	6
Farnsworth	5	4	3	2	1
Grettle	60	58	56	54	52
Howe	6	6	6	6	6
Irwin	18	17	14	10	6
Jarvis	12	11	10	9	8
Koch	13	13	12	12	12
Leaton	11	10	10	9	9
Martin	5	4	3	3	3
Nash	90	80	70	65	60
Otter	12	11	11	10	10

   a. Calculate the market demand curve for good X.

   b. What is the price elasticity of demand when
      (1) the price is between $1 and $2 per pound?
      (2) the price is between $3 and $4 per pound?

6. Suppose that Jeremy Jarvis considers Geritol of supreme importance and that he spends all of his income on Geritol.

   a. To this consumer, what is the price elasticity of demand for Geritol?

   b. What is the income elasticity of demand for Geritol?

   c. What is the cross elasticity of demand between Geritol and any other good?

7. Which of the following are likely to have a positive cross elasticity of demand:
   a. automobiles and oil?
   b. wood tennis rackets and metal tennis rackets?
   c. gin and tonic?
   d. fishing poles and fishing licenses?
   e. a Harvard education and a Stanford education?
8. Show that, if the Engel curve for a good is a straight line through the origin, the income elasticity of demand for the good is 1.
9. According to Richard Tennant, "The consumption of cigarettes is . . . [relatively] insensitive to changes in price. . . . In contrast, the demand for individual brands is highly elastic in its response to price. . . . In 1918, for example, Lucky Strike was sold for a short time at a higher retail price than Camel or Chesterfield and rapidly lost half its business." Explain why the demand for a particular brand is more elastic than the demand for all cigarettes. If Lucky Strike raised its price by 1 percent in 1918, was the price elasticity of demand for its product greater than 2?
10. *a.* According to S. Sackrin of the U S. Department of Agriculture, the price elasticity of demand for cigarettes is between 0.3 and 0.4, and the income elasticity of demand is about 0.5. Suppose the federal government, influenced by findings that link cigarettes and cancer, were to impose a tax on cigarettes that increased their price by 10 percent. What effect would this have on cigarette consumption?
    b. Suppose a brokerage house advised you to buy cigarette stocks because, if incomes rose by 50 percent in the next decade, cigarette sales would be bound to double. What would be your reaction to this advice?

Key Concepts for Review

Market	Marginal revenue
Market demand curve	Marginal revenue curve
Price elasticity of demand	Industry demand curve
Total revenue	Firm demand curve
Income elasticity of demand	Perfect competition
Engel's law	Identification problem
Cross elasticity of demand	Direct experimentation
Substitute	Consumer clinics
Complement	Horizontal demand curve

ANSWERS

Case Study: The Demand for Oranges

a. The price elasticity of demand for Florida Indian River oranges seems to be 3.1; the price elasticity of demand for Florida Interior oranges seems to be 3.0; and the price elasticity of demand for California oranges seems to be 2.8.

b. The cross elasticities (η_{xy}) are as follows:

		X	
Y	*Florida Indian River*	*Florida Interior*	*California*
Florida Indian River	—	1.6	0.01
Florida Interior	1.2	—	0.1
California	0.2	0.1	—

c. Clearly, Florida Indian River and Florida Interior oranges are closer substitutes than the Florida and California oranges.

d. The fact presented in part (*c*) is of obvious use to orange growers in both parts of the country.

e. The study is limited, of course, by the fact that it pertains to only one city at only one relatively short period of time.

Completion Questions

1. zero; zero
2. substitutes
3. 1 percent
4. high
5. price elastic
6. $-\frac{\Delta Q}{Q} \div \frac{\Delta P}{P}$
7. the income elasticity of food is low
8. horizontal
9. higher
10. total revenue
11. marginal revenue
12. horizontal
13. risky

True or False

1. True	2. False	3. True	4. False	5. False	6. False
7. False	8. False	9. False	10. False	11. False	12. False
13. False	14. True				

Multiple Choice

1. *b*	2. *b*	3. *a*	4. *c*	5. *b*	6. *b*

Review Questions

1. It will result in an increase in the amount of money spent on steel.

2. A market is a group of firms and individuals that are in touch with each other in order to buy or sell some good. Basically, all markets consist primarily of buyers and sellers, although third parties like brokers and agents may be present as well. The market demand curve is simply the horizontal summation of the individual demand curves of all the consumers in the market.

3. If ΔP is very small, we can compute the point elasticity of demand, which is $-(\Delta Q/Q) \div (\Delta P/P)$.

   If we have data concerning only large changes in price, we can compute the arc elasticity of demand which is $-\Delta Q(P_1 + P_2)/\Delta P(Q_1 + Q_2)$.

4.

$$-\frac{\Delta Q\,(P_1+P_2)}{\Delta P\,(Q_1+Q_2)} = \frac{1\,(1+2)}{1\,(8+7)} = \frac{3}{15} = .20$$

$$-\frac{\Delta Q\,(P_1+P_2)}{\Delta P\,(Q_1+Q_2)} = \frac{1\,(2+3)}{1\,(7+6)} = \frac{5}{13} = .38$$

$$-\frac{\Delta Q\,(P_1+P_2)}{\Delta P\,(Q_1+Q_2)} = \frac{1\,(4+5)}{1\,(5+4)} = \frac{9}{9} = 1.00$$

5. First, and foremost, the price elasticity of demand for a commodity depends on the number and closeness of the substitutes that are available. If a commodity has many close substitutes, its demand is likely to be price elastic.

   The extent to which a commodity has close substitutes depends on how narrowly it is defined. In general, one would expect that, as the definition of the market becomes narrower and more specific, the product will have more close substitutes and its demand will become more price elastic.

   Second, it is sometimes asserted that the price elasticity of demand for a commodity is likely to depend on the importance of the commodity in consumers' budgets.

   Third, the price elasticity of demand for a commodity is likely to depend on the length of the period to which the demand curve pertains.

6. The income elasticity of demand is $(\Delta Q/Q) \div (\Delta I/I)$, where Q is quantity demanded and I is income.

   One would expect the income elasticity of demand generally to be higher for luxuries than for necessities.

   Engel's law states that better-off nations spend a smaller proportion of their incomes on food than do poorer nations, the income elasticity of demand for food being quite low.

7. The cross elasticity of demand is the percentage change in the quantity demanded of good X resulting from a 1 percent change in the price of good Y. The cross elasticity of demand is positive if goods X and Y are substitutes and negative if goods X and Y are complements.

8. The following table derives marginal revenue:

Quantity	*Price (dollars)*	*Total revenue (dollars)*	*Marginal revenue (dollars)*
1	10	10	
2	9	18	18 - 10 = 8
3	8	24	24 - 18 = 6
4	7	28	28 - 24 = 4
5	6	30	30 - 28 = 2
6	5	30	30 - 30 = 0
7	4	28	28 - 30 = -2
8	3	24	24 - 28 = -4
9	2	18	18 - 24= -6

The marginal revenue curve is as follows:

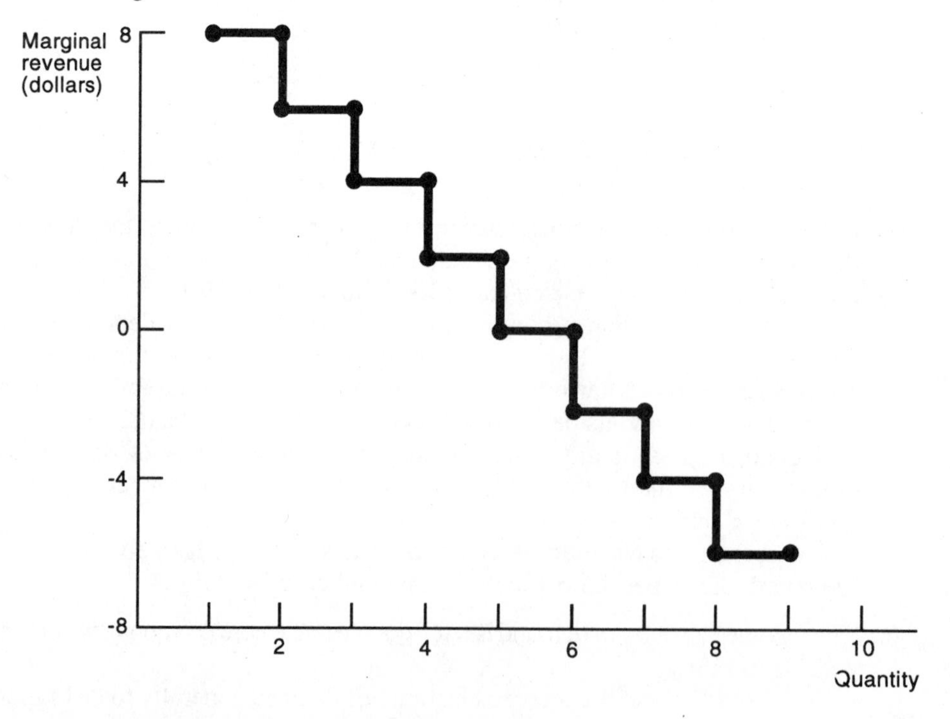

9. The firm's demand curve under perfect competition is a horizontal line.
   This is caused by the fact that the firm can sell all it wants without influencing the price. A very small decrease in price would result in an indefinitely large increase in the quantity it could sell, and a very small increase in its price would result in its selling nothing.

10. For students with a grasp of elementary calculus, the derivation is

$$MR = p + q\frac{dp}{dq}$$
$$= p\left(1 + \frac{q}{p}\frac{dp}{dq}\right)$$
$$= p\left(1 - \frac{1}{\eta}\right)$$

11. One method is direct market experimentation; another is to interview consumers and administer questionnaires concerning buying habits; another is to use statistical methods to extract information from past market data. Each of these methods has its problems, but the difficulties, although sometimes formidable, are not insoluble.

12. It would indicate the extent to which fare increases would decrease subway travel. For example, William Vickrey's data seem to indicate that fare increases would increase total revenue since demand is price inelastic. Obviously this is an important fact.

13. It might indicate that both the companies and the public might be better off if steel prices were lowered somewhat, since the quantity demanded might expand more than heretofore expected.

14. Other factors—notably the general level of prices and incomes and the quality of the students—have not been held constant. Holding these factors—and the tuition rates at other universities—constant, it is almost surely untrue that large increases in tuition at this university would not cut down on the number of students demanding admission to the university.

15. There have been changes in the price of servants, the distribution of income, tastes, and other relevant factors. Taking these changes into account, there is no contradiction.

16. *a.* There would be a 3.6 percent increase in quantity demanded.
    b. There would be a 6 percent increase in quantity demanded.

17. All other things equal, you could expect that a 1 percent increase in income would result in about a 1 percent increase in the total quantity of liquor demanded.

Problems

1. *a.* Statistical and econometric analyses, as well as experiments.
   b. Yes.
   c. Yes.

2. Since $MR = P(1 - 1/\eta)$, it follows that $P = MR \div (1 - 1/\eta)$. In this case, $MR = \$2$ and $\eta = 2$; thus, $P = \$2 \div (1 - 1/2)$, or \$4. (*MR* equals marginal revenue, *P* equals price, and η equals the price elasticity of demand.)

3. The price elasticity of demand is $-(dQ/dP)(P/Q)$. If the demand curve is D_1,

$$\frac{dQ}{dP} = -1$$

Thus,

$$-\frac{dQ}{dP}\frac{P}{Q} = \frac{P}{Q} = \frac{P}{10-P}$$

If the demand curve is D_2,

$$\frac{dQ}{dP} = \frac{-1}{2}$$

Thus,

$$-\frac{dQ}{dP}\frac{P}{Q} = \frac{P}{2Q} = \frac{P}{10-P}$$

Since the price elasticity of demand equals $P/(10 - P)$ in each case, it must be the same for both demand curves if the price is the same. (Note that if the demand curve is D_1, $Q = 10 - P$; and if it is D_2, $Q = 5 - ½P$. This is clear from the diagram in the question.)

4. *a.* If P equals 1, the demand for the firm's product is inelastic. To see that this is the case, note that, if $P = 1$, $Q = 5 - 1 = 4$; thus, the firm's total revenue is \$4 million per day. If $P = 2$, $Q = 5 - 2 = 3$; thus, the firm's total revenue is \$6 million per day. Since an increase in price (from \$1 to \$2) increases total revenue, demand must be price inelastic. It is unwise for a firm to operate at a point where the demand for its product is price inelastic. By increasing its price, it can increase its total revenue and (because it generally costs less to produce less output) reduce its total cost. Thus, it can increase its total profit. Thus, the Somalia Manufacturing Company should consider a price hike.

   b. Let's compute the arc elasticity of demand when price is between \$1.00 and \$1.01. When $P = 1$, $Q = 4$. If $P = 1.01$, $Q = 5 - 1.01 = 3.99$. Thus

$$\eta = -\frac{4.00 - 3.99}{(4.00 + 3.99)/2} \div \frac{1.00 - 1.01}{(1.00 + 1.01)/2} = \frac{0.01}{3.995} \div \frac{0.01}{1.005} = 0.25$$

   Thus, the price elasticity of demand is about 0.25, not 1.0. The marketing specialist is wrong.

5. *a.* Summing up the quantities demanded by the individuals at each price, the market demand curve is:

Price (dollars)	*Quantity*
1	407
2	372
3	336
4	304
5	274

   b. When the price is between \$1 and \$2,

$$\eta = -\frac{407 - 372}{(407 + 372)/2} \div \frac{1 - 2}{(1 + 2)/2} = \frac{35}{389.5} \div \frac{1}{1.5} = 0.13$$

When the price is between $3 and $4,

$$\eta = -\frac{336 - 304}{(336 + 304)/2} \div \frac{3 - 4}{(3 + 4)/2} = \frac{32}{320} \div \frac{1}{3.5} = 0.35$$

6. *a.* Holding his income constant, the total amount he spends on Geritol is constant too; that is

$$PQ = I$$

where P is the price of Geritol, Q is the quantity demanded by Mr. Jarvis, and I is his income. Thus

$$Q = \frac{I}{P}$$

Since I is held constant, this demand curve is a rectangular hyperbola, and the price elasticity of demand equals 1.

b. Since $Q = I/P$, it follows that a 1 percent increase in I will result in a 1 percent increase in Q, when P is held constant. Thus the income elasticity of demand equals 1.

c. Since Q does not depend on the price of any other good, the cross elasticity of demand equals zero.

7. Substitutes have a positive cross elasticity of demand. Thus cases (*b*) and (*e*) are likely to have a positive cross elasticity of demand.

8. The Engel curve shows the relationship between money income and the amount consumed of a particular commodity. If this relationship is a straight line through the origin, it follows that the amount consumed of this commodity is *proportional* to the consumer's money income. Thus, a 1 percent increase in the consumer's money income results in a 1 percent increase in the amount consumed of this commodity. Consequently, the income elasticity of demand for this commodity equals 1.

9. Because there are lots of very close substitutes for a particular brand, but not for cigarettes as a whole.
   Yes.

10. *a.* It would reduce the quantity demanded by about 3 to 4 percent.
    b. Since the income elasticity is 0.5, cigarette sales would increase only about 25 percent; they would not double. Also, there are many factors other than income to consider. For example, many groups concerned with health care are interested in discouraging cigarette consumption.

Case 1 The Standard Oil Company: British Petroleum Loses Patience[1]

J. David Hunger

It was April 14, 1987 and the members of the special committee of Standard Oil's board of directors were in a quandary. The committee had only nine days to decide what to recommend to the minority stockholders regarding the offer by British Petroleum (BP) to buy the 45% of Standard's shares BP did not already own. British Petroleum was offering $70 a share, but the committee worried that the price was too low. A report to the special committee prepared by First Boston Corporation argued that the shares not owned by BP had an acquisition value of at least $85 per share—significantly higher than BP's offer.

The special committee, composed of the seven outside members of Standard's board of directors, had been unable to decide upon a recommendation. Because of differing valuations of the company, the committee had proposed that Standard's board not take a position on British Petroleum's offer until April 23—only five days before BP's offer expired. Douglas Danforth, committee chairman, knew that the matter could be delayed no longer. What should the committee recommend?

BACKGROUND

Standard Oil of Ohio was officially established as an independent firm in 1911 when the United States Supreme Court ordered the giant Standard Oil Trust to divest itself of holdings in 33 other companies. The largest, Standard of New Jersey, eventually became the successful Exxon Corporation. Standard Oil of Ohio, in contrast, was established as a one-state marketer with obsolete and inadequate refining capacity. Even though it continued to operate under the original Standard Oil charter signed by J. D. Rockefeller in 1870, the company owned no crude oil and no pipelines. Its assets in 1911 were only $6.6 million, consisting of Rockefeller's original Cleveland refinery, some storage tanks, and wagons. Reduced to the status of buying and selling other companies' oil, Standard of Ohio concentrated on marketing gasoline to the developing automobile market. Its red, while, and blue SOHIO signs shone from attractive service stations located on the most desirable intersections throughout Ohio. It also marketed in contiguous states under the name "Boron." Standard's marketing expertise made the firm a very strong competitor. Nevertheless, until 1970, Standard Oil or Sohio, as it referred to itself, was a very minor player in the global petroleum industry.

In 1970, Sohio exchanged a special stock interest equal to about a 25% stake in Standard Oil to acquire British Petroleum's U.S. interests which included large Alaskan holdings where rich deposits of oil were believed to be located. This gave

1. This case was prepared as a basis for class discussion rather than to illustrate either effective or ineffective handling of a managerial situation. Originally presented at a workshop of the Midwest Society for Case Research, July 1987. Revised 1989.

Sohio a 50% share of what became the 1.5 million barrel-a-day Prudhoe Bay oil field. The company also took over management of the Sinclair/BP operation which ranged from Texas to New England. In return, the British company obtained two seats on Sohio's 15-person board of directors, a voice in Sohio's spending plans, a healthy infusion of cash in the form of dividends from Sohio, and an experienced U.S. marketer for its Alaskan oil. Based on a formula tied to Prudhoe Bay crude oil production, British Petroleum (BP) gained a 53% majority interest in Standard and a third seat on its board of directors by 1978.

The huge cost of environmental battles plus inflation increased the price of the Prudhoe Bay oil pipeline from its initial 1971 estimate of $900 million to $9.3 billion at completion in 1978, thus forcing Sohio to borrow approximately $4.6 billion to pay its share of construction costs. Fortunately, world oil prices began to soar with the 1973 OPEC oil embargo. The bold gamble paid off when the oil started flowing from Alaska. Sohio suddenly found itself in 1978 among the major oil companies, ranking ninth in total assets. Prudhoe Bay, however, accounted for 80% or its assets, about 97% of its oil production, and more than 85% of its profits. Standard Oil of Ohio became the holder of one of the largest U.S. oil reserves, as well as one of the largest companies in the United States. By 1981, Sohio's earnings of $1.95 billion were more than 46 times its profits of 10 years earlier.

Realizing that the flow of oil coming from Prudhoe Bay was likely to decline in the late 1980s, and with it the huge annual cash profits, the top management of Sohio set in motion in 1978 a two-pronged strategy of (1) expanding oil and gas exploration and (2) diversifying into energy-related industries. Spending billions of dollars for exploration rights, the company began actively searching for oil in the lower 48 states and the Gulf of Mexico. From 1977 when Sohio spent only $20 million on exploration to 1982 when its exploration/production capital expenditure budget was $1.96 billion, Standard attempted to find enough oil to reduce its heavy dependence on Prudhoe Bay. It increased its exploration staff from 60 in 1977 to nearly 1,000 by the end of 1983. Seeking to expand beyond its current operations, Standard acquired in 1984 Truck-stops Corporation of America with its facilities along interstate highways in 15 states and Gulf Oil Corporation's refining and marketing properties in the 8 southeastern states of Kentucky, Tennessee, Alabama, Mississippi, Georgia, Florida, South Carolina, and North Carolina. The company also spent money on a number of unprofitable ventures as well as some that may be profitable in the long run.

Disappointing Kennecott Acquisition

By far, however, the company's two largest investments, Kennecott and Mukluk, were unsuccessful. In 1981, Sohio purchased the nation's largest copper producer, Kennecott Corporation, for $1.77 billion. Unfortunately for Sohio, the price of copper soon began to drop precipitously as foreign copper producers began to oversupply the world market. Attempting to stem Kennecott's losses, Standard acted to close Kennecott's abrasive operations (part of Carborundum, which Kennecott had earlier purchased) as well as to cut back and modernize copper operations. Even with these cost-cutting moves, by end-1985 Kennecott lost a total of $702 million (in operating losses) since its purchase by Sohio.

The 12 major U.S. copper producers were operating in 1985 at 59% of capacity and were in danger of going out of business. By keeping output high, Third World competitors, such as Chile (where ore was twice as rich in copper, as most U.S. mines) kept prices around 67 cents per pound in 1985 (about half the 1980 prices) versus the 82 cents per pound average cost of U.S. production. Nevertheless, in 1985, Sohio's top management decided to spend $400 million on a three-year modernization program of Kennecott's Bingham Canyon mine in Utah. The stated goal was to make Kennecott profitable by 1989 by reducing the costs of producing copper from 75 cents to around 45 cents a pound. Sohio's top management may have had little choice given the estimated costs of closing the Bingham mine at anywhere from $500 million to $5 billion depending on environmental and severance pay requirements.(1)

Costly Mukluk Exploration

Realizing that it needed a big oil strike to replace the $6 billion in annual revenues from Prudhoe, Sohio invested heavily in a prospect called *Mukluk* in the Beaufort Sea. After an estimated cost of $1.7 billion for leases and drilling 14 miles off Alaska's north coast, the exploratory well was found on December 3, 1983 to contain water, not oil. Even though this was the most expensive "dry well" in history, Sohio's top management continued to push for continued exploration both in Alaska and in the lower 48. "Because we want to replace Prudhoe, we need to do some elephant-hunting," said Richard Bray, head of Sohio's exploration and production unit. "But because most times you don't bag one, you also need to go after rabbits, squirrels, anything that's out there."(2)

Unfortunately, after spending nearly $5 billion on exploration and leases from 1980 to 1985, the company had replaced only a little more than a third of the oil it had produced during that time. By early 1986, management estimated that output from Prudhoe Bay would begin to decline at a rate of 10 to 12 percent a year beginning in 1988. "If they had gone out three or four years ago (1982–1983) and bought an independent oil company, they would have been in a substantially better position," stated Lawrence Tween, an oil analyst for Kidder Peabody and Company.(3) During the fourth quarter of 1985, Sohio finally cut its aggressive annual oil exploration budget by $200 million and reduced its exploration personnel by 600 employees. The 1986 and 1987 exploration budgets were subsequently cut even further.

Sohio Becomes Standard Oil

In February 1986, Sohio changed its name to "The Standard Oil Company." This appeared to reflect its desire to become more like the company John D. Rockefeller had founded 116 years before. In their February 19, 1986 letter to shareholders and employees, Alton Whitehouse, chief executive officer and chairman of the board, and John Miller, president and chief operating officer, stated:

> We are the original Standard Oil Company, founded by John D. Rockefeller in Cleveland in 1870. All other Standard Oil Companies that were a part of The Standard Oil Trust have either disappeared entirely or have changed their corporate names. Standard Oil also says plainly that we are primarily in energy, a field where we have attained a major stature and where we intend to remain and continue to grow. It is a strong corporate name with a proud heritage and we plan to use it widely. (4)

In a speech to security analysts in early February 1986, Whitehouse reported that he had succeeded in streamlining the company's work force and asserted that "we're ready to handle the worst."(5) Taking over from Charles Spahr in 1978 as chief executive officer, Alton W. Whitehouse, Jr. had worked hard to transform Sohio from a small regional marketer/refiner into a major integrated oil company. Coming to the company in 1968 as the firm's first chief legal counsel, Whitehouse represented Sohio in the negotiations with British Petroleum. He subsequently became Sohio's president in 1970 and its vice-chairman in 1977. Admitting his lack of technical knowledge of the oil industry, Mr. Whitehouse divided the company into business groups and delegated most operating responsibility to the unit heads.

Explaining why the company's net income had dropped from $1.5 billion in 1984 to $308 million in 1985, Whitehouse stated that income had been reduced by $1.86 billion before tax ($1.15 billion after tax) in the fourth quarter of 1985. The charges against earnings reflected a reevaluation of assets, the cost of the continued shutdown of the Utah Copper Division during modernization, as well as expenses related to staff reductions (some 1,300 positions had been eliminated). "In 1985, total operating income before special charges was almost equal to 1984, as the combination of higher sales volumes of Alaskan crude oil and significant gains in refining and marketing income balanced the effects of lower crude oil prices and higher exploration expenses," reported Whitehouse and Miller.(6)

British Petroleum Takes Charge

With almost no warning, however, Standard Oil's parent company, British Petroleum decided to assert itself in the affairs of Whitehouse's company. Since BP had acquired a significant interest in the company in 1970, it had allowed Standard to operate independently of BP. Under the 1970 pact, BP had been limited initially to two, then to three members of Sohio's Board of Directors. By 1986, however, Sir Peter Walters, chief executive of British Petroleum Company and his top aides decided that BP's hands-off policy had to change. The precipitous drop in oil prices, coupled with Standard's $1.8 billion pretax write-off in 1985 of Kennecott and coal assets, triggered the action. "There were a number of disappointments in the past few years," said Sir Peter. "Looking back at those, and the testing time ahead for oil prices, we felt we had to strengthen management for all shareholders' benefits." (7) Informed by Sir Peter of BP's plan to take over managerial control of Standard Oil, Alton Whitehouse reluctantly resigned his position as CEO at Standard's February 27, 1986 meeting of its board of directors.

Robert B. Horton, an executive with British Petroleum and a Standard Oil board member, succeeded Whitehouse as chairman and chief executive officer of Standard Oil on April 1, 1986. Frank Mosier, Standard's executive vice-president, succeeded John Miller as president and chief operating officer. The functions of chief vice-president/chief financial officer and of executive vice-president for engineered products; chemicals, metallurgical products and technology were filled by BP executives John Browne and Colin Webster, respectively, brought to Cleveland from London. Described as a workaholic, CEO Horton had experience at turning around two problem BP units; a $2.8 billion chemical subsidiary and a tanker company. Concerned with Standard's failure to reinvest successfully the cash from Prudhoe Bay, Horton was faced with the prospect of recommending to the board either that Standard continue trying to replace Prudhoe Bay oil or that the company be shrunk. "If I have somewhat of a reputation as a hatchet man, it's perhaps because I've never been reluctant to admit it when I made a mistake," asserted Horton. "It's really best at that point to cut your losses and move on."(8)

THE U.S. OIL INDUSTRY

1986 had been another difficult year for the U.S oil industry. United States oil demand of 16 million barrels a day was still far below peak demand of 19 million barrels a day in 1978. The "spot" price for West Texas Intermediate, the barometer of U.S. oil prices, dropped from $30 per barrel in October I985 to $10 per barrel in March 1986 and stayed below $18 for the rest of the year. Standard Oil's 1986 budget had been based on oil priced at $18 a barrel. The company's total cost of producing and transporting Alaskan crude oil to refineries ranged between $11 and $12 per barrel.(9) Other oil companies faced the same problem. An industry composite of the 29 largest fuel companies in the United States by *Business Week* revealed a 26% decline in dollar sales and a 24% decline in dollar profits in 1986 from 1985.(10) The National Petroleum Council estimated that approximately 150,000 jobs, roughly 25% of the total, were lost in the oil industry in 1986 as U.S. oil production dropped by 700,000 barrels a day, or about 6%, even though demand was rising nearly 3 percent. "America's oil industry has been devastated by this price skid, and the devastation continues," reported Fred Hartley, chairman of Unocal Corporation.(11)

Problems with Diversification

A number of the major integrated oil companies have found the industrywide problems to be aggravated by their experiences in diversifying outside the energy-related field. Standard Oil's poor experience with its Kennecott acquisition was paralleled by Mobil's problems with its cash-hungry Montgomery Ward subsidiary, Atlantic Richfield's difficulties with its Anaconda mining unit, Ashland Oil's unsuccessful move into insurance, and Exxon's string of bad investments. In particular, Exxon's purchase of Reliance Electric in 1979 for $1.2 billion had been selected by *Fortune* as one of the decade's seven worst mergers. The list of oil firms diversifying out of oil during the 1970s had become quite a long one by 1980. Nevertheless by 1986, most of these non-oil related acquisitions had been sold or were in the process of being sold as companies began to concentrate once again on the oil/energy business.

Threat of Takeovers

Adding to the pressures facing the major oil firms in the 1980s was the increasing threat of being taken over by another firm. The acquisition of Conoco by DuPont, Cities Service by Occidental Petroleum, Getty Oil by Texaco, Superior Oil by Mobil, and Gulf Oil by Chevron (Standard Oil of California) shook the industry. In addition, Royal Dutch/Shell acquired the remainder of its 64.9% owned U.S. subsidiary after a lengthy stockholder law suit. Pennzoil's dissatisfaction with having lost Getty Oil to Texaco led to a successful law suit against the winner and to Texaco's surprising declaration of Chapter 11 bankruptcy in April 1987.

Also in early 1987, Amoco (Standard Oil of Indiana) outbid both Exxon (Standard

Oil of New Jersey) and TransCanada Pipelines for Canada's Dome Petroleum. Canadian Energy Minister Marcel Masse was concerned with Amoco's purchase, given the Canadian government's commitment to its goal of increasing Canadian ownership in the oil and gas sector from its then current 48% level to 50%. The purchase of Dome by Amoco, the fifth largest U.S. oil company in terms of revenue, reduced Canadian ownership in the industry to about 40%.(12)

Forces leading to the intense merger activity included undervalued stock prices, high cash positions, and declining U.S. oil reserves. With growth in world demand for refined oil products projected at a rate of only 1 to 2% annually for the rest of the 1980s, stocks of the big oil firms were trading at substantial discounts to the value of the wealth they had amassed to date. The huge cash flows coming for the production and sales of oil and gas reserves were generally far larger than could be prudently spent on new drilling prospects. After noting the bad experiences of Mobil, Sohio, Ashland Oil, and Exxon, among others, in investing outside of the energy industry, and considering the depressed nature of coal and nuclear energy, almost all of the large firms had been raising their dividends and buying back their own stock. Purchasing one's own stock served several functions. Assets were acquired with excess cash at low risk without a takeover premium. By reducing stock outstanding, earnings per share was increased. The company's stock price was also supported.

With the average finding cost of around $12 per barrel, U.S. oil companies reduced their exploration spending by 40% in 1987 from 1986. Many U.S. oil operators insisted in April 1987 that they would not resume much domestic exploration until the price of oil increased from its current $18 per barrel to $25 per barrel.(13) (See Exhibit 1 for a summary of current reserves of representative companies.)

It therefore made sense for oil firms to supplement their exploration activities by buying other companies' reserves. Noting that the industry was not heavily concentrated by Federal Trade Commission and Justice Department standards, the U.S. Congress in March 1984 had rejected a proposed 11-month moratorium on further oil industry mergers. A green light was thus given to further merger activity for the foreseeable future.

STANDARD OIL'S OPERATIONS

The Standard Oil Company ended 1986 with a loss. In his first official letter in February 1987 to the shareholders and employees, Robert Horton reported:

> It was a terrible year; we reported a net loss of $345 million after special and extraordinary charges of $844 million net of tax. Even before these charges, our earnings fell some 66 percent to $499 million, largely because our average oil price was only $13.83 per barrel in 1986, compared with $26.43 per barrel in the previous year. And things would have been worse if our downstream oil and non-oil businesses had not generally put in good performances.(14)

(Financial and operational data for Standard Oil are presented in Exhibits 2 through 4.) Having spent a year of pruning the exploration portfolio, cutting capital expenditures, reducing overhead, selling off some businesses, and modernizing others, Horton next presented in the February 1987 letter his plan for surviving oil prices he felt would average only $15 per barrel:

Our strategy for Standard Oil has four simple rules:

- First and foremost, we are an oil company.
- Diversify in moderation only. Nonoil must fit, be competitive, and be profitable.
- Keep our financial position strong.
- Manage for profitability, not for size or growth for its own sake.(15)

Business Segments

The Standard Oil Company in 1987 was composed of seven business segments: Exploration and Production, Refining and Marketing, Chemicals, Metals Mining, Coal, QIT, and Other Businesses. (Exhibit 5 presents information on these business units over a five-year period.)

Exploration and Production was responsible for finding and developing crude oil and natural gas for the company. These activities were typically referred to in the industry as "upstream" operations. At year-end 1986 Standard's proved developed and undeveloped reserves of crude oil, condensate, and natural gas liquids were 2.4 billion barrels plus 7.3 trillion cubic feet of natural gas. According to *Value Line*, the after-tax present value of Standard's combined oil and gas reserves using low year-end 1986 prices was $3.4 billion. (16) Weak prices continued to rule out building the costly pipelines needed to bring the natural gas to market. Exploration spending during 1986 was $331 million, about one-third that spent during 1985, but $31 million more than that planned for 1987. Exploration activities were to focus on northern Alaska, the Gulf of Mexico, and Oklahoma's Anadarko Basin. Top management's new strategy was to increase operating efficiency and to reduce costs of the upstream business. Based on its forecast that crude oil sales prices would average $15 per barrel in 1986 dollars for the rest of the decade, top management wanted to ensure the company's profitability while maintaining its ability to take advantage of attractive acquisitions.

Refining and Marketing was responsible for all "downstream" operations such as transporting and trading crude oil, refining the oil into gasoline and diesel fuel, and marketing products and services to the general public. Since wholesale and retail product prices fell less rapidly in 1986 than did crude oil prices, operating income was $436 million in 1986, up from $351 million in 1985. About 29% of 1986 operating income, however, was contributed by the former Gulf Oil properties. Standard's four refineries had an above-average ability to maximize production for the more profitable light products, such as gasoline and diesel fuel They operated at 92% capacity in 1986, compared to the industry average of 84%. Seeking more operating efficiency, management cut the number of company-operated stations in Ohio by 16% from 1983 through 1986 while increasing retail gasoline sales by 10%. To make up for the reduction of full-service stations, Standard introduced PROCARE automotive service centers in Ohio metropolitan areas. Management planned to expand the number of PROCARE centers in Ohio and western Pennsylvania from 72 in 1986 to 100 by 1988. The company also planned to add to its 1986 total of 41 truckstops in 19 states. Management hoped to expand its presence in the Southeastern states and was "particularly interested in acquiring a presence on the West Coast."(17)

Chemicals, in comparison with other Standard businesses, was relatively small, but profitable. The unit produced and marketed two lines of commodity chemicals, acrylonitrile (used to create plastics and fibers) and nitriles, as well as nitrogen products and benzene. Operating income equaled $57 million in 1986, $2 million less than the previous year once unusual charges were excluded from the 1985 figures. Since more than 90% of worldwide acrylonitrile capacity used technology licensed from Standard Oil, the company planned to emphasize the licensing of this technology to developing countries. The company thus seemed to have a competitive advantage in a small segment within the very competitive, but growing world chemical industry.

Metals Mining was composed of the Bingham Canyon copper mine in Utah, the last operating remnant of Kennecott Corporation. One of Robert Horton's first acts as Standard's chairman and CEO was to tie the $400 million modernization program to wage concessions by the United Steelworkers. Although the Bingham Canyon mine was closed in March 1985, operations began again piecemeal in September 1986, as modernization proceeded. Standard sold its Ray Mines Division in Arizona and Chino Mines Company in New Mexico as well as some other properties in November 1986, for approximately $160 million. Although metals mining recorded a loss of $342 million in 1986, if unusual charges associated with the sale of the Ray and Chino copper mines were excluded, operating income would have been $ 11 million. When the modernization of Bingham Canyon was completed in 1988, it was to have the capacity to produce 185,000 tons of copper, 260,000 ounces of gold, 2 million ounces of silver, and 8 million pounds of molybdenum annually. Its production costs were expected to be the lowest in the United States and in the lowest quartile among free-world producers. Nevertheless, the market demand for copper was expected to remain poor to moderate, although prices could rise as mines continued to be closed in the United States and third-world producers stopped adding to mining capacity.

Coal was composed of the Old Ben Coal Company, a business which recorded an operating loss of $255 million in 1986. This loss included, however, $300 million worth of unusual charges related to the disposal of undeveloped coal properties and to the closing of the high-cost Kitt Mine in West Virginia. As a result of asset sales and writedowns, net coal assets were $200 million (706 million tons) at the end of 1986. As a part of implementing the unit's plans to reduce costs and improve productivity, coal headquarters was moved from Lexington, Kentucky, to Standard headquarters in Cleveland and its staff reduced 37%. The unit's overall productivity improved from 17.8 tons per man shift in 1985 to 21.8 tons in 1986. Nevertheless, the situation facing the coal industry continued to be one of worldwide overcapacity and declining prices.

QIT was the shortened form of the Canadian-based "Qit-Fer et Titane, Inc." As the leading producer of titanium dioxide slag, QIT's operating income was $110 million in 1986, up $30 million from 1985. Titanium dioxide slag was used to make white pigments for paint, paper, plastics, ceramics, and textiles. With demand consistently increasing over the past 10 years, titanium dioxide pigments had replaced nearly all other white pigments in the world. Since QIT was the low cost producer of this slag (stemming from its technological expertise and its access to abundant raw materials) the demand for QIT's slag had grown faster than had overall pigment demand. A modernization and construction program was in process to expand the company's slag production capacity by 25% by 1988. The company also was the major supplier of high-purity iron (generated as a coproduct in QIT's proprietary process of making titanium dioxide slag) to the ductile-iron casting industry.

Other Businesses of Standard Oil were its Engineered Materials Company and its Chase Brass and Copper Company. This segment recorded an operating loss of $151 million in 1986 compared with a loss of $98 million in 1985. Results for 1986 included unusual charges of $140 million for various divestitures and losses totaling $12 million from the businesses divested. Operating income from Engineered Materials increased from $30 million in 1985 to $36 million in 1986. Its principal products were high-temperature ceramic fiber insulation, advanced refractories, polyester resins and panels, and molten metal pumping systems. The outlook for structural and electronic ceramics appeared to be very bright. Chase Brass and Copper, in contrast, reported an operating loss of $10 million in 1986 compared to a $4 million loss in 1985. Chase's new copper and brass narrow-strip mill was just completed in time for declining prices in brass and copper items.

Strategic Managers

The board of directors of Standard Oil was composed of 15 members in 1987 (see Exhibit 6) Butler, Hartigan, and Keep were representatives of British Petroleum's 55% interest in the company. (The percentage had increased from 53% as a result of Standard's purchases of its own stock from 1983 to 1985.) Browne, Horton, and Webster were executives of Standard Oil who had previously served in managerial positions with British Petroleum. Mosier and Bray were also executives of Standard Oil. Of these eight inside or management directors, Mosier had served seven consecutive years on the board. Horton had served on the board since 1983. Bray and Webster had served one previous year. For the other four inside directors, 1986 was their first year on the board.

The remaining seven directors were outside or nonmanagement related members of the board. D. W. Buchanan, Jr., was a retired president of Old Ben Coal Company. The other six outside directors with no past connections with the company were Bailey, Danforth, De Lancey, Gorman, Hangen, and Knight. Of the seven, Bailey, Buchanan, Hangen, and Knight had served for at least the previous four years. These seven outside directors formed a Special Committee established by the board in April 1986. This committee was given the task of monitoring Standard Oil's relationship with British Petroleum and advising the board with regard to that relationship and all joint Standard-BP transactions and ventures.

Standard Oil's top management was composed of five people in 1987. Robert B. Horton, chairman of the board and chief executive officer, was 47 years old. Before 1986 he had served as a managing director (similar to executive vice-president) of British Petroleum with responsibility for finance, planning, and the company's operations in the Western Hemisphere. Frank E. Mosier, president and chief operating officer, was 56 years old. Before 1986, he had served as Standard Oil's executive vice-president overseeing downstream petroleum, metals mining, and corporate planning. He had joined Sohio in 1953 and had been primarily involved with the downstream petroleum business. Richard A. Bray, executive vice-president with responsibility for exploration and production, coal, and external affairs, was 55 years old. Since joining Standard Oil in 1982, he had been in charge of exploration and production. Prior to that time, he had held a variety of exploration and production offices during 24 years with Exxon. E. John P. Browne, executive vice-president and chief financial officer, was 39 years old. Before 1986, he had served as group treasurer of British Petroleum and chief executive of BP Finance International. J. Colin E. Webster, executive vice-president responsible for chemicals, metals, mining, industrial products companies, research and development, and ventures, was 50 years old. Before 1986, he had served as president of BP North America, Inc.

THE BRITISH PETROLEUM COMPANY p.l.c.

Registered originally in the United Kingdom on April 14, 1909, as the Anglo-Persian Oil Company, Ltd., to develop newly discovered oil resources in Persia (now Iran), British Petroleum, or BP as it was commonly known, grew to become one of the most successful and powerful multinational oil companies. Along with Exxon, Royal Dutch/Shell, Mobil, Texaco, Chevron, and Gulf, British Petroleum was one of the famed "Seven Sisters"—the major integrated multinationals which engaged in every stage of petroleum operation throughout the world and controlled around 70% of the world's oil trade until the OPEC oil embargo in 1973. Led by Sir Peter Walters as chairman of the board since 1981, BP worked hard to overcome one of he most outmoded refining and marketing networks in the industry. According to *Business Week*, Sir Peter's concentration on quiet profitability during the 1980s transformed BP into a pacesetter for the industry. (18) In contrast to Standard Oil, BP's top management had not hesitated to get rid of businesses that failed to perform profitably.

BP Operations and Ownership

Besides its widespread oil and gas operations in 1987, BP had interests in chemicals and plastics, animal feeds and agricultural seeds (including its 1986 purchases of the U.S. firms Purina Mills and Edward J. Funk and Sons), minerals mining, coal, marine shipping, detergents, and computer software. Until 1977, 68.3% of British Petroleum stock had been owned by the British Government and the Bank of England. Over time, various sales of stock by the government reduced the percentage to 31.6% (578,496,892 shares) by end-1986. On March 18, 1987, the British government announced plans to sell its remaining shares, valued at $8.51 billion, by March 31, 1988.(19) BP shares were listed on the stock exchanges in the United Kingdom, the United States, Canada, Switzerland, France, West Germany, and the Netherlands.

Following the pattern of other oil companies in 1986, BP's profits of 817 million pounds sterling ($1,201 million) were 49% less than those in 1985, due primarily to the dramatic drop in oil prices. (See Exhibits 7 and 8 for BP's financial data.) Oil production averaged 718,000 barrels a day in 1986 compared with 694,000 barrels in 1985. BP management estimated, however, that oil production would fall to 620,000 barrels a day by 1990. Estimated net proved reserves of crude oil stood at 2,252 million barrels in 1987 compared to Standard Oil's 2,406 million barrels. Oil from the United Kingdom (primarily the North Sea), which comprised 55% of BP's total proved reserves, had showed a steady decline since 1984. This was a significant problem for

British Petroleum since the oil and gas business accounted for 84% of BP's operating profit in 1986 and 95% in 1985 (when oil prices were higher). Exploration in offshore China and Brazil during the 1980s had proved disappointing, although some oil had been found in New Guinea and in Ecuador.

BP's Bid for Standard Oil

Only eight days after the British government had announced its intention to sell its remaining shares in British Petroleum, the top management of BP announced a 28-day tender offer to begin April 1, 1987 for the 45% of Standard Oil it did not already own. It offered to pay $70 for each of the approximately 105.8 million shares for a total of $7.4 billion. The bid was the largest ever by a British company and the largest in the United States since Chevron's $13.23 billion acquisition of Gulf in 1984. If it succeeded, the offer would make BP the third largest oil company in the world behind Exxon and Royal Dutch/Shell.

At a London news conference on March 26, Sir Peter Walters said that BP's management wanted to eliminate the minority holding to make expansion easier in the United States—"an important focal point of our future strategy." In the past, Sir Peter added, BP had been hampered by a cumbersome dual board structure and potential competitive conflicts with Standard in planning U.S. opportunities. David Simon, chief financial managing director of BP, said that a combination of the two companies "cleared the ground for" major acquisitions of oil reserves or downstream operations in the United States. "Much of what BP can do, we feel it can do in the United States," added Sir Peter Walters. Walters stressed that his goal in taking full ownership of Standard was not to make BP larger but to make it more efficient and profitable by focusing on low-cost production, Referring to past relations with Standard Oil, he pointed out that "we were either missing tricks because our joint overheads were too high, or we were pursuing different paths."(20) The last point was a reference to the tendency of Standard's previous CEO, Alton Whitehouse, to act independently of BP and to ignore the parent company's advise. For example, when Standard purchased a concession in 1985 to prospect for oil in Qatar, a Middle East state where BP had considerable exploration experience, BP's top managers first learned of the purchase when they read it in the newspapers.(21)

According to BP's David Simon, the company planned to fund about one-third of the stock purchase from its own cash with another third coming from commercial paper and the remainder from bank borrowings. "We would intend to (repay the bank loans) from the joint cash flow of the combined companies," stated Simon. Rodney Chase, BP group treasurer, said that BP's debt to total capital would increase from 33 to 44%, but would be back in the high 30s within a year of the purchase. The price for Standard's shares had been set at $70 based upon top management's assumption of an inflation-adjusted crude oil price of $18 a barrel through the end of the century. This was an increase over the $15 a barrel price for the same time period which both BP and Standard had used during 1986 for planning purposes, said Simon.(22)

Upon BP's announcement on March 26, the price of Standard Oil stock rose $6.50 on the New York Stock Exchange to $71.375 on a volume of 5.5 million shares. Just hours after BP's bid was announced, a class-action suit was filed in Cleveland on behalf of Standard's minority stockholders. The suit argued that BP was offering "a fraudulently low and unfair price" for the stock. This action was reminiscent of Royal Dutch/Shell's attempt in 1985 to buy the remaining 30.5% of the U.S. Shell stock it didn't own. In that instance, the parent company's offer of $55 a share faced a similar class-action suit. Royal Dutch/Shell was forced to raise the bid twice to $60 a share because of protests from Shell's minority shareholders and outside directors. After much delay, the terms were finally approved in a Delaware court.(23)

In response to a question asking if BP might increase its offer, Sir Peter Walters uttered a flat "no" and said that unlike Royal Dutch/Shell; BP wasn't proposing a merger. British Petroleum was simply buying Standard's stock in the open market and only had to follow full disclosure rules. Therefore, Standard's board did not need to place an independent valuation on the company. In addition, Sir Peter stated that under BP's interpretation of Ohio corporate law, British Petroleum could force minority shareholders to accept the $70 offer if BP's current 55% stake in Standard Oil reached 90%. The shareholders of British Petroleum were to vote on BP's action at a special April 22 meeting.

STANDARD OIL'S RESPONSE

British Petroleum had presented its plans to purchase Standard Oil's remaining stock at a special Standard Oil board meeting on March 9, 1987. The special committee of outside directors established to monitor Standard-BP relations gave BP a guarded response. The committee's chairman, Douglas Danforth, asked Sir Peter to postpone the offer for one or two months to give the committee adequate time to review the offer. BP replied negatively. The committee hired First Boston Corporation to examine the offer. In an April 3 letter to the committee, First Boston contended that BP's offer did not include a premium over Standard's current stock market price since Standard's stock would have been trading at $70 even without BP's bid. It concluded that Standard Oil stock was worth at least $85 a share. After first announcing that it would issue a statement on BP's offer on April 14, the special committee of outside board members decided to postpone its recommendation until April 23. The committee did, however, ask shareholders to wait until the committee made its report to Standard's board before tendering their shares to BP. The tender offer was due to expire on April 28, unless it was extended by BP.

The wide gap between First Boston's valuation of Standard Oil and that of Goldman, Sachs, & Company, which had prepared the BP offer, created a war of words between the two investment banking firms. In addition to differences on the amount of probable reserves in Prudhoe Bay and on the future operating income of Standard's refining and marketing business, the two valuations differed on the future price of oil. BP's offer was based on a forecasted 1988 price of $18 per barrel for West Texas Intermediate crude oil adjusted for an expected 5% annual rate of inflation through the end of the century. First Boston, in contrast, took into account three possible oil-price scenarios, ranging from a starting price of $17.25 a barrel with modest price escalation to $20 a barrel with rapid escalation.(24) British Petroleum labeled First Boston's calculations "ill-founded" and "seriously flawed."

The eight inside directors of Standard Oil were in a rather difficult position. The three representatives of BP on the board obviously had to support BP's offer. The three top executives of Standard Oil who had been placed there by BP were caught between the two sides. In Horton's case, as an ex-managing director of British Petroleum and a likely candidate to run BP some day, anything the chairman of Standard Oil did was bound to provoke criticism. Directors Mosier and Bray had no past connections with BP, but obviously depended on Horton and BP executives for their future with Standard Oil. The level of discussions which look place during April between the committee of seven outside directors and top management of British Petroleum suggested that the outside directors were not willing to go along with BP's $70 per share.

What should the board recommend to the minority stockholders of Standard Oil? A failure by BP to obtain the stock or a drawn-out legal battle could hurt the British government's plan to sell its 32% stake in BP in the fall of 1987. Given that one of the members of BP's 13-member board of directors was appointed by the British government this was no small concern to BP's top management This situation worked to the benefit of Standard's special board committee as it deliberated on what to recommend.

THE FUTURE PRICE OF OIL

Every oil company and industry analyst appeared to have a different estimate of the future price of oil. Many believed that the price was bound to increase significantly. Exxon, for example, assumed in 1987 that the price of oil would stay flat in constant dollar terms for a few years and then rise faster than the rate of inflation. In a speech delivered to the Australian Petroleum Exploration Association on March 22, 1987, Donald McIvor, an Exxon senior vice president, said that "the price of oil must rise significantly again before too long." He noted that the world had been using oil at rates of 20 to 25 billion barrels a year since 1970, while discoveries had been accumulating at only 10 to 15 billion barrels a year. "Most major consuming countries, including those with substantial production of their own, will ultimately come to depend heavily on Middle East oil."(25) In general agreement with this view, the U.S. Department of Energy forecast a $33 a barrel price in constant dollars by the year 2000. Data Resources, Inc., an economic forecaster, estimated a $32 constant dollar price ($64 with likely inflation) by the end of the century.

A case could be made, however, for a reasonably constant oil price. For example, Arlon Tussing, an energy economist, predicted that oil prices in constant dollars were "likely to remain within a range of $10 a barrel and $20 a barrel for the rest of this century." As for the longer term, he expected technology to push energy prices not up but down. Tussing argued that increases in the price of oil would cause large switches to natural gas and coal as substitutes. Unlike the situation in the early 1970s, oil was "no longer the indispensable fuel," said Tussing. For example, many industrial energy users in 1987 were equipped with dual or triple fuel capacity and could choose oil, gas, or coal, depending on the price. At prices much above $20, Tussing said, oil loses almost the entire global bulk-fuels market to other energy sources (26)

A report issued in February 1987 by the National Petroleum Council proposed two possible trends for future oil prices. One trend began at $18 per barrel in 1986 with growth at a constant dollar rate of 5% per year to $36 in the year 2000. The second trend started at $12 per barrel in 1986 and grew at a constant dollar rate of 4% per year to $21 by the year 2000. The report also warned that the percentage of U.S.

consumption that came from imports could increase from 33% in 1986 to 48% by 1990 if crude oil prices remained near 1986 levels and could even rise to 66% of consumption by the end of the century. The report further warned that the United States could face a return to the oil crises of the 1970s, when political problems in the Mideast combined with high American dependence on imported oil to create shortages.(27)

These varying estimates regarding the future price of oil were reflected in the different valuations of Standard Oil by Goldman, Sachs and Company (BP's financial adviser) and First Boston Corporation (financial adviser to the special board committee of Standard Oil). According to Goldman, Sachs and Company, a $1 change in the per barrel price of oil had the effect of changing the value of Standard Oil's proved oil and gas reserves by an amount equivalent to approximately $4 per share of common stock.(28)

QUESTIONS

1. If you were a member of the special committee of Standard Oil board members, what use would you make of the following microeconomic concepts: (a) demand curve, (b) supply curve, (c) price elasticity of demand, (d) income elasticity of demand, and (e) cross elasticity of demand?
2. Evaluate Donald McIvor's forecasts of the price of oil. Do you agree with his reasoning? Why or why not? Has he been correct thus far?
3. Evaluate Arlon Tussing's forecasts of the price of oil. Do you agree with his reasoning? Why or why not? Has he been correct thus far?
4. Suppose that the market demand for oil is such that $Q_D = 22 - 0.1P$, where Q_D is the quantity demanded of oil (in billions of barrels) and P is the price of oil (in dollars per barrel). Is this in accord with Mr. Tussing's argument? Why or why not?
5. In 1981, Standard Oil bought Kennecott Copper. During the late 1970s, there was a decrease in the rate of growth of electricity generation in many countries, and during the 1980s aluminum and fiber optics became more important as a substitute for copper. Did these events tend to shift the demand curve for copper? If so, in what direction?
6. The price of copper fell from about $1 per pound in 1980 to about 60 cents in 1986. Is this consistent with your answer to the previous question? Why or why not?
7. If you had been a member of Standard Oil's board in 1985, would you have recommended that it sell Kennecott? Why or why not?

NOTES

1. A. Sullivan and J. Valentine, "Copper Industry Is Ill and Getting Sicker," *Wall Street Journal*, June 18, 1985, p. 6; G. Stricharchuk, "Sohio Predicts It Can Weather Oil-Price Plunge," *Wall Street Journal*, February 6, 1986, p. 11; and, D. Cook and W. Glasgall, "Is Sohio Getting In Shape for a Buyout?" *Business Week*, December 16, 1985, pp. 28-29.

2. G. Brooks, "After Mukluk Fiasco, Sohio Strives to Find or Perhaps to Buy Oil," *Wall Street Journal*, April 19, 1984, p. 22.

3 G. Putka, R.E. Winter, and G. Stricharchuk, "How and Why BP Put Its Own Commanders at Standard Oil Helm," *Wall Street Journal*, March 7, 1986, p. 8.

4. *1985 Annual Report*, Standard Oil Company, p. 4.

5. G. Stricharchuk, "Sohio Predicts It Can Weather Oil-Price Plunge."

6. A. W. Whitehouse and J. R. Miller, "Letter to Shareholders and Employees," *1985 Annual Report*, Standard Oil Company, p. 2.

7. Putka et al., "How and Why BP Put Its Own Commanders at Standard Oil Helm," p. 1.

8. D. Cook, "Will Horton Have to Take a Hatchet to Standard Oil?" *Business Week*, May 23, 1986, p. 79

9. G. Stricharchuk, "Sohio Predicts It Can Weather Oil-Price Plunge."

10. "The Top 1,000 U.S. Companies Ranked by Industry," *Business Week*, April 17, 1987, p. 134.

11. M. Potts, "Concern Grows over Rise in U.S. Oil Imports," *Washington Post*, March 8, 1987, p. H3.

12. B. Richards, J. McNish, and L. Zehr, "Amoco Agrees to Buy Dome Petroleum Ltd.," *Wall Street Journal*, April 20, 1987, p. 2.

13. J. Tanner and Y. Ibrahim, "Price Stability Encourages Oil Industry," *Wall Street Journal*, April 28, 1987, p. 6.

14. *1986 Annual Report*, Standard Oil Company, p. 2.

15. Ibid.

16. "Petroleum (Integrated) Industry," *Value Line Investment Survey*, Part 3: Ratings and Reports, Edition 3, April 10, 1987, p. 402.

17. *1986 Annual Report*, Standard Oil Company, p. 15.

18. S. Miller and D. Cook, "Why BP Is Going All Out for All of Standard Oil," *Business Week*, April 13, 1987, p. 50.

19. "The British Petroleum Company p.l.c.," *Moody's Industrial Manual*, vol. 1, A-1, 1986, p. 1072.

20. Putka and Winter.

21. Putka, et al., p. 8.

22. Putka and Winter, pp. 2 and 16.

23. G Anders, "Standard Holders File Suit to Block BP Bid in Effort to Replay Shell Case," *Wall Street Journal*, March 27, 1987, p. 2.

24. M. W. Miller and R. E. Winter, "British Petroleum's Standard Bid Spurs Price Dispute," *Wall Street Journal*, April 7, 1987, pp. 3 and 28.

25. J. Tanner, "Exxon Official Sees a Significant Rise in Price of Oil as Inevitable Before Long," *Wall Street Journal*, March 23, 1987, p.7.

26. A. Bayless, "A Bear in the Oil Patch," *Wall Street Journal*, April 28, 1987, p. 34.

27. Potts, "Concern Grows over Rise in U.S. Oil Imports."

28. *The Standard Oil Company*, special report prepared by Goldman, Sachs and Company for British Petroleum, April 21, 1987.

EXHIBITS

EXHIBIT 1

YEAR-END OIL AND NATURAL GAS RESERVES OF REPRESENTATIVE U.S INTEGRATED PETROLEUM CORPORATIONS

	Oil Volumes (Millions of barrels)		Gas Volumes (Billions of cubic feet)	
	1985	1986	1985	1986
Amerada	692	458	1,882	1,936
Amoco	2,769	2,424	15,137	15,375
Atlantic Richfield	2,931	2,927	7,085	6,895
Chevron	3,831	3,513	9,994	10,081
Exxon	6,733	6,512	29,723	29,430
Mobil	2,366	2,460	20,687	20,479
Occidental	963	752	2,703	3,258
Pennzoil	111	98	961	900
Phillips	901	718	4,883	5,144
Standard Oil	2,648	2,406	7,219	7,308
Sun	848	796	3,174	917
Texaco	3,333	3,225	8,869	8,165
Unocal	751	752	6,189	6,073

Source:
Value Line Investment Survey, Part 3: Ratings and Reports, Edition 3, vol. 42, no. 29, (April 10, 1987), p. 402.

EXHIBIT 2

STATEMENT OF INCOME — STANDARD OIL COMPANY

$ Millions, Except Per-Share Amounts Year Ended December 31	1986	1985	1984
Revenues			
Sales and operating revenue	$9,219	$13,002	$11,692
Excise taxes	803	816	559
	10,022	13,818	12,251
Costs and Expenses			
Costs of products sold and operating expenses	4,903	6,156	5,406
Taxes other than income taxes	1,368	1,817	1,579
Depreciation, depletion and amortization	1,158	927	796
Oil and gas exploration expenses, including amortization of unproved properties	926	1,101	704
Selling, general and administrative expenses	954	943	728
Unusual items (write down or disposal of properties in coal, metals, and exploration)	1,079	1,699	90
	10,388	12,643	9,303
Income (Loss) before Interest, Income Taxes, and Extraordinary Item	(366)	1,175	2,948
Interest expense	(335)	(396)	(347)
Interest income	93	97	132
Income (Loss) before Income Taxes and Extraordinary Item	(608)	876	2,706
Income taxes	297	(568)	(1,218)
Income (Loss) before Extraordinary Item	(311)	308	1,488
Extraordinary Item—loss on early payment of some debt, net of income taxes	(34)	—	—
Net Income (Loss)	$(345)	$308	$1,488
Per Share of Common Stock			
Income (loss) before extraordinary item	$(1.32)	$1.31	$6.14
Extraordinary item	(.15)	—	—
Net income (Loss)	$(1.47)	$1.31	$6.14
Cash dividends	$2.80	$2.80	$2.65
Average Number of Common and Equivalent Shares Outstanding (millions)	235	235	242
STATEMENT OF RETAINED EARNINGS			
Balance at beginning of year	$7,628	$7,977	$7,128
Net Income (loss)	(345)	308	1,488
Cash dividends			
Common	(305)	(305)	(306)
Special	(352)	(352)	(333)
Balance at end of year	$6,626	$7,628	$7,977

Source: *1986 Annual Report*, Standard Oil Company, p. 41.

EXHIBIT 3

BALANCE SHEET — STANDARD OIL COMPANY

$ Millions—December 31	1986	1985
Assets		
Current Assets		
Cash, including time deposits	$ 138	$ 120
Marketable securities at cost, which approximates market	275	235
Accounts receivable, less allowances	861	1,611
Refundable federal income tax	771	—
Inventories	1,200	1,437
Net investment in operations to be divested	127	—
Prepaid expenses and deferred charges	83	92
	3,455	3,495
Property, Plant, and Equipment		
Petroleum		
Exploration and production (successful efforts accounting method)	13,733	13,502
Refining and marketing	2,130	1,998
Coal	501	962
Metals mining	1,396	2,114
Chemicals	468	461
QIT	279	219
Other business	290	420
Corporate and other	454	460
	19,251	20,136
Less accumulated depreciation, depletion, and amortization	7,434	7,001
	11,817	13,135
Other Noncurrent Assets		
Investments in unconsolidated affiliates	218	291
Receivables	333	390
Prepaid expenses and deferred charges	132	197
	683	878
	$15,955	$17,508

EXHIBIT 3 (continued)

BALANCE SHEET — STANDARD OIL COMPANY

$ Millions—December 31	1986	1985
Liabilities and Shareholders' Equity		
Current Liabilities		
Notes payable	$ 318	$ 72
Current maturities of long-term obligations	101	328
Accounts payable	974	1,568
Accrued income and other taxes	365	413
Accrued interest	134	113
Other	554	733
	2,446	3,227
Long-Term Obligations and Accruals		
Long-term debt	2,951	2,962
Capital lease obligations	325	343
Accruals and reserves	1,152	1,300
	4,428	4,605
Deferred Income Taxes	2,061	1,658
Shareholders' Equity		
Capital stock		
Common—$1.25 stated value, 300 million shares authorized, shares issued—122,498,893 and 122,337,553	154	154
Special—stated value 1,000 shares authorized and issued	25	25
	179	179
Additional paid-in capital	822	818
Retained earnings	6,626	7,628
Common stock in treasury, at cost—13,626,248 shares and 13,623,050 shares	(607)	(607)
	7,020	8,018
	$15,955	$17,508

Source: *1986 Annual Report*, Standard Oil Company, pp. 42–43.

EXHIBIT 4

OPERATING AND OTHER STATISTICS — STANDARD OIL COMPANY

	1986	1985	1984	1983	1982
Petroleum					
Crude oil and natural gas produced (net) bbl. / day					
Alaska	706,400	699,700	617,900	594,800	676,700
Lower 48 states	20,200	20,000	16,500	17,200	18,200
Foreign	—	—	—	—	—
	726,600	719,700	634,400	612,000	694,900
Produced natural gas sold (net)—thousands of cu. ft./ day	154,400	110,100	87,600	95,600	90,700
Refinery runs—bbl. / day	622,800	597,500	405,700	393,700	360,100
Refinery capacity (year end) bbl. / calendar day	656,000	656,000	456,000	456,000	456,000
Refined petroleum products sold—bbl. / day	644,500	604,200	410,800	404,700	370,900
Marketing retail outlets[2]	8,100	8,200	3,050	3,175	3,550
Nonpetroleum					
Acrylonitrile produced—millions of pounds	840	830	760	660	590
Ilmenite ore shipped—thousands of tons	3,000	2,780	2,040	1,700	2,030
Coal sold—thousands of tons	15,400	13,900	14,400	10,700	10,900
Produced copper sold—thousands of tons	181	189	310	316	268
Operating Results					
Revenues (millions of dollars)	$ 10,022	$ 13,818	$ 12,251	$ 11,958	$ 13,490
Income (loss) before extraordinary item (millions of dollars)	(311)	308	1,488	1,512	1,879
Net income (loss) (millions of dollars)	(345)	308	1,488	1,512	1,879
Return on average capital employed	(1.3%)	3.8%	12.4%	13.2%	17.2%
Ratio of earnings to fixed charges[3]		2.6	6.0	5.7	5.3

2. Includes outlets supplied by jobbers, automobile dealers, marine dealers, etc.
3. Earnings for 1986 were inadequate to cover fixed charges. The amount of the deficiency in total adjusted earnings was $679 million.

EXHIBIT 4 (continued)
OPERATING AND OTHER STATISTICS — STANDARD OIL COMPANY

	1986	1985	1984	1983	1982
Per Share of Common Stock					
Income (loss) before extraordinary item	$ (1.32)	$ 1.31	$ 6.14	$ 6.14	$ 7.63
Net income (loss)	(1.47)	1.31	6.14	6.14	7.63
Dividends paid	2.80	2.80	2.65	2.60	2.55
Market price, high–low	52–40	56–40	51–40	59–35	42–26
Other Data					
Average number shares outstanding (millions)	235	235	242	246	246
Shareholders of record of common stock	52,100	55,300	59,900	62,800	63,300
Employees	39,700	42,100	44,200	44,000	49,800
Wages, salaries, employee benefits (millions of dollars)	$ 1,419	$ 1,627	$ 1,593	$ 1,510	$ 1,612
R&D expense (millions of dollars)[4]	125	158	148	135	96
Oil and gas exploration expenses (millions of dollars)	926	1,101	704	834	486

Source: *1986 Annual Report*, Standard Oil Company, p. 63.

4. Includes research and development expense funded at both the corporate and business segment levels, and expenses associated with synthetic fuels and other alternate-energy development projects.

EXHIBIT 5

BUSINESS SEGMENT DATA — STANDARD OIL COMPANY

$ Millions	1986	1985	1984	1983	1982
Business Segment Information					
Revenues [5]					
Petroleum	$ 7,959	$11,425	$ 9,480	$ 9,438	$10,875
Coal	465	437	465	361	346
Metals mining	178	270	482	619	557
Chemicals	475	551	643	516	554
QIT	337	303	257	198	213
Other businesses	708	857	909	827	953
Corporate and other	(28)	23	15	(1)	(8)
Intersegment eliminations	(72)	(48)	—	—	—
	$10,022	$13,818	$12,251	$11,958	$13,490
Income (Loss) Before Interest, Income Taxes and Extraordinary Item					
Petroleum					
Exploration / production	$ 13	$ 2,442	$ 2,954	$ 2,848	$ 3,647
Refining / marketing	436	351	200	453	390
	449	2,793	3,154	3,301	4,037
Coal	(255)	(518)	61	21	(1)
Metals mining	(342)	(851)	(160)	(91)	(187)
Chemicals	57	39	45	(7)	(22)
QIT	110	80	41	16	34
Other businesses	(151)	(98)	(33)	(173)	(32)
Corporate and other	(234)	(270)	(160)	(158)	(132)
	$ (336)	$ 1,175	$ 2,948	$ 2,909	$ 3,697
Assets					
Petroleum					
Exploration / production	$ 8,677	$ 9,608	$ 9,366	$ 8,536	$ 8,023
Refining / marketing	2,469	2,949	2,079	1,803	1,568
	11,146	12,557	11,445	10,339	9,591
Coal	376	679	1,131	1,103	1,113
Metals mining	1,289	1,729	2,317	2,174	2,240
Chemicals	347	402	495	536	576
QIT	458	365	318	341	359
Other businesses	390	736	715	713	874
Corporate and other	1,949	1,040	1,066	1,156	1,263
	$15,955	$17,508	$17,487	$16,362	$16,016

5. Petroleum revenues include exploration/production, refining, and marketing.

EXHIBIT 5 (continued)

BUSINESS SEGMENT DATA — STANDARD OIL COMPANY

$ Millions	1986	1985	1984	1983	1982
Capital Expenditures					
Petroleum					
Exploration / production	$ 1,121	$ 1,535	$ 1,633	$ 1,795	$ 1,996
Refining / marketing	185	489	247	151	120
	1,306	2,024	1,880	1,946	2,116
Coal	53	74	51	23	168
Metals mining	101	98	176	156	202
Chemicals	10	13	23	20	69
QIT	68	80	19	12	6
Other businesses	42	81	58	58	65
Corporate and other	38	114	122	83	82
	$ 1,618	$ 2,484	$ 2,329	$ 2,298	$ 2,708

Source: *1986 Annual Report*, Standard Oil Company, p. 62.

EXHIBIT 6

BOARD OF DIRECTORS — STANDARD OIL COMPANY

E. E. Bailey [2, 3, 5]
Dean, Graduate School of Industrial Administration, Carnegie-Melllon University

R. A. Bray
Executive vice-president

E. J. Browne
Executive vice-president and chief financial officer

D. W. Buchanan Jr. [2, 3]
Retired president, Old Ben Coal Company

B. R. R. Butler [1]
Managing director, The British Petroleum Company p. l. c. (International Oil Company) and chairman BP Exploration Company Limited (subsidiary of the British Petroleum Company p. l. c.)

D. D. Danforth [2, 4]
Chairman of the board and chief executive officer, Westinghouse Electric Corporation (Diversified Electrical Company)

W. J. De Lancey [1, 2, 4]
Retired Chairman of the board and chief executive officer, Republic Steel Corporation (Steel Manufacturer)

J. T. Gorman [2, 4]
President and chief operating officer, TRW, Inc. (Diversified High-Technology Company)

J. J. Hangen [1, 2, 3]
Retired chairman of the board and chief executive officer, Appleton Papers, Inc. (Paper Manufacturer)

I. G. S. Hartigan [1]
President, BP North America, Inc. (subsidiary of the British Petroleum Company p. l. c.)

R. B. Horton [1]
Chairman of the board and chief executive officer

K. R. Keep [5]
Director technical, BP Exploration Company Limited (subsidiary of the British Petroleum Company p. l. c.)

C. F. Knight [2, 4]
Chairman of the board and chief executive officer, Emerson Electric Co. (Electrical and Electronic Products and Systems Manufacturing)

F. E. Mosier
President and chief operating officer

J. C. E. Webster [5]
Executive vice-president

Committee Memberships as of January 31, 1987

1. Member of Executive Committee
2. Member of Special Committee
3. Member of Audit Committee
4. Member of Compensation Committee
5. Member of Contribution Committee

Source: *1986 Annual Report*, Standard Oil Company, p. 64.

EXHIBIT 7

SUMMARIZED GROUP INCOME STATEMENTS
BRITISH PETROLEUM COMPANY

$ Millions	1986	1985	1984	1983	1982
Turnover	39,941	53,281	50,830	49,219	51,300
Operating expenses	37,229	48,551	46,440	45,020	47,418
	2,712	4,730	4,390	4,199	3,882
Other income	1,155	1,039	837	800	1,246
Replacement cost operating profit	3,867	5,769	5,227	4,999	5,128
Realized stock holding gain (loss)	(1,724)	(323)	162	(217)	120
Historical cost operating profit	2,143	5,446	5,389	4,782	5,248
Interest expense	735	749	760	841	1,214
Profit before taxation	1,408	4,697	4,629	3,941	4,034
Taxation	62	1,797	1,911	1,845	1,930
Profit after taxation	1,346	2,900	2,718	2,096	2,104
Minority shareholders interest	145	823	840	780	851
Profit before extraordinary items	1,201	2,077	1,878	1,316	1,253
Extraordinary items	(467)	(1,207)	(309)	251	(7)
Profit for the year	734	870	1,479	1,567	1,246
Distribution to shareholders	944	809	734	666	647
Retained profit / (deficit) for the year	(210)	61	745	901	599
Earnings per ordinary share	$0.66	$1.14	$1.03	$0.72	$0.69
Replacement cost profit					
Historical cost profit before extraordinary items	1,201	2,077	1,878	1,316	1,253
Realized stock holding gain (loss) less minority interest	1,414	284	(185)	158	(156)
Replacement cost profit before extraordinary items	2,615	2,361	1,693	1,474	1,097

Source: *1986 Annual Report*, British Petroleum Company, p. 57.

EXHIBIT 8

SUMMARIZED GROUP BALANCE SHEETS
BRITISH PETROLEUM COMPANY

$ Millions	1986	1985	1984	1983	1982
Fixed assets	26,464	26,442	24,636	24,857	25,110
Stocks and debtors	11,788	14,181	12,367	13,218	14,878
Liquid resources	3,743	3,198	2,685	1,327	2,560
Total assets	41,995	43,821	39,688	39,402	42,548
Creditors and provisions					
excluding finance debt	14,615	17,121	13,048	12,394	42,548
Capital employed	27,380	26,700	26,640	27,008	29,390
Financed by:					
Finance debt	7,506	7,438	8,187	7,928	10,588
Minority shareholders' interest	5,115	4,895	5,063	5,105	4,795
BP shareholders' interest	14,759	14,267	13,390	13,975	14,007
	27,380	26,700	26,640	27,008	29,390

Source: *1986 Annual Report*, British Petroleum Company, p. 59.

Part 3 The Firm: Its Technology and Costs

Chapter 7 The Firm and Its Technology

Case Study: Japanese Manufacturing Techniques

No topic has received more attention among policy-makers in Washington and elsewhere than Japanese competition. In industries like consumer electronics, steel and autos, the Japanese have increased their share of the American market, and, according to many experts, have achieved superior productivity and quality. An engineering comparison of a compact-car plant in Japan and the United States was as follows:

	United States	*Japan*
Parts stamped per hour	325	550
Manpower per press line	7–13	1
Time needed to change dies	4–6 hours	5 minutes
Average production run	10 days	2 days
Time needed per small car	59.9 hours	30.8 hours
Number of quality inspectors	1 per 7 workers	1 per 30 workers

a. One of the concepts at the core of Japanese production management is "just-in-time" production, which calls for goods being produced and delivered just in time to be sold, subassemblies to be produced and delivered just in time to be assembled into finished goods, and fabricated parts to be produced and delivered just in time to go into subassemblies. What are the advantages of this system?

b. What are the disadvantages of this system?

c. Another major Japanese production concept is "total quality control" or "quality at the source," which means that errors should be found and corrected *by the people performing the work.* In the West, inspection often is performed by statistical sampling *after* a lot of goods is produced. What are the advantages of the Japanese system?

d. Does microeconomics play an important role in determining whether these production techniques should be adopted in the United States? If so, how?

Completion Questions

1. At the Tangway Corporation, the average product of labor equals $3L$, where L is the number of units of labor employed per day. The total output produced per day if 4 units of labor are employed per day is ___________. The total output produced per day if 5 units of labor are employed per day is___________. The marginal product of the fifth unit of labor employed per day is _________.

2. In the ________________, all inputs are variable.
3. The ______________ production function can be written $Q = AL^{\alpha_1} K^{\alpha_2} M^{\alpha_3}$
4. A fixed input is ________________.
5. A variable input is ________________.
6. In both the short run and the long run, a firm's productive processes generally permit substantial ______________ in the proportions in which inputs are used.
7. The average product of an input is total product divided by ______________.
8. The marginal product of an input is the addition to total output due to _______.
9. Underlying the law of diminishing returns is the assumption that technology remains ________________.
10. Two isoquants can never ________________.

True or False

____ 1. If the average product of labor equals $10/L$, where L is the number of units of labor employed per day, total output is the same regardless of how much labor is used per day.

____ 2. The law of diminishing marginal returns is inconsistent with increasing returns to scale.

____ 3. An isoquant is analogous to the budget line in the theory of consumer demand.

____ 4. The marginal rate of technical substitution equals minus one times the slope of the isoquant.

____ 5. Isoquants are concave.

____ 6. All production functions exhibit constant return to scale.

____ 7. Increasing returns to scale can occur because of the difficulty of coordinating a large enterprise.

____ 8. Whether there are increasing, decreasing, or constant returns to scale in a particular case is an empirical question.

____ 9. Statistical studies of production functions are hampered by the fact that available data do not always represent technically efficient combinations of inputs and outputs.

____ 10. The only goal of any firm is to maximize profits.

____ 11. The concept of profit maximization is well defined in an uncertain world.

____ 12. The production function is not closely related to a firm's or industry's technology.

____ 13. The law of diminishing marginal returns applies to cases where there is a proportional increase in all inputs.

Multiple Choice

1. At the La Roche Company, the average product of labor equals $5 \div \sqrt{L}$, where L is the amount of labor employed per day. Thus,
   a. labor always is subject to diminishing marginal returns.
   b. labor is subject to diminishing marginal returns only when L is greater than 5.
   c. labor always is not subject to diminishing marginal returns.
   d. labor always is not subject to diminishing marginal returns when L is greater than 5.
   e. none of the above.

2. Suppose that the production function is as follows:

Quantity of output per year	*Quantity of input per year*
2	1
5	2
9	3
12	4
14	5
15	6
15	7
14	8

   The average product of the input when 7 units of the input are used is
   a. 7.
   b. 15.
   c. $2\frac{1}{7}$.
   d. 7/15.
   e. none of the above.

3. If the production function is as given in Question 2, the marginal product of the input when between 1 and 2 units of the input is used is
   a. 2.
   b. 5.
   c. 3.
   d. 4.
   e. none of the above.

4. If the production function is as given in Question 2, the marginal product of the input begins to decline
   a. after 3 units of input are used.
   b. after 2 units of input are used.
   c. after 4 units of input are used.
   d. after 7 units of input are used.
   e. none of the above.

5. If the production function is as given in Question 2, the marginal product of the input is negative when more than

a. 7 units of input are used.
b. 6 units of input are used.
c. 5 units of input are used.
d. 4 units of input are used.
e. none of the above.

6. The marginal product equals the average product when the latter is
   a. ½ of its maximum value.
   b. ¼ of its maximum value.
   c. equal to its maximum value.
   d. 1½ times its maximum value.
   e. none of the above.

Review Questions

1. Discuss the limitations of profit maximization as an assumption concerning the motivation of the firm. Why does profit maximization remain the principal assumption made by economists?
2. Discuss the nature of technology and the constraints it imposes on the firm's behavior. What are fixed inputs? Variable inputs?
3. Discuss the meaning of the production function. What is the short run? The long run? How does the production function in the short run differ from that in the long run?
4. *a.* Suppose the production function for a cigarette factory is as given below, there being only one input.

Amount of input (units per year)	*Amount of output (units per year)*
1	7
2	14.5
3	22
4	29
5	35
6	39
7	39

Plot the average product curve for the input in the graph below.

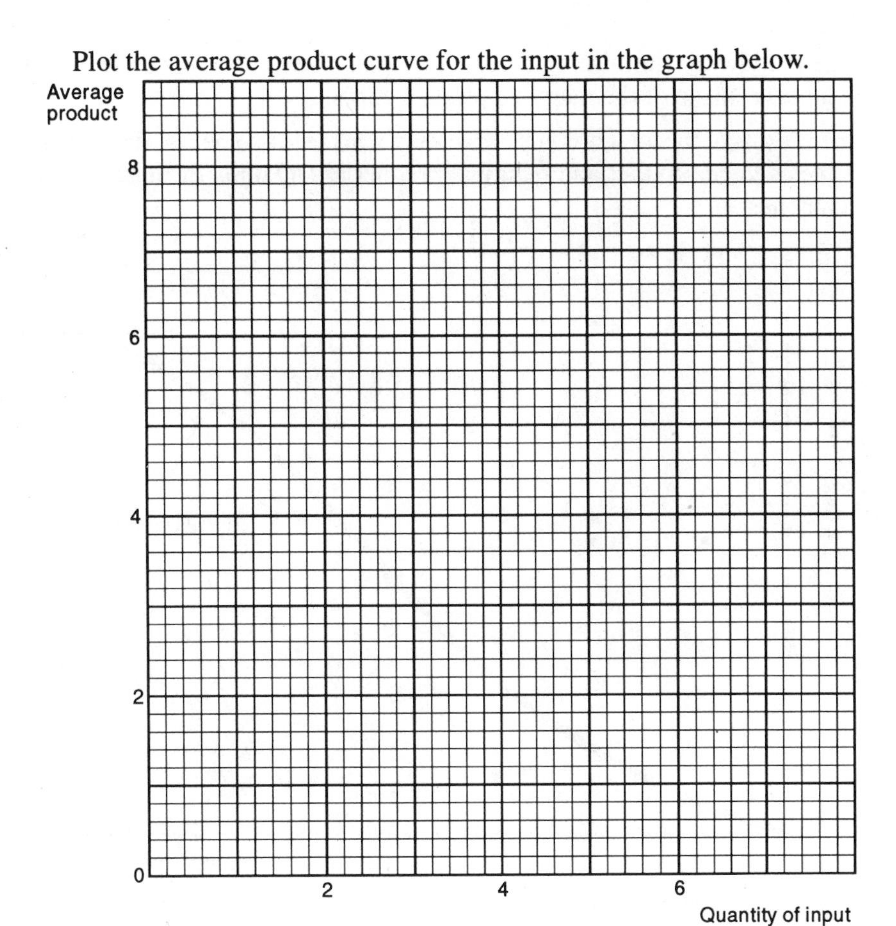

b. On the basis of the production function given in the first part of this question, plot the marginal product curve of the input.

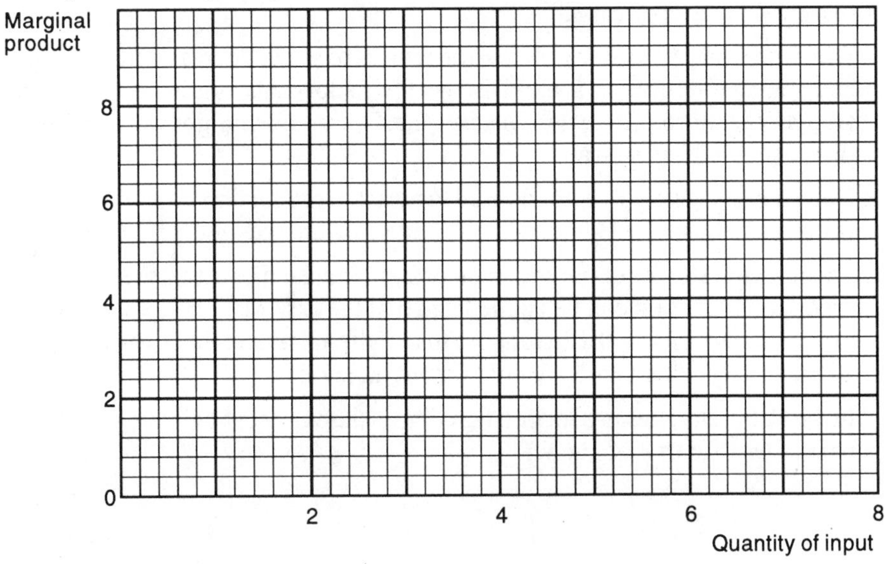

5. State the law of diminishing marginal returns. Indicate how this law is used to deduce the shape of the marginal product curve.
6. On the basis of the total product curve shown in the graph below, derive a measure of the average product and marginal product at *OQ* units of input by graphical techniques.

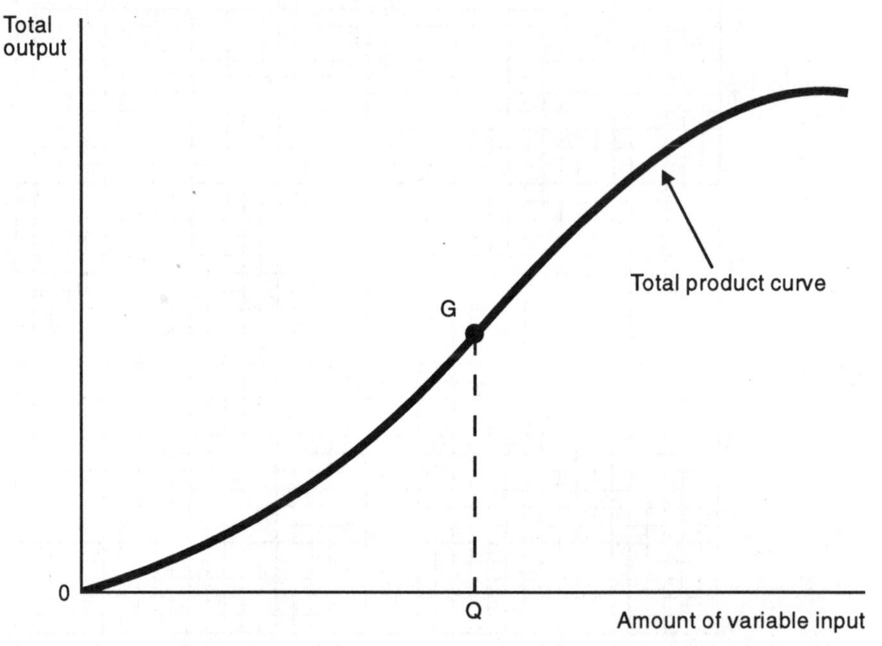

7. Describe an isoquant. Define the marginal rate of technical substitution.
8. (Advanced) Use calculus to show that the marginal rate of technical substitution of labor for capital is equal to the ratio of the marginal product of labor to the marginal product of capital.
9. Describe what is meant by increasing returns to scale, decreasing returns to scale, and constant returns to scale. Discuss the factors that might be responsible for increasing returns to scale. Discuss the factors that might be responsible for decreasing returns to scale.
10. Describe the principal methods used to measure production functions in particular firms and industries. What are their limitations and advantages?
11. Econometric studies of the cotton industry in India indicate that the Cobb-Douglas production function can be applied and that the exponent of labor is 0.92 and the exponent of capital is 0.12. Suppose that both capital and labor were increased by 1 percent. By what percent would output increase?
12. In the Cobb-Douglas production function, is the exponent of labor generally larger or smaller than that of capital?

Problems

1. The production function for the Omni Laser Company is
$$Q = 10L^{.5} K^{.3} M^{.3}$$
where Q is the number of lasers produced per week, L is the amount of labor used per week, K is the amount of capital used per week, and M is the quantity of raw materials used per week.
   a. Does this production function exhibit diminishing marginal returns?
   b. Does this production function exhibit decreasing returns to scale?
   c. Does the average product of labor depend on the amount of the other inputs used? Explain.

2. According to data derived by Earl Heady and others on the basis of an experiment carried out at the Animal Husbandry Department of Iowa State University,[1] the following combinations of hay and grain consumption per lamb will result in a 25-pound gain on a lamb:

Pounds of hay	*Pounds of grain*
40	130.9
50	125.1
60	120.1
70	115.7
80	111.8
90	108.3
110	102.3
130	97.4
150	93.8

   a. You are asked by a local farmer to estimate the marginal product of a pound of grain in producing lamb. Can you provide such an estimate on the basis of these data?
   b. The local farmer is convinced that constant returns to scale prevail in lamb production. If this is true, and if hay and grain consumption per lamb are the only inputs, how much gain will accrue if the hay consumption per lamb is 100 pounds and the grain consumption per lamb is 250.2 pounds?
   c. What is the marginal rate of technical substitution of hay for grain when between 40 and 50 pounds of hay (and between 130.9 and 125.1 pounds of grain) are consumed per lamb?
   d. Suppose that a major advance in technology occurs which allows farmers to produce a 25-pound gain on a lamb with less hay and grain than the above table indicates. If the marginal rate of technical substitution (at each rate of consumption of each input) is the same after the technological advance as before, can you draw the new isoquant corresponding to a 25-pound gain on a lamb?

1. See E. Heady, *Economics of Agricultural Production and Resource Use.*

3. Suppose that an entrepreneur's utility depends on the size of her firm (as measured by its output) and its profits. In particular, the indifference curves are as follows:

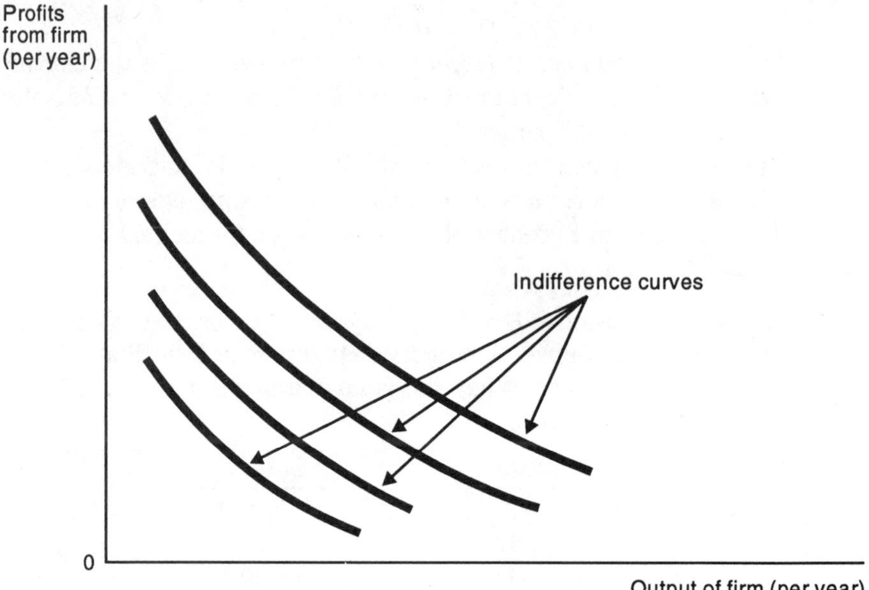

   a. Will the entrepreneur maximize profit?
   b. Draw a graph to indicate how one can determine the output she will choose.

4. At the Amarillo Piano Company, the average product of labor equals 5, regardless of how much labor is used.
   a. What is the marginal product of the first unit of labor?
   b. What is the marginal product of the eightieth unit of labor?
   c. By how much will output increase if labor is increased by 100 units?
   d. By how much will output fall if labor is reduced by 50 units?
   e. Does this case conform to the law of diminishing marginal returns? Why or why not?
   f. Does this case seem realistic? Why or why not?

5. In Canada, suppose that the relationship between a cow's total output of milk and the amount of grain it is fed is as follows:

Amount of grain (pounds)	*Amount of milk (pounds)*
1,200	5,917
1,800	7,200
2,400	8,379
3,000	9,371

   (This relationship assumes that forage input is fixed at 6,500 pounds of hay.)
   a. Calculate the average product of grain when each amount is used.

b. Estimate the marginal product of grain when between 1,200 ant 1,800 pounds are fed, when between 1,800 and 2,400 pounds are fed, and when between 2,400 and 3,000 pounds are fed.

c. Does this production function exhibit diminishing marginal returns?

6. The Antioch Corporation, a hypothetical producer of paper napkins, claims that in 1994 it has the following production function:

$$Q = 3 + 4L + 2P$$

where Q is the number of paper napkins it produces per year, L is the number of hours of labor per year, and P is the number of pounds of paper used per year.

a. Does this production function seem to include all of the relevant inputs? Explain.

b. Does this production function seem reasonable, if it is applied to all possible values of L and P? Explain.

c. Does this production function exhibit diminishing marginal returns?

7. Earl Heady of Iowa State University estimated a Cobb-Douglas production function for six types of farms, there being 5 inputs in the production function: (1) land, (2) labor, (3) equipment, (4) livestock and feed, and (5) other resource services. The exponent of each input was as shown.

	Exponent				
Farm Type	*Land*	*Labor*	*Equipment*	*Livestock and feed*	*Other resource services*
Crop farms	0.24	0.07	0.08	0.53	0.02
Hog farms	0.07	0.02	0.10	0.74	0.03
Dairy farms	0.10	0.01	0.06	0.63	0.02
General farms	0.17	0.12	0.16	0.46	0.03
Large farms	0.28	0.01	0.11	0.53	0.03
Small farms	0.21	0.05	0.08	0.43	0.03

Source: Heady, *Economics of Agricultural Production and Resource Use.*

a. Do there appear to be increasing returns to scale in any of these six types of farms?

b. In what type of farm does a 1 percent increase in labor have the largest percentage effect on output?

c. Based on these results, would you expect that output would increase if many of the farms included in Heady's sample were merged?

8. Fill in the blanks in the following table:

Number of units of variable input	*Total output (number of units)*	*Marginal product* of variable input*	*Average product of variable input*
3	___	Unknown	30
4	___	20	___
5	130	___	___
6	___	5	___
7	___	___	19½

* These figures pertain to the interval between the indicated amount of the variable input and one unit less than the indicated amount of the variable input.

9. As the quantity of a variable input increases, explain why the point where *marginal* product begins to decline is encountered before the point where average product begins to decline. Explain too why the point where *average* product begins to decline is encountered before the point where *total* product begins to decline.

10. The following graph shows the combinations of quantities of grain and protein that must be used to produce 150 pounds of pork. Curve *A* assumes that no Aureomycin is added, while curve *B* assumes that some of it is added.

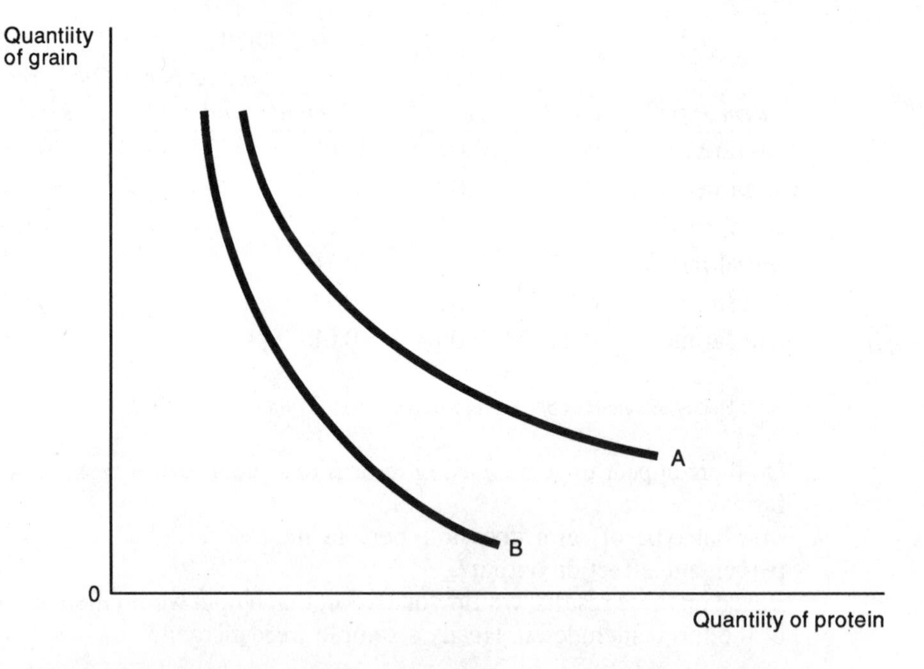

a. If Aureomycin can be obtained free, should pork producers add it?

b. Does the addition of Aureomycin affect the marginal rate of technical substitution? If so, how?

11. In the early 1880s, cigarette-rolling machinery became available. The new machines cut the cost of fabricating a cigarette in half. Assuming that capital and labor are the only inputs, what effect did this innovation have on the production

function for cigarettes? Once the new machines were available, how did the optimal input combination differ from what it was before their introduction?

12. According to some estimates, there are important economies of scale in automobile production up to a plant size equaling about 5-10 percent of U.S. auto output. What are the implications of this fact for the number of auto companies that we are likely to find in the United States? In Chile? How would you go about obtaining data concerning economies of scale in some other industry?

13. If Q is the number of cars washed per hour and L is the number of persons employed, a study of an auto laundry found the following short-run relationship:
$$Q = -0.8 + 4.5L - 0.3L^2$$
Do there appear to be diminishing marginal returns?

14. Suppose that in a chemical plant $Q = AL^a K^b$ where Q is the output rate, L is the rate of labor input, and K is the rate of capital input. Statistical analysis indicates that $a = 0.8$ and $b = 0.3$. The owner of the plant claims that there are increasing returns to scale in the plant. Is he right?

15. (Advanced) Suppose you are assured by the owner of an aluminum plant that his plant is subject to constant returns to scale, labor and capital being the only inputs. He claims that output per worker in his plant is a function of capital per worker only. Is he right?

Key Concepts for Review

Profit
Profit maximization
Technology
Input
Fixed input
Variable input
Short run
Long run
Production function
Marginal product
Average product
Law of diminishing marginal returns
Isoquants
Economic region of production
Marginal rate of technical substitution
Increasing returns to scale
Decreasing returns to scale
Constant returns to scale
Cobb-Douglas production function

ANSWERS

Case Study: Japanese Manufacturing Techniques

a. Because of its hand-to-mouth nature, just-in-time production means that a firm holds less inventories of parts, subassemblies, and finished goods. Because it is expensive to hold inventories, this is an advantage. Also, defects tend to be discovered more quickly and their causes may be nipped in the bud.

b. If goods, subassemblies, and parts are to be produced and delivered just in time for use or sale, they must be produced in small lots, which means that the equipment used to produce them must be set up more often. Each setup is expensive; it often involves moving heavy dies into place, making adjustments, and inspecting the results until the settings are right. To overcome this disadvantage, the Japanese have worked very hard to reduce setup costs. For example, Toyota has reduced the time to set up 800-ton presses used in forming auto hoods and fenders from an hour in 1971 to about 10 minutes or less.

c. Higher quality of product, less waste of materials, a heightened awareness of the causes of defects on the part of the people doing the work, and fewer inspectors.

d. Yes. To see whether they should be adopted in the United States, engineering principles alone do not suffice. It is essential that microeconomic concepts be applied to find out whether these techniques really would lower costs and increase efficiency.[2]

Completion Questions

1. 48; 75; 27
2. long run
3. Cobb-Douglas
4. fixed in quantity
5. variable in quantity
6. variation
7. the quantity of the input
8. an extra unit of the input
9. constant
10. intersect

True or False

1. True	2. False	3. False	4. True	5. False	6. False
7. False	8. True	9. True	10. False	11. False	12. False
13. False					

Multiple Choice

1. *a*	2. *c*	3. *c*	4. *a*	5. *a*	6. *c*

Review Questions

1. First, the making of profits generally requires time and energy, and, if the owners of the firm are the managers as well, they may decide that it is preferable to sacrifice profits for leisure. Second, in an uncertain world, the concept of maximum profit is not clearly defined. Third, it is often claimed that firms pursue other objectives than profits: e.g., better social conditions, a good image, higher market share, and so forth.

   Profit maximization remains the standard assumption in microeconomics because it is a close enough approximation for many important purposes, because alternative theories sometimes require unavailable data, and because it provides rules of rational behavior for firms that do want to maximize profits.

2. See R. Schonberger, *Japanese Manufacturing Techniques* New York: The Free Press, 1982 and Y. Tsurumi, "Japan's Challenge to the U.S.", *Columbia Journal of World Business*, Summer 1982.

2. Technology is the sum total of society's pool of knowledge concerning the industrial and agricultural arts. How much can be produced with a given set of inputs depends upon the level of technology.

   A fixed input is an input whose quantity cannot be changed during the period of time under consideration. A variable input is an input whose quantity can be changed during the relevant period.

3. The production function is the relationship between the quantities of various inputs used per period of time and the maximum quantity of the output that can be produced per period of time.

   The short run is that period of time in which some of the firm's inputs are fixed.

   The long run is that period of time in which all inputs are variable.

   In the short run, certain inputs are fixed in quantity whereas this is not the case in the long run.

4. *a.* The average product curve is as follows:

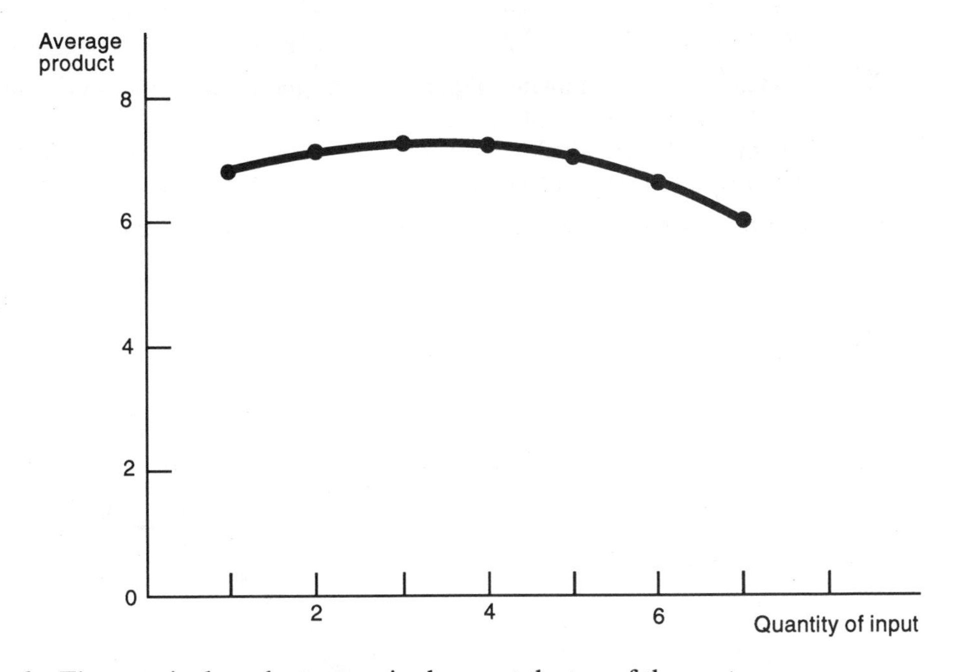

   b. The marginal product curve is shown at the top of the next page.

5. The law of diminishing marginal returns states that, if equal increments of an input are added, the quantities of other inputs being held constant, the resulting increments in product will decrease beyond some point.

   Since marginal product decreases beyond some point, the marginal product curve must turn down beyond some point.

6. The average product of *OQ* units of input equals the slope of *OG*, a straight line that connects *O* and *G*. The marginal product at *OQ* units of input equals the slope of VV^1, the tangent at *G*. See graph at the bottom of the next page.

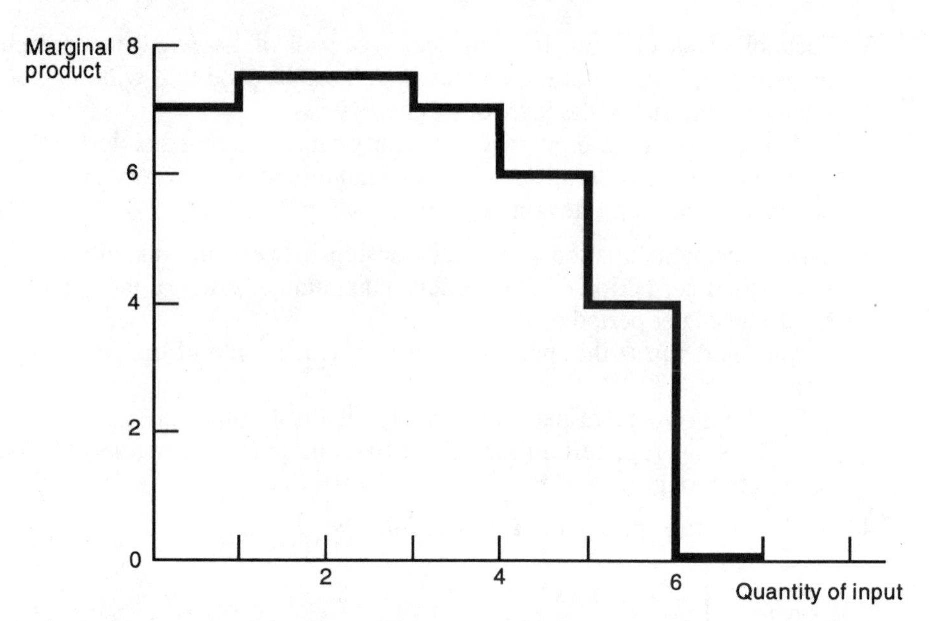

7. An isoquant is a curve showing all possible (efficient) combinations of inputs that can produce a certain quantity of output.

   The marginal rate of technical substitution is the rate at which one input can be substituted for another to maintain a constant output rate.

8. Letting Q be output, x_1 be the amount of labor, and x_2 be the amount of capital,

$$dQ = \frac{\partial f}{\partial x_1}\, dx_1 + \frac{\partial f}{\partial x_2}\, dx_2$$

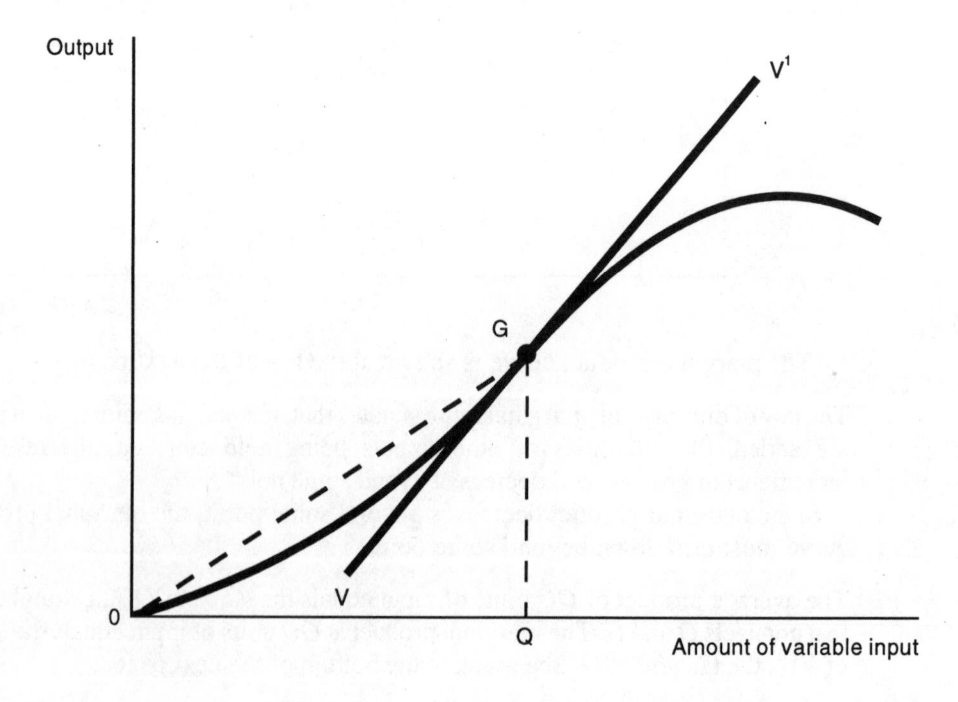

where f is the production function. Since $dQ = 0$, the marginal rate of technical substitution is

$$\frac{-dx_2}{dx_1} = \frac{\partial f}{\partial x_1} \div \frac{\partial f}{\partial x_2}$$

9. If all inputs are increased by a certain percentage, with the result that output increases by more than this percentage, this is a case of increasing returns to scale.

   If all inputs are increased by a certain percentage, with the result that output increases by less than this percentage, this is a case of decreasing returns to scale.

   If all inputs are increased by a certain percentage, with the result that output increases by the same percentage, this is a case of constant returns to scale.

   Increasing returns to scale may be due to indivisibilities, various geometrical relations, greater specialization, probabilistic considerations, and so forth.

   Decreasing returns to scale may result from the difficulties of managing a large enterprise.

10. The three principal methods have been time-series analysis, cross-section studies, and methods based on information supplied by engineers or agricultural scientists. One problem is that the data may not pertain to efficient combinations of inputs. Another problem is the difficulty of measuring capital. Despite these and many other problems, the resulting estimates have been of considerable interest and use.

11. 1.04 percent.

12. It is generally larger than the coefficient of capital.

Problems

1. *a.* Yes.
   b. No.
   c. Yes.

2. *a.* No.
   b. 50 pounds, since half of these amounts (that is, 50 pounds of hay and 125.1 pounds of grain) results in a 25-pound gain.
   c. 0.58.
   d. No, because it is impossible to tell (from the information given in the question) how much hay and grain can be used to produce a 25-pound gain after the advance in technology.

3. *a.* No.
   b. She will choose point A in the graph on the next page. Thus, output will be Oq. See Question 7.1 in the text.

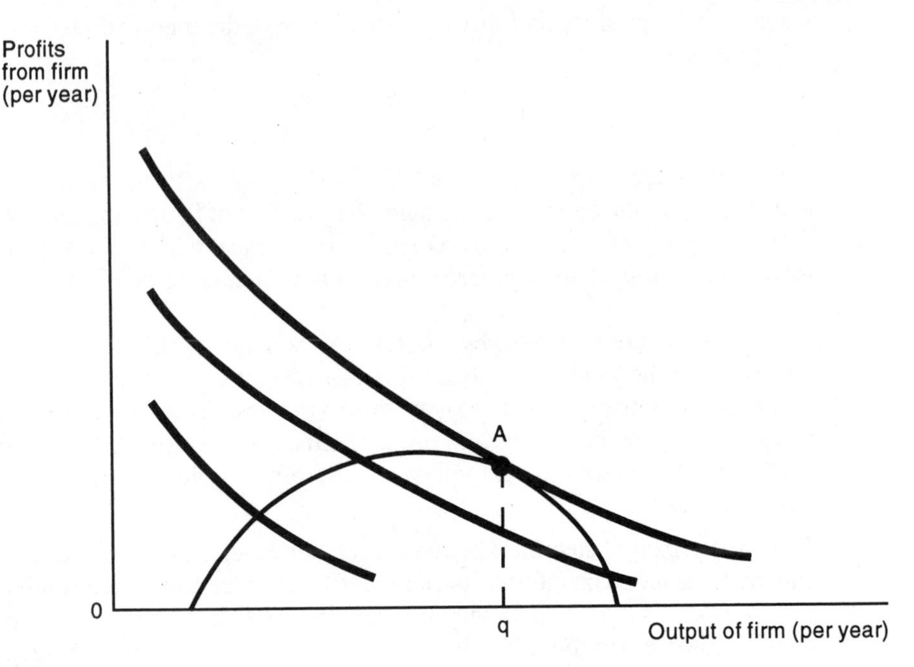

4. *a.* 5.
   b. 5.
   c 500.
   d. 250.
   e. No.
   f. No.

5. a. and b. The average and marginal products of grain when each amount is used are calculated as follows:

Amount of grain	*Average product*	*Marginal product*
1,200	$5{,}917 \div 1{,}200 = 4.93$	
		$\frac{7{,}200 - 5{,}917}{1{,}800 - 1{,}200} = 2.22$
1,800	$7{,}250 \div 1{,}800 = 4.03$	
		$\frac{8{,}379 - 7{,}250}{2{,}400 - 1{,}800} = 1.88$
2,400	$8{,}379 \div 2{,}400 = 3.49$	
		$\frac{9{,}371 - 8{,}379}{3{,}000 - 2{,}400} = 1.65$
3,000	$9{,}371 \div 3{,}000 = 3.12$	

   c. Yes. The marginal product of grain decreases as more of it is used.

6. *a.* No. It seems likely that some capital and land are used.
   b. No. It says that an extra 1/4 hour of labor, or an extra 1/2 pound of paper, will result in an extra paper napkin regardless of the number of paper napkins produced. Beyond some output level, one would expect the marginal product of labor and of paper to fall.
   c. No.

7. *a.* No.
   b. General farms.
   c. No.

8. The complete table is:

Number of units of variable input	*Total output (number of units)*	*Marginal product of variable input*	*Average product of variable input*
3	90	Unknown	30
4	110	20	27½
5	130	20	26
6	135	5	22½
7	136½	1½	19½

9. Because of the law of diminishing marginal returns, the marginal product begins to decline at some point. If the marginal product exceeds the average product at that point, the marginal product can fall to some extent without reducing the average product. Only when it falls below zero will the total product begin to decrease.

10. *a.* Yes, since it seems to reduce the amount of grain and protein that must be used to produce 150 pounds of pork. However, this assumes that it has no negative effect on the quality of the pork and that it does not increase the amount of other inputs that must be used.
    b. At each quantity of protein, curve *B* is steeper than curve *A*. In other words, the absolute value of its slope is greater. Thus, the marginal rate of technical substitution of protein for grain is greater when Aureomycin is added than when it is not.

11. It resulted in more output being obtainable from a fixed amount of capital and labor than before. Once the new machines were available, the optimal input combination was one where the capital-labor ratio was higher than before.

12. It is clear that we are unlikely to find more than about 10-20 auto firms in the United States. Since Chile produces far fewer autos, we would expect to find far fewer auto firms than this number. Engineering estimates and statistical analysis of costs are some important techniques to estimate economies of scale.

13. Yes. The marginal product of labor equals the slope of this relationship. If you plot the relationship, you will find that the slope decreases as L increases.

14. Yes.

15. Yes.

Chapter 8 Optimal Input Combinations and Cost Functions

Case Study: Cost Functions In the Railroad Industry

Edwin Mansfield and Harold Wein estimated the relationship between total cost and output in a railroad freight yard.[1] All yards contain sets of tracks. In large yards, the tracks are generally of three types: receiving tracks where incoming freight cars are stored, classification tracks where cars are switched, and outbound tracks where cars that are situated on a classification track are stored until a locomotive hauls them away as a train. Freight yards switch cars; that is, they sort incoming cars by putting them on the appropriate classification tracks, and in this way they break up incoming trains to form new trains. Most yards also deliver and pick up cars. Engines are assigned to deliver cars to industrial sidings and other yards and to pick them up there.

For the particular yard studied by these authors, the relationship between total cost and output is

$$C = 4{,}914 + 0.42S + 2.44D$$

where C is daily cost (in dollars), S is the number of cuts switched per day, and D is the number of cars delivered per day. A cut is a group of cars that rolls as a unit on to the same classification track; it is often used as a measure of switching output.

a. If this yard delivers no cars and switches 1,000 cuts on a given day, what is the average total cost of switching each cut?

b. If this yard switches no cuts and delivers 1,000 cars on a given day, what is the average total cost of delivering a car?

c. What is the marginal cost of switching a cut? Based on the above equation, does the marginal cost of switching a cut depend on the number of cars delivered?

d. What is the marginal cost of delivering a car? Based on the above equation, does the marginal cost of delivering a car depend on the number of cuts switched?

e. Do you think that the above equation holds for any value of S or D, no matter how large? Explain.

Completion Questions

1. If the average fixed cost of producing 10 units of output at the Rothschild Manufacturing Company is $10, the average fixed cost of producing 20 units is _______________. If the marginal cost of each of the first 20 units of output is $5, the average variable cost of producing 20 units is _______________. And the average total cost of producing 20 units is _______________.

2. If it minimizes cost, a firm will produce at a point where the isocost curve is _______________ to the isoquant.

1. E. Mansfield, and H. Wein, "A Regression Control Chart for Costs," *Applied Statistics*.

3. Plant and equipment of a firm are ______________ in the short run.
4. Total cost equals ______________ plus variable cost.
5. Average cost must equal marginal cost at the point where average cost is a ______________.
6. The long-run total cost equals output times ______________.
7. ______________ include opportunity costs of resources owned and used by the firm's owner.
8. An important criticism of cross-section studies of cost functions is that they sometimes are subject to the ______________.
9. Some determinants of the shape of the long-run average cost curve are ________.
10. The total variable cost curve turns up beyond some output level because of the ______________.
11. The marginal cost curve turns up beyond some output level because of the ______________.
12. Average variable cost equals the price of the variable input divided by ______________, if the price of the variable input is constant.
13. Marginal cost equals the price of the variable input divided by ___________, if the price of the variable input is constant.

True or False

____ 1. If average variable cost always equals \$20 when output is less than 100 units, marginal cost is less than \$20 when output is in this range.

____ 2. Long-run marginal cost can never exceed short-run marginal cost.

____ 3. A firm always can vary the quantity of labor inputs in the short run.

____ 4. Costs that have already been incurred are important factors in making production decisions.

____ 5. The opportunity cost doctrine says that the production of one good may reduce the cost of another good.

____ 6. The marginal rate of technical substitution of labor for capital is the marginal product of capital divided by the marginal product of labor.

____ 7. When the firm has constructed the scale of plant that is optimal for producing a given level of output, long-run marginal cost will equal short-run marginal cost at that output.

____ 8. The shape of the long-run average cost function is due primarily to the law of diminishing marginal returns.

____ 9. Average cost must exceed marginal cost at the point where average cost is a minimum.

____ 10. The break-even point lies well above the output level that must be reached if the firm is to avoid losses.

___ 11. Whether or not an industry is a natural monopoly depends on the long-run average cost curve and the industry demand curve.

___ 12. Empirical studies often indicate the short-run average cost curve is S-shaped.

Multiple Choice

1. Firm X's average total cost per month equals $5 x *Q*, where *Q* is the number of units of output produced per month. The marginal cost of the third unit of output produced per month is
   a. $15.
   b. $20.
   c. $25.
   d. $30.
   e. none of the above.

2. The curve in the graph below has the shape of
   a. marginal cost curve.
   b. average variable cost curve.
   c. average fixed cost curve.
   d. all of the above.
   e. none of the above.

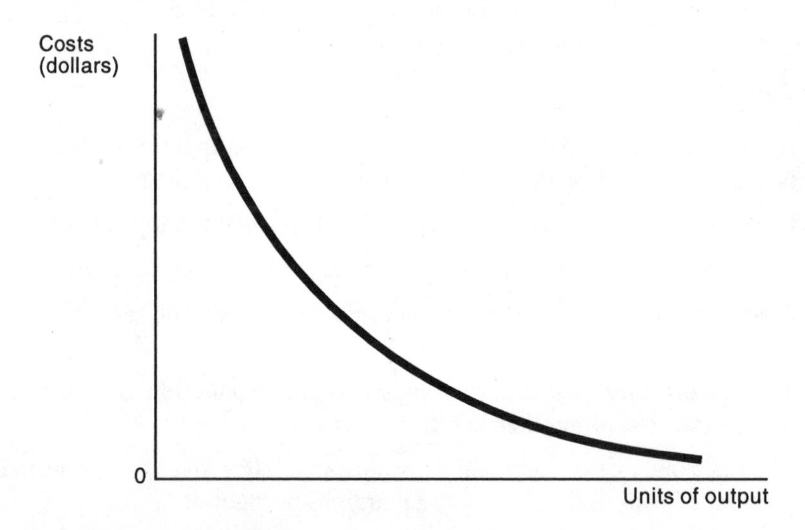

3. The firm's cost functions are determined by
   a. the price of its product.
   b. its advertising agency.
   c. its production function.
   d. the age of the firm.
   e. none of the above.

4. The following industry often is a natural monopoly;
   a. cigarette industry.
   b. publishing industry.

c. drug industry.
d. electric power industry.
e. none of the above.

5. The curve shown below is
   a. an isoquant.
   b. an isocost curve.
   c. an average cost curve.
   d. a marginal cost curve.
   e. none of the above.

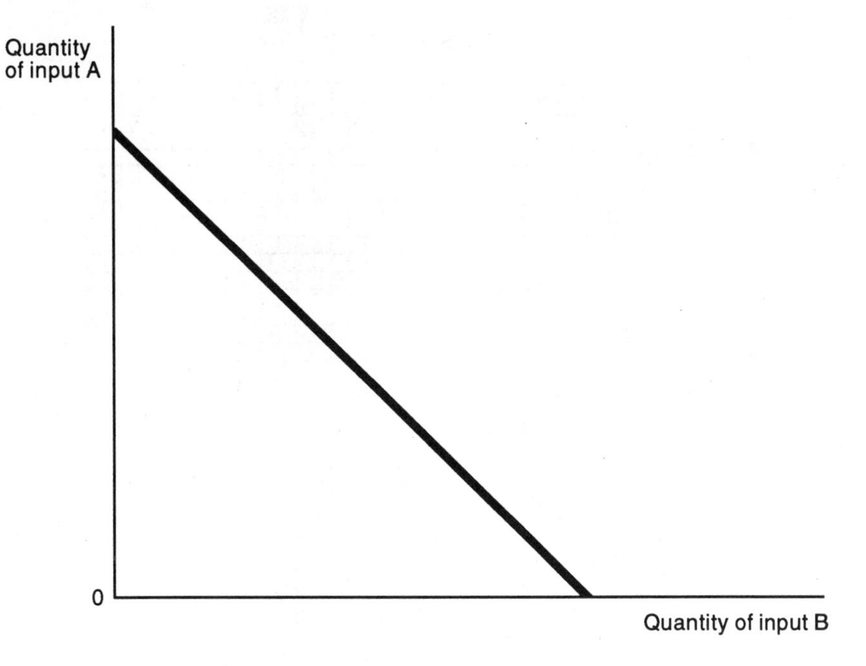

Review Questions

1. Show that a firm will maximize output—for a given outlay—by distributing its expenditures among various inputs in such a way that the marginal product of a dollar's worth of any input is equal to the marginal product of a dollar's worth of any other input that is used.
2. Suppose that capital and labor are the only inputs used by a printing plant and that capital costs $1 a unit and labor costs $2 a unit. Draw the isocost curves corresponding to an outlay of $200 and $300.

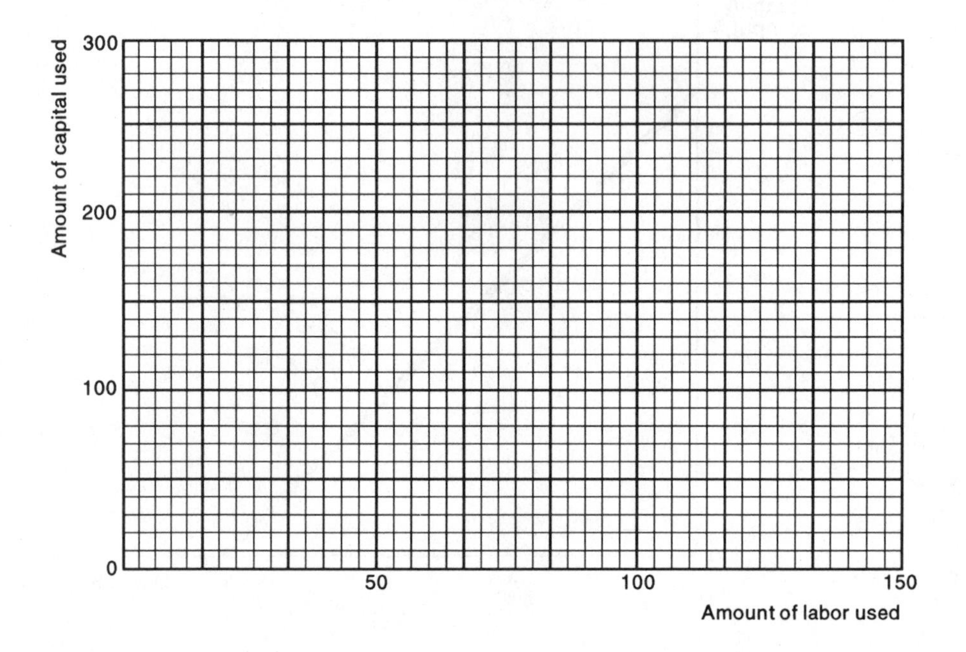

3. Discuss the reasons why the firm, if it is to minimize the cost of producing a given output, must equate the marginal rate of technical substitution and the input-price ratio.
4. Discuss the nature and importance of the opportunity cost or alternative cost doctrine.
5. What are the differences between private and social costs? Illustrate your answer with cases of environmental pollution.
6. What are the differences between explicit and implicit costs? Why do economists bother with implicit costs?
7. What is the difference between the short run and the long run? What is the difference between fixed inputs and variable inputs?
8. *a.* Suppose that a small textile firm's costs are as shown on the next page.

Units of output	*Total Fixed cost (dollars)*	*Total variable cost (dollars)*
0	500	0
1	500	50
2	500	90
3	500	140
4	500	200
5	500	270
6	500	350
7	500	450
8	500	600

Draw the firm's total fixed cost function below.

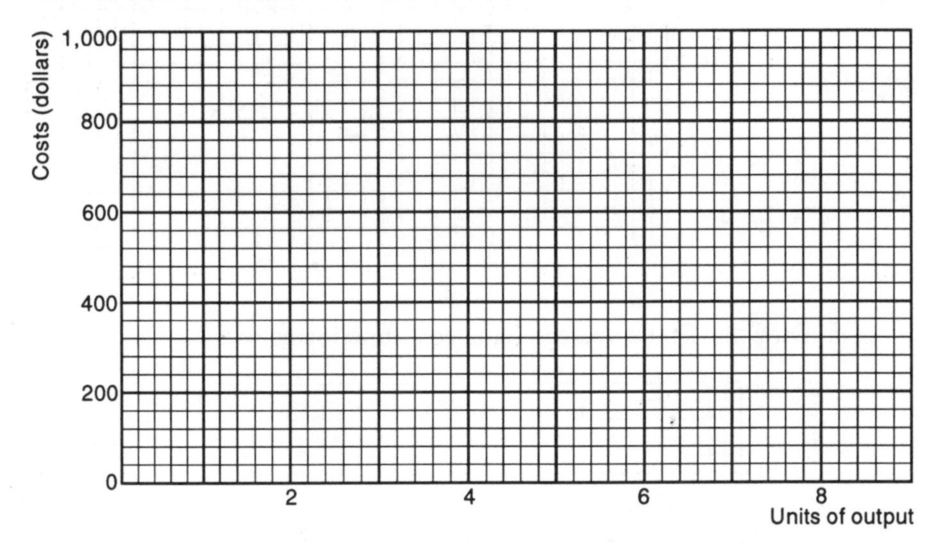

b. Draw the firm's total variable cost function below.

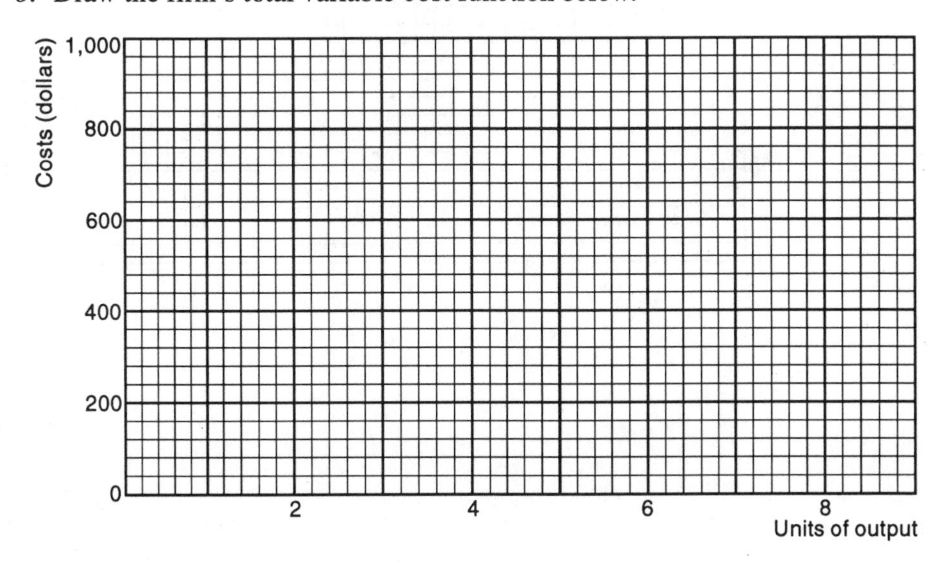

c. Draw the firm's total cost function below.

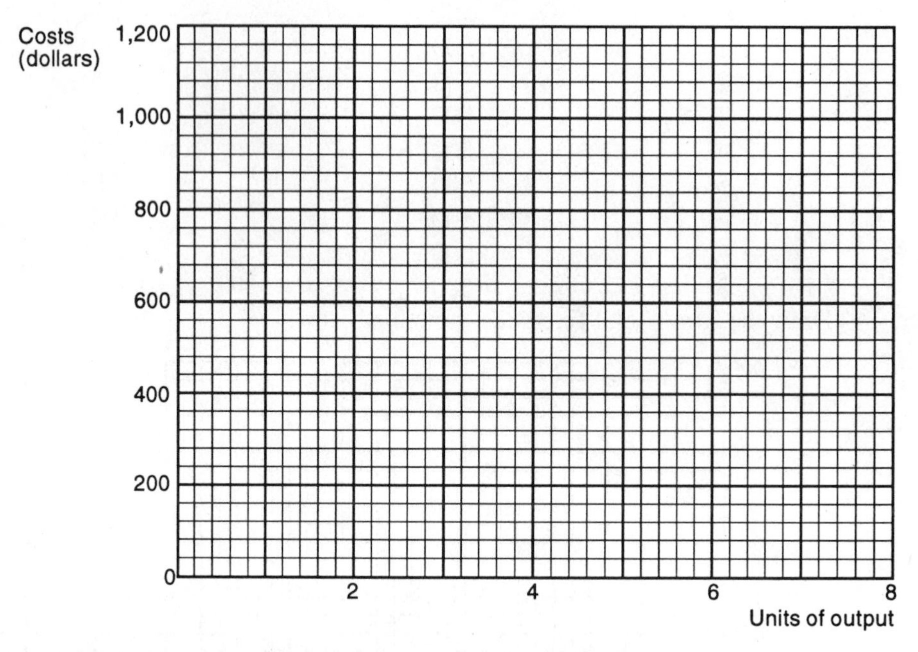

d. Draw the firm's average fixed cost function below.

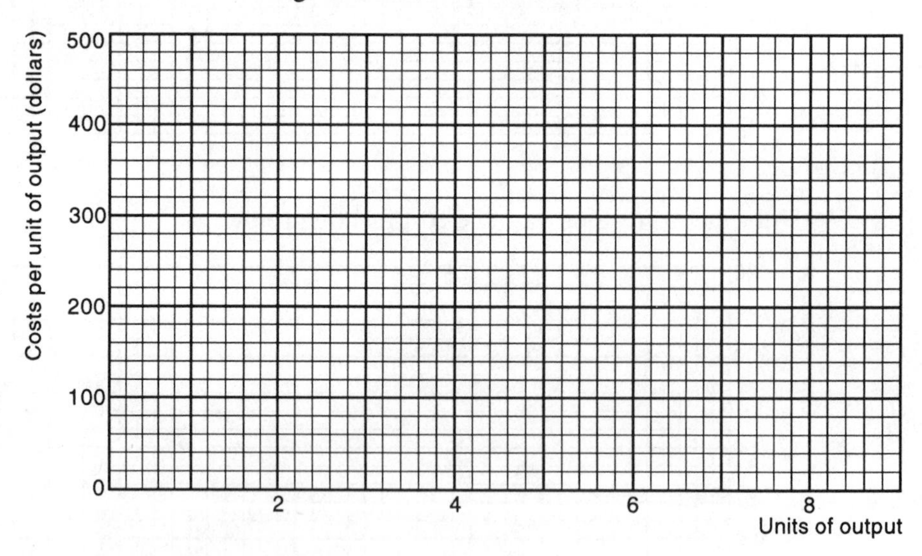

e. Draw the firm's average variable cost function below.

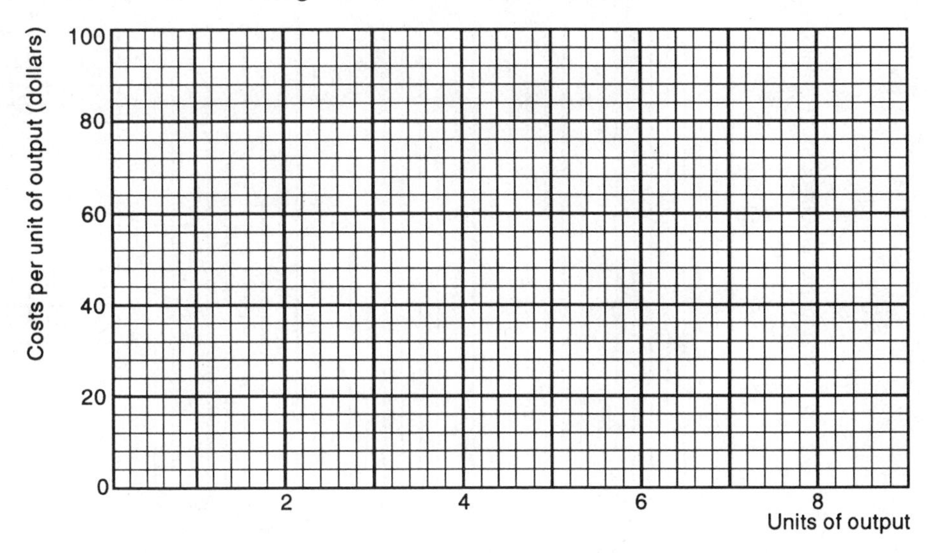

f. Draw the firm's average total cost function below.

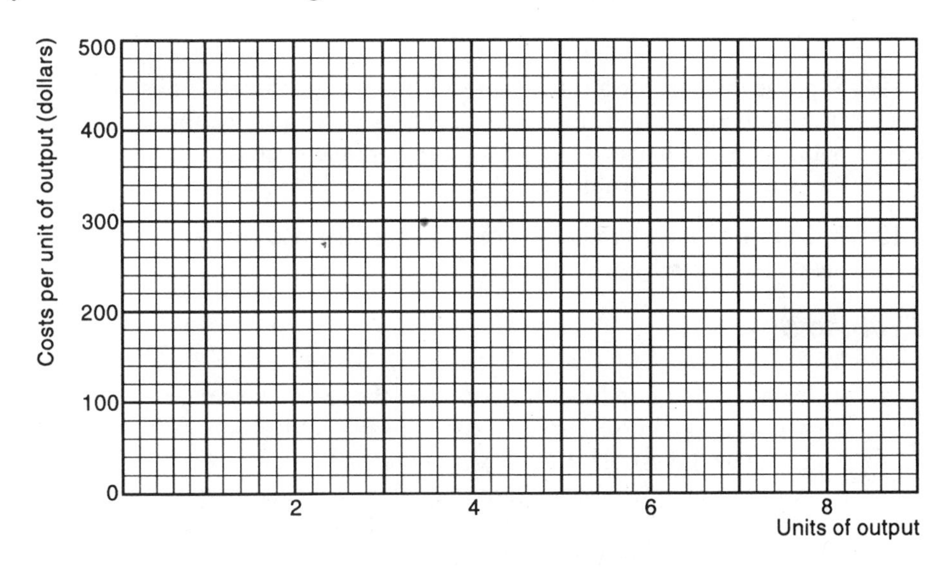

g. Draw the firm's marginal cost function below.

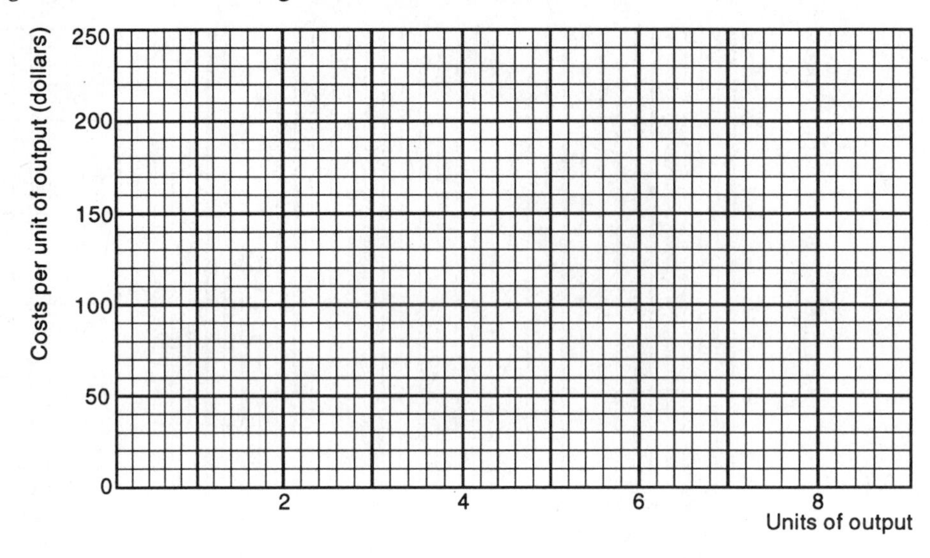

9. Suppose that you are a consultant to a firm that publishes books. Suppose that the firm is about to publish a book that will sell for $10 a copy. The fixed costs of publishing the books are $5,000; the variable cost is $5 a copy. What is the break-even point for this book?

10. If the price were $8 rather than $10 in the previous problem, what would be the break-even point?

Problems

1. *a.* Suppose that two firms have exactly the same marginal cost curve, but their average fixed cost curve is not the same. Will their average variable cost curve be the same? Why or why not?
   b. A firm's marginal product of capital is twice its marginal product of labor; the price of labor is $6 and the price of capital is $3. Is the firm minimizing cost? If not, how can it reduce its costs? Explain.
   c. Will your answer to part (*b*) differ if the firm is a monopolist rather than a perfectly competitive firm? Explain.

2. *a.* Explain the determinants of optimal lot size.
   b. With respect to lot size, how do Japanese manufacturing methods tend to differ from those in the United States?

3. Elizabeth Winter's farm is a profit maximizing, perfectly competitive producer of tomatoes. It produces tomatoes using one acre of land (price = $2,000) and varying amounts of labor (price = $500 per month of labor). The production function is as follows:

Number of months of labor (per month)	*Output of tomatoes per month (in truckloads)*
0	0
1	1
3	2
7	3
12	4
18	5
25	6

Show that Ms. Winter's farm is subject to increasing marginal cost as output increases.

4. John Merriwether, vice-president of sales of a textile firm, is 47 years old, makes $250,000 a year, and is an executive on the rise. Besides his accomplishments in business, Mr. Merriwether is an artist with some talent but no training. He talks wistfully of spending a year or two at an art school but concludes ruefully that "It would cost too much." Mrs. Merriwether, who has never had a job, also has some artistic talent and enrolls at a local art school when her youngest child enters college.
   a. If Mr. Merriwether feels that a year or two of art school would cost too much, why does he encourage his wife to enroll?
   b Is the cost the same to Mrs. Merriwether as to her husband?
   c. What is the cost to her husband of a year at art school?
   d. Indicate how this case helps to explain why the young, rather than the middle-aged, go to college.

5. According to data (which will be analyzed further in Chapter 9 of the text) published by the U.S. Department of Agriculture, 8,500 pounds of milk can be produced by a cow fed the following combinations of hay and grain:

Quantity of hay (pounds)	*Quantity of grain (pounds)*
5,000	6,154
5,500	5,454
6,000	4,892
6,500	4,423
7,000	4,029
7,500	3,694

   a. A cost-minimizing farmer has a cow producing 8,500 pounds of milk, and he is feeding the cow 7,000 pounds of hay and 4,029 pounds of grain. What is the lowest possible value for the ratio of the price of grain to the price of hay? Why?
   b. Under the conditions described in part (*a*), what is the highest possible value of this ratio? Why?
   c. If the price of a pound of hay equals one-half the price of a pound of grain (which equals P), what is the cost of each input combination?

d. Under the conditions described in part (*c*), what is the minimum-cost input combination (of those shown above)?

6. The Miracle Manufacturing Company's short-run average cost function in 1994 is
$$AC = 3 + 4Q$$
where AC is the firm's average cost (in dollars per pound of the product), and Q is its output rate.
   a. Obtain an equation for the firm's short-run total cost function.
   b. Does the firm have any fixed costs? Explain.
   c. If the price of the Miracle Manufacturing Company's product (per pound) is $2, is the firm making profits or losses? Explain .

7. Fill in the blanks in the table below.

Output	*Total cost*	*Total fixed cost*	*Total variable cost*	*Average fixed cost*	*Average variable cost*
0	50	___	___	___	___
1	70	___	___	___	___
2	100	___	___	___	___
3	120	___	___	___	___
4	135	___	___	___	___
5	150	___	___	___	___
6	160	___	___	___	___
7	165	___	___	___	___

8. In recent decades, there has been considerable pressure from consumer advocates and other groups for more and better safety devices in automobiles. What effect do you think the adoption of these devices has on the total cost function of an automobile manufacturer? Does it affect total fixed costs? Marginal costs? If so, how?

9. Based on results obtained by Harold Cohen, the total cost of operating a hospital can be approximated by $4,700,000 + $.00013X^2$, where X is the number of patient days (a crude measure of output.)[2] Derive an expression for the relationship between the cost per patient day and the number of patient days. How big must a hospital be (in terms of patient days) to minimize the cost per patient day?

10. J. A. Nordin found the following relationship between an electric light and power plant's fuel costs (C) and its eight-hour output as a percent of capacity (Q):
$$C = 16.68 + 0.125Q + 0.00439Q^2$$
When Q increases from 50 to 51, what is the increase in the cost of fuel for this electric plant? Of what use might this result be to the plant's managers?

11. *a.* Suppose that a steel plant's production function is $Q = 5LK$, where Q is its output rate, L is the amount of labor it uses per period of time, and K is the amount of capital it uses per period of time. Suppose that the price of labor

2. H. Cohen, "Hospital Cost Curves with Emphasis on Measuring Patient Care Output," *Empirical Studies in Health Economics*, ed. H. Klarman.

is $1 a unit and the price of capital is $2 a unit. The firm's vice-president for manufacturing hires you to figure out what combination of inputs the plant should use to produce 20 units of output per period. What advice would you give him?

b. Suppose that the price of labor increases to $2 per unit. What effect will this have on output per unit of labor?

12. *a.* According to Joel Dean's classic study of a hosiery mill, total cost equaled $\$2{,}936 + 1.998Q$, where Q is output. How does marginal cost behave, according to this finding?
    b. How does average cost behave, according to this finding?
    c. What factor could account for Dean's finding that marginal cost does not increase as output rises?
13. According to Frederick Moore, engineers sometimes rely on the so-called "0.6 rule" which states that the increase in cost is given by the increase in capacity raised to the 0.6 power; that is,

$$C_2 = C_1\left(\frac{X_2}{X_1}\right)^{0.6}$$

where C_1 and C_2 are the costs of two pieces of equipment and X_1 and X_2 are their respective capacities. Does the 0.6 rule suggest economies of scale?

14. According to many econometric studies of long-run average cost in various industries, the long-run average cost curve tends to be L-shaped. Does this mean that there are constant returns to scale at all levels of output?

Key Concepts for Review

Isocost curve
Alternative cost
Opportunity Cost
Social cost
Private cost
Explicit cost
Implicit cost
Cost function
Total fixed cost
Total variable cost
Total cost
Average fixed cost
Average variable cost
Average cost
Marginal cost
Break-even chart
Long-run average cost curve
Long-run marginal cost curve
Statistical cost functions

ANSWERS

Case Study: Cost Functions in the Railroad Industry

a. Total cost equals $4,914.00 + $420.00 = $5,334.00, so average cost equals about $5.33.

b. Total cost equals $4,914.00 + $2,440.00 = $7,354.00, so average cost equals about $7.35.

c. 42 cents. No.

d. $2.44. No.

e. No. Eventually, as more cuts are switched, the marginal cost of switching a cut is bound to rise. Eventually, as more cars are delivered, the marginal cost of delivering a car is bound to rise.

Completion Questions

1. \$5; \$5; \$10
2. tangent
3. fixed
4. fixed cost
5. minimum
6. long-run average cost
7. implicit costs
8. regression fallacy
9. economies and diseconomies of scale
10. law of diminishing marginal returns
11. law of diminishing marginal returns
12. average variable product
13. marginal product

True or False

1. False	2. False	3. False	4. False	5. False	6. False
7. True	8. False	9. False	10. False	11. True	12. False

Multiple Choice

1. *c*	2. *c*	3. *c*	4. *d*	5. *b*

Review Questions

1. Draw the firm's isoquants, as shown in the graph on the next page. Also draw the isocost curve corresponding to the given outlay. Clearly, point P is the input combination that maximizes output for this outlay. Since the firm's isoquant is tangent to the isocost curve at point P, the slope of the isocost curve (which equals -1 times the price of input 1 $\div$ price of input 2) must equal the slope of the isoquant (which equals -1 times the marginal product of input 1 $\div$ marginal product of input 2). Thus, at point P, the ratio of the marginal product to the price of each input must be the same.

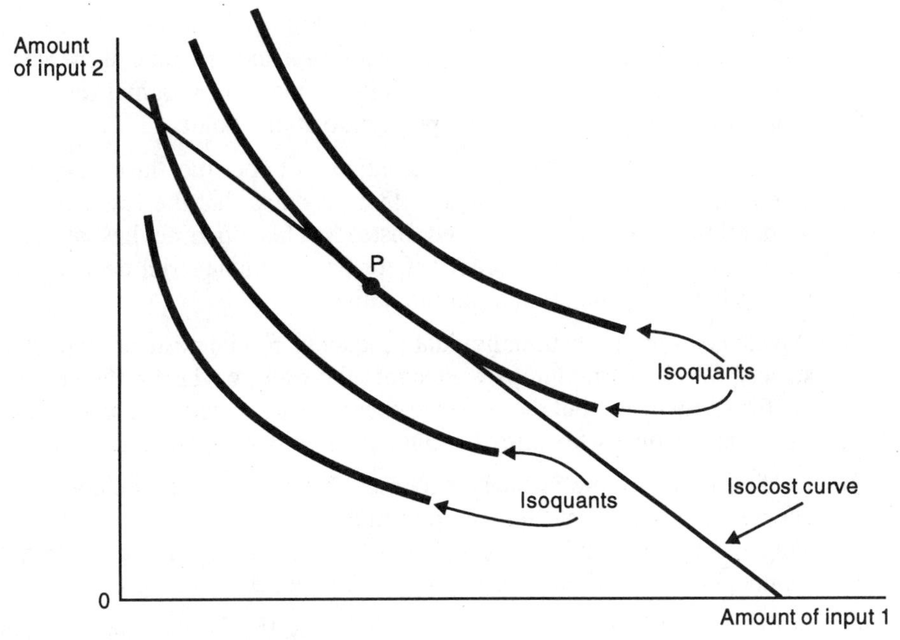

2. The isocost curves are as follows:

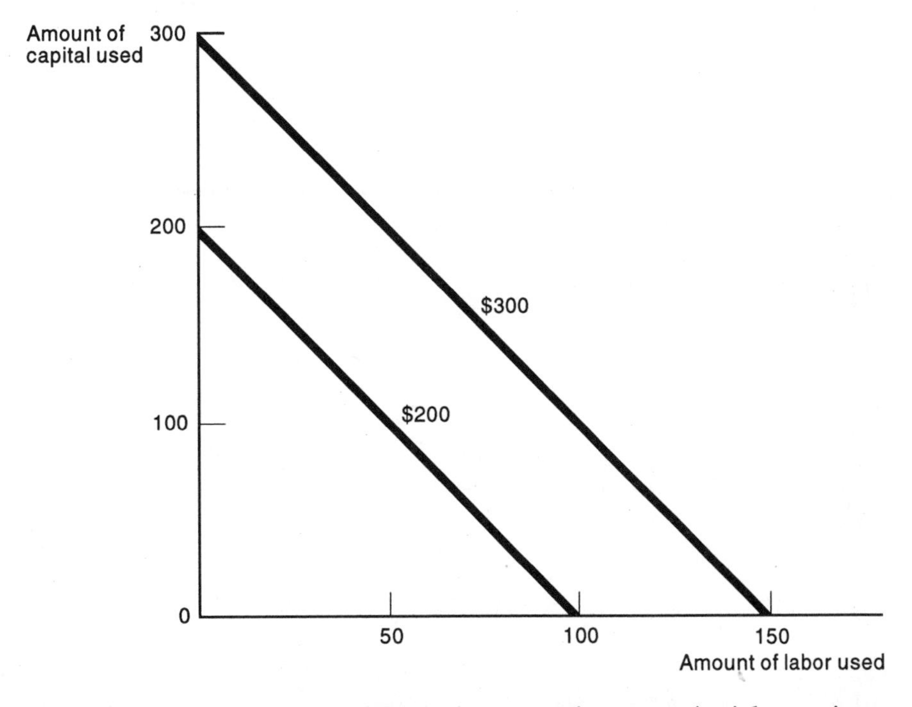

3. If the firm minimizes the cost of producing a certain output, it picks a point on the isoquant (corresponding to this output) that is on the lowest isocost curve. This means that the optimal point is a point of tangency between this isoquant and an isocost curve. But if it is a point of tangency, the slope of the isoquant must equal

the slope of the isocost curve at this point. Since the slope of the isoquant is –1 times the marginal rate of technical substitution and the slope of the isocost curve is –1 times the input price ratio, it follows that the marginal rate of technical substitution must equal the input price ratio at this point.

4. According to the opportunity or alternative cost doctrine, the cost of producing a certain product is the value of the other products that the resources used in its production could have produced instead. This doctrine lies at the heart of economic analysis and is important for proper managerial decision-making as well as for the formulation of public policy.

5. Private costs are costs to individual producers. Social costs are the total costs to society. When a firm dumps wastes into the water or the air, the private costs to the firm may be nil, but the costs to other parts of society—drinkers of the water, fishermen, people who enjoy boating, etc.—may be very great.

6. Explicit costs are the ordinary expenses that accountants include as the firm's expenses. Implicit costs are the opportunity costs of the labor and capital owned and used by the firm's owners. Unless implicit costs are considered, the firm cannot determine whether or not it is making an economic profit.

7. In the short run, some inputs, particularly the firm's plant and equipment, are fixed. In the long run, no inputs are fixed. A fixed input is one that is fixed in quantity. A variable input is one that is not fixed in quantity.

8. *a.*

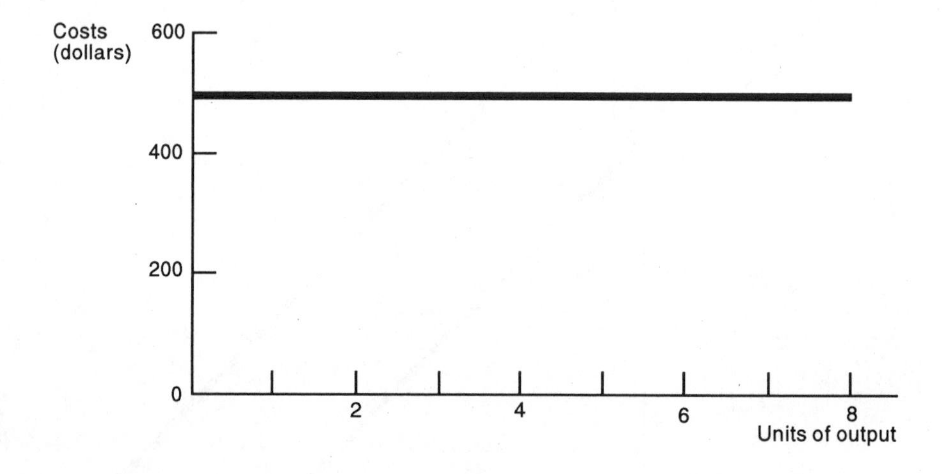

b.

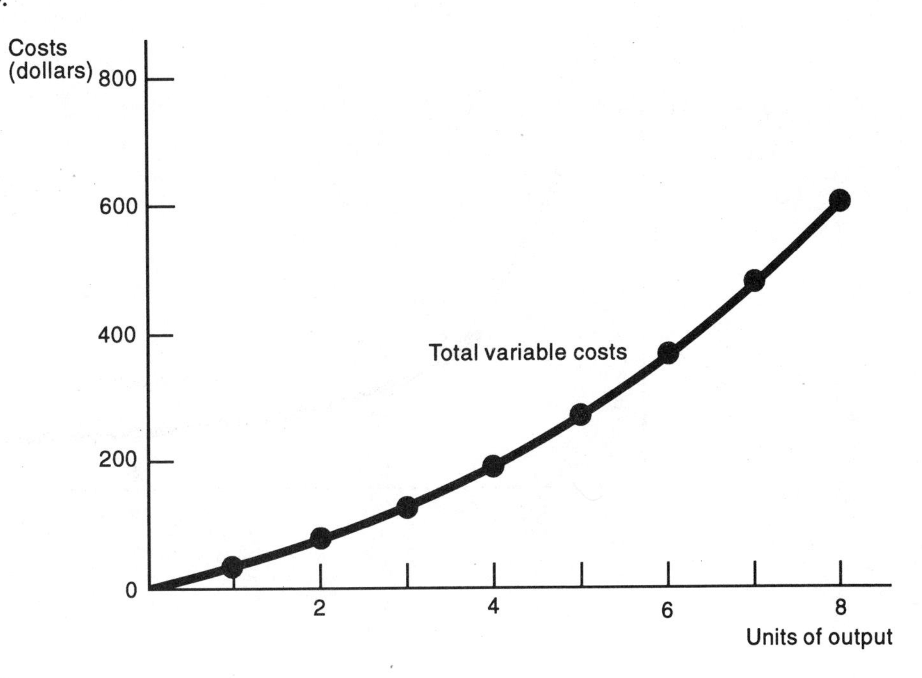
Costs
(dollars)
800
600
400
200
0
Total variable costs
2
4
6
8
Units of output

c.

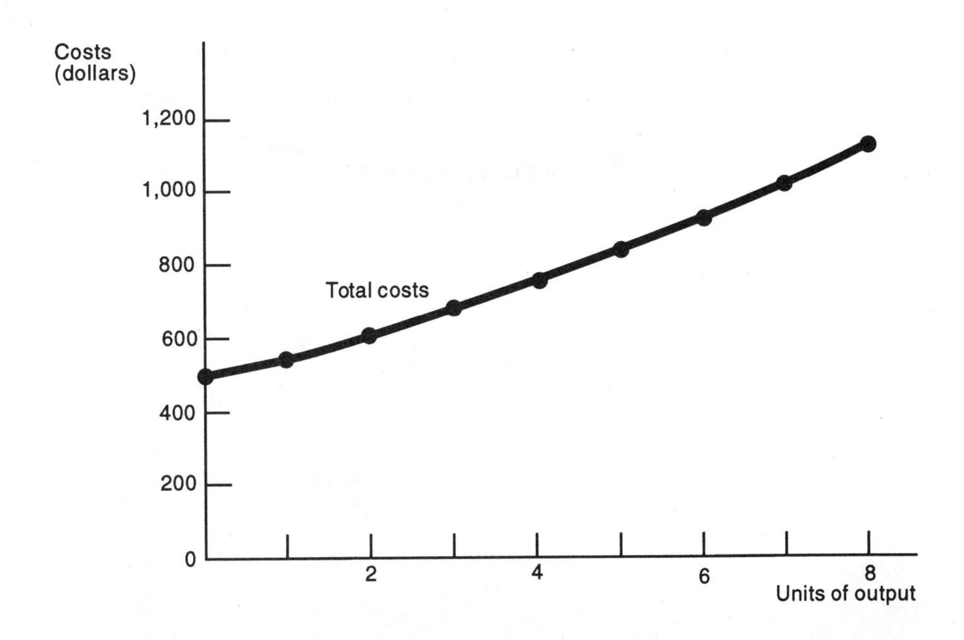
Costs
(dollars)
1,200
1,000
800
600
400
200
0
Total costs
2
4
6
8
Units of output

d.

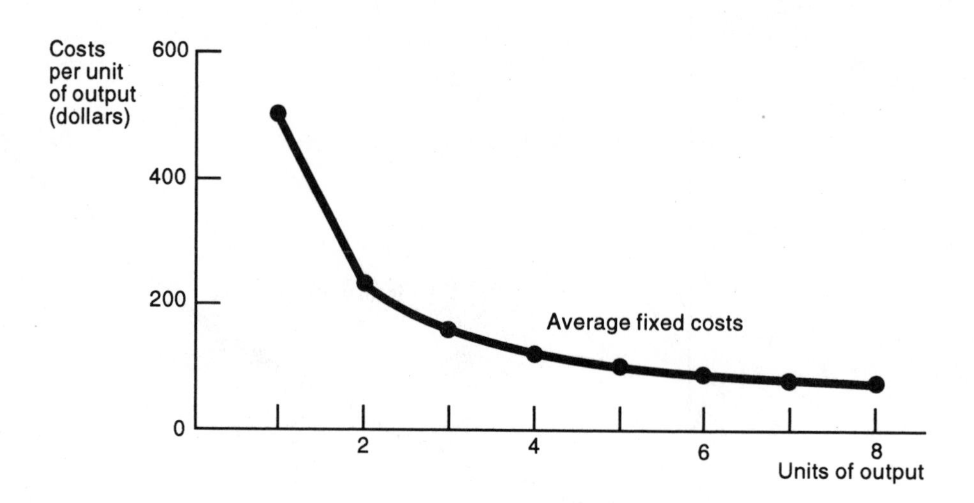
Costs
per unit
of output
(dollars)
600
400
200
0
Average fixed costs
2
4
6
8
Units of output

e.

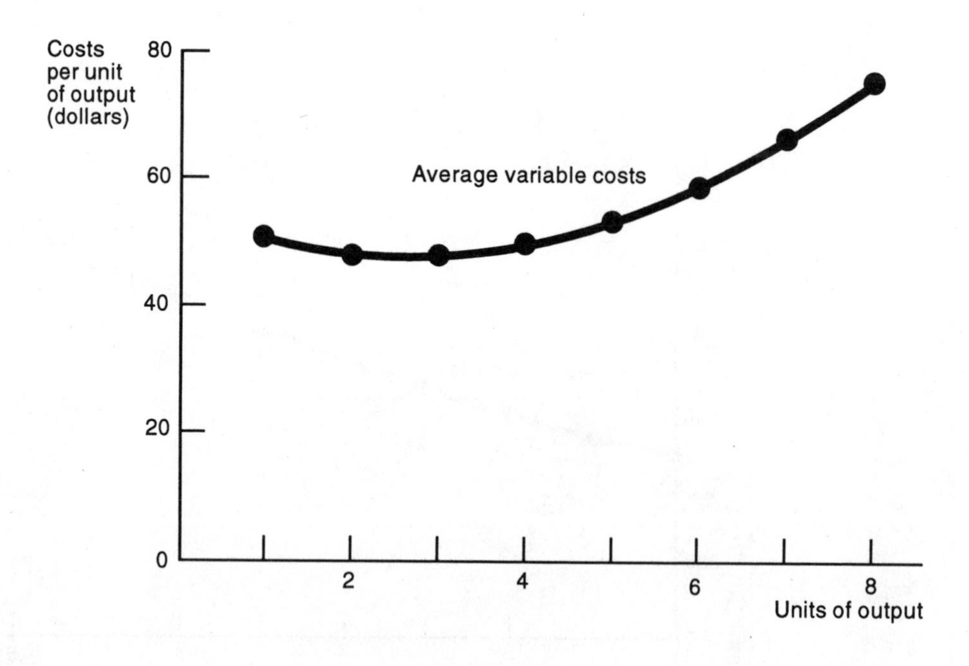
Costs
per unit
of output
(dollars)
80
60
40
20
0
Average variable costs
2
4
6
8
Units of output

f.

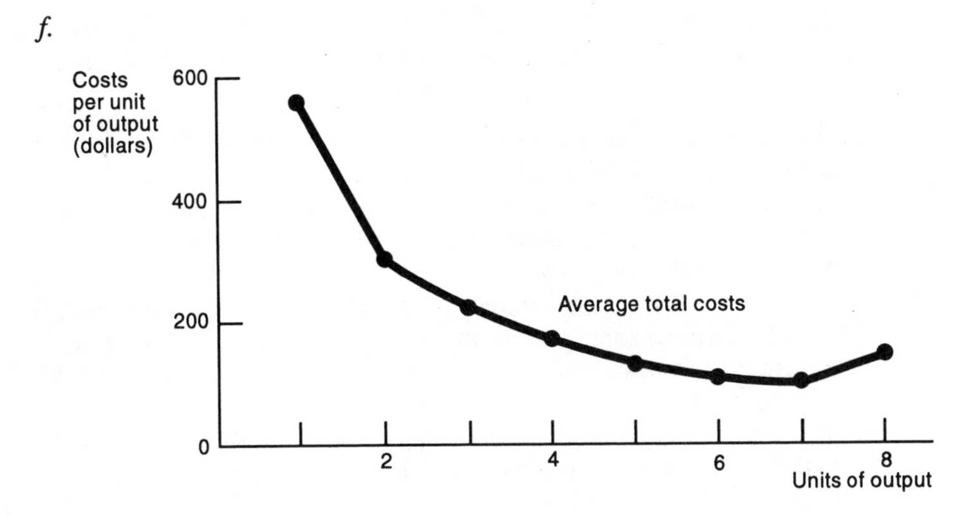

g.

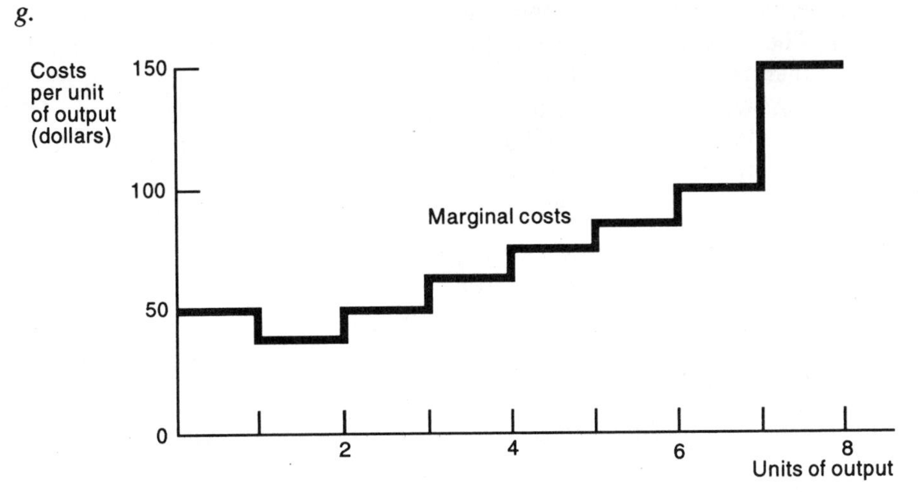

9. 1,000 copies sold.
10. 1,667 copies sold.

Problems

1. *a.* Yes.
   b. No. Since (marginal product of capital ÷ price of capital) exceeds (marginal product of labor ÷ price of labor), the firm can lower its costs by substituting capital for labor.
   c. No.
2. *a.* Optimal lot size is determined by the cost per setup, the total annual requirement of the relevant part, and the annual cost of holding each part of this sort in inventory for a year.
   b. Japanese firms tend to produce goods in smaller lots than do American firms.
3. The *extra* labor needed to produce the first, second, third, fourth, fifth, and sixth

truckloads of tomatoes are 1, 2, 4, 5, 6, and 7 months of labor. Thus, the marginal cost of the first, second, third, fourth, fifth, and sixth truckloads of tomatoes are \$500, \$1,000, \$2,000, \$2,500, \$3,000, and \$3,500.

4. *a.* One important reason is that the cost of Mrs. Merriwether's going to art school is likely to be much less than the cost of his going, if the cost is measured correctly. This cost includes not only the amount spent on tuition and books, but also the alternative cost of the time it takes.

   b. No, because the alternative cost of her time is less than the alternative cost of his time. If he spends a year at art school, they will lose \$250,000 in forgone earnings (and perhaps more, because it may jeopardize his income in future years), whereas if she spends a year at art school, their earnings will not decline.

   c. The cost is the amount spent on tuition and books (and whatever other expenses are associated with attendance at art school) plus the alternative cost of his time. The alternative cost of his time is the amount he could have earned if he had not gone to school.

   d. The costs of going to school tend to be much higher for the middle-aged than for the young, because the alternative cost of a middle-aged person's time tends to be higher than that of a young person (because the middle-aged tend to earn more per hour than the young). In addition, of course, the benefits from going to school may be less for the middle-aged than for the young because there are fewer years ahead to reap such rewards.

5. *a.* If P_G is the price of a pound of grain and P_H is the price of a pound of hay,

$$7{,}000P_H + 4{,}029P_G < 7{,}500P_H + 3{,}694P_G$$
$$7{,}000P_H + 4{,}029P_G < 6{,}500P_H + 4{,}423P_G$$

   These inequalities hold because the chosen input combination is known to have lower cost than any of the others. Thus

$$(4{,}029 - 3{,}694)\ P_G < (7{,}500 - 7{,}000)\ P_H$$

or

$$\frac{4{,}029 - 3{,}694}{7{,}500 - 7{,}000} = \frac{335}{500} = 0.67 < \frac{P_H}{P_G}$$

and

$$(7{,}000 - 6{,}500)\ P_H < (4{,}423 - 4{,}029)\ P_G$$

or

$$\frac{7{,}000 - 6{,}500}{4{,}423 - 4{,}029} = \frac{500}{394} = 1.269 < \frac{P_G}{P_H}$$

   Thus, the lowest possible ratio of the price of grain to the price of hay is 1.269.

   b. In part (*a*), we showed that $P_H/P_G > 335/500$. Thus

$$\frac{P_G}{P_H} < \frac{500}{335} = 1.492$$

   Consequently, the highest possible ratio of the price of grain to the price of hay is 1.492.

   c. The cost of each input combination (in the order they appear in the table in the question) is

8,654*P*
8,204*P*
7,892*P*
7,673*P*
7,529*P*
7,444*P*

d. The minimum-cost combination (of those considered here) is 7,500 pounds of hay and 3,694 pounds of grain.

6. *a.* Since total cost equals average cost times output, the firm's total cost function is

$$C = AC \times Q = 3Q + 4Q^2$$

b. No, since total cost equals zero when $Q = 0$.

c. If the price is \$2, total revenue (R) equals $2Q$. Thus, the firm's profit equals

$$\pi = R - C = 2Q - (3Q + 4Q^2) = -Q - 4Q^2$$

If Q is greater than zero, π must be negative, and the firm is incurring losses. If the firm is producing nothing, it is incurring neither profits nor losses. Thus, the firm is better off to produce nothing.

7. The table is as follows.

Total fixed cost	*Total variable cost*	*Average fixed cost*	*Average variable cost*
50	0	—	—
50	20	50	20
50	50	25	25
50	70	16⅔	23⅓
50	85	12½	21¼
50	100	10	20
50	110	8⅓	18⅓
50	115	7⅐	16 3/7

8. It will probably shift the total cost function upward; that is, the total cost of producing a particular number of automobiles will increase. To the extent that automobile manufacturers must increase their plant and equipment to produce and assemble such safety devices, fixed costs will rise. Marginal cost will probably increase because the addition of the safety devices may increase the cost of producing an extra car.

9. Cost per patient day is

$$\frac{4{,}700{,}000}{X} + .00013X$$

which is a minimum when X is approximately equal to 190,000 patient days.

10. When $Q = 50$, $C = 16.68 + (0.125)(50) + (0.00439)(2500)$, which equals 33.905. When $Q = 51$, $C = 16.68 + (0.125)(51) + (0.00439)(2601)$, which equals 34.473.

Thus, the increase in fuel cost is 0.568.

This result might be of use to the managers in determining whether it would be profitable to increase output.

11. *a.* The isocost curve is: Cost = $1L + 2K$, or $L = \text{cost} - 2K$. The relevant isoquant is $20 = 5LK$, or $L = 4 \div K$. The point on this isoquant that is on the lowest isocost curve is $K = \sqrt{2}$ and $L = 2\sqrt{2}$.
    b. If the price of labor is \$2 per unit, the optimal value of K is 2 and the optimal value of L is 2. Thus, output per unit of labor is $20 \div 2$, or 10, whereas it formerly was $20 \div 2\sqrt{2}$, or $10 \div \sqrt{2}$. Thus output per unit of labor has risen.
12. *a.* It is constant: \$1.998.
    b. It declines as Q increases. Specifically, it equals $1.998 + 2936/Q$.
    c. The data do not cover the range of output near and at the plant's capacity.
13. Yes.
14. No.

Case 2 The Dairy Industry

Richard F. Fallert
Don P. Blayney
James J. Miller

STRUCTURE OF THE DAIRY INDUSTRY

Dairy products account for about 13 percent of total cash receipts from all farm commodities. In 1988, cash receipts from dairy products totaled $17.7 billion, ranking second only to cattle and calves with $36.3 billion. Soybeans and corn followed dairy products in cash receipts, with $12.4 and $10.1 billion, respectively.

Milk, which is bulky, highly perishable, and subject to bacterial and other contamination, must be produced and handled under sanitary conditions and marketed quickly, either for drinking or for manufacture into storable dairy products. Price is the fundamental coordinator of activities in milk production, assembly, processing, and distribution. Prices, even though influenced by government programs, allocate raw milk supplies among competing demands and provide production and marketing signals to dairy farmers and processing and marketing firms.

The ability of market prices to efficiently coordinate economic activities depends in part on the inherent characteristics of milk and its products. Government involvement attempts to overcome certain market deficiencies created by these characteristics. These factors are not unique to milk; but, in combination, they create unique conditions and problems. These characteristics include:

- Extreme perishability of the raw product, with a high potential for transmitting diseases, requiring rapid product movement, refrigeration, and heat treatment;
- Highly inelastic demand—low quantity response to price changes;
- Bulkiness due to its high water content (87 percent);
- Production through a continuous biological process, creating (among other effects) a need for skilled workers every day;
- Unsynchronized seasonality of production and demand;
- Biological lags in output (about 36 months from the time a cow is bred until the heifer enters the milking herd); and
- Joint assembly and hauling of milk for most dairy farmers.

Milk Production

Although milk is produced and processed in every state, over half total 1988 U.S. milk production came from Wisconsin, California, New York, Minnesota, and Pennsylvania. Over two-thirds the total milk supply was produced in 10 states. Large drylot dairy farms with 1,000 to 2,000 cows are common in Florida and the Southwest (southern and central California, Arizona, and New Mexico), but dairy operations of this type are rare elsewhere.

Structure

The number of farms with milk cows declined from 2.8 million in 1955 to about 205,000 in 1989 (Table 1). The number of milk cows declined from 21 million in 1955 to 11.1 million in 1975, and 10.1 million in 1989. A 144 percent increase in milk production per cow enabled production to more than keep pace with commercial needs over the 1955 to 1989 period.

Along with the aggregate structural changes, regional shifts in milk production from the more traditional dairy areas of the Upper Midwest and Northeast to the West and Southwest have been observed. The shift began about three decades ago and has accelerated in the last 20 years. Wisconsin is still far ahead as the number one milk producing state, but California is closing the gap. Population is shifting from the "frostbelt" to the "sunbelt" and may explain part of the milk production shifts. However, other factors, such as a milder climate requiring less overhead in buildings, better control of hay and forage quality, and specialization in strictly milking and managing cows, may be important factors. In addition, the large drylot operations of 1,000 cows or more seem to show economies of specialization allowing more intensive use of facilities and thereby reducing overhead costs.

The size distribution of dairy farms has changed over the last three decades (Table 2). In 1959, 86 percent of the farms with milk cows had fewer than 20 cows. By 1987, only 33 percent fell in this category and they had only 3 percent of the milk cows. In contrast, only 7,172 farms (0.4 percent) had 100 or more cows in 1959, but in 1987, about 10 percent of the herds were in this category and had 42 percent of the milk

TABLE 1

DAIRY INDUSTRY CHANGES, 1955–89

Item	1955	1975	1989	Change per year 1955–75	Change per year 1975–89
		Thousand		Percent [1]	
Cows	21,044	11,139	10,127	–3.1	–0.7
Farms with milk cows	2,763	444	205 [2]	–8.7	–5.4
		Number			
Average cows per farm	8	25	49	5.9	4.9
		Pounds			
Milk per cow (annual)	5,842	10,360	14,244	2.9	2.3
		Million Pounds			
Total milk production	122,945	115,398	145,252	–0.3	1.7

1. Compound annual rate.
2. Commercial dairy farms (farms with 10 or more milk cows) are estimated at around 160,00 in 1989 with an average of around 65 cows per farm.

Source: U.S. Department of Agriculture.

cows. The average herd size on all farms with milk cows was 50 in 1987. The average herd size on farms with 5 or more cows was 63.

If only herds with 5 or more milk cows are considered as commercial dairy farms, 57 percent of the commercial dairy farms had between 5 and 50 milk cows and had 26 percent of the total commercial dairy cow herd in 1987. In contrast, commercial dairy herds with 200 or more milk cows represented about 3 percent of the total commercial herds, but had 24 percent of the commercial dairy cows.

In the Southwest (Arizona and California), 28 percent of the commercial herds had 500 or more cows and accounted for 64 percent of the total cows in commercial dairy herds. In contrast, only 3 percent of the cows in herds with 5 or more cows were in

TABLE 2

FARMS REPORTING MILK COWS, BY HERD SIZE, SELECTED YEARS[1]

Herd size (cows)	1959	1964	1969	1974	1978	1982	1987
Farms reporting milk cows (number):							
1–19	1,706,395	947,236	402,022	224,277	167,840	166,078	65,678
Percent	85.9	77.2	64.1	55.5	50.3	41.8	32.5
20–49	242,733	228,911	171,996	118,706	101,195	88,548	67,622
Percent	12.2	18.7	27.4	29.4	30.4	31.9	33.5
50–99	30,018	40,549	42,426	46,266	48,138	53,334	48,310
Percent	1.5	3.3	6.7	11.5	14.4	19.2	23.9
100 plus	7,172	9,622	11,059	14,505	16,312	19,650	20,335
Percent	.4	.8	1.8	3.6	4.9	4.1	10.1
Total	1,986,318	1,226,318	627,503	403,754	333,485	277,610	201,945
Percent	100.0	100.0	100.0	100.0	100.0	100.0	100.0
Milk cow numbers (1,000 cows):							
1–19	NA	4,489	2,165	1.072	735	538	347
Percent		28.7	17.6	10.1	7.1	5.0	3.4
20–49	[2]13,831	6,832	5,315	3,793	3,300	2,949	2,301
Percent	[2]82.2	43.6	43.2	35.6	31.9	27.2	22.9
50–99	1,785	2,571	2,700	2,973	3,121	3,474	3,169
Percent	10.6	16.4	22.0	27.9	30.1	32.0	31.5
100 plus	1,208	1,768	2,112	2,817	3,199	3,875	4,254
Percent	7.2	11.3	17.2	26.4	30.9	35.8	42.2
Total	16,824	15,660	12,292	10,655	10,355	10,836	10,071
Percent	100.0	100.0	100.0	100.0	100.0	100.0	100.0

NA= not available

1. Does not include Alaska and Hawaii
2. Herd size 1–49

Source: Derived from published U.S. Census of Agriculture data.

commercial herds of 200 or more cows in the Lake States (Minnesota and Wisconsin), while 82 percent were in the 20–99 category.

Herd size reflects only the size of the dairy enterprise, not the size of the whole farm operation. In the Southwest, for example, most farms specialize only in milking cows. Most feed (both forage and concentrate) is purchased, with much of the forage in the region produced under irrigation on specialized hay-producing farms. In other regions, where herds are smaller, a larger proportion of the feed is grown on the farm and other farm enterprises are important to the overall farm operation. Some dairy farmers in these regions expand herd size and specialize the dairy enterprise by shifting from grain to forage production and purchasing more of their concentrates.

Supply Adjustment

Major expansion of the milk supply is a long-term process, mainly because of biology. It takes an average of 27 months from birth until a heifer enters the milking herd (Fig. 1). Contraction of milk supply is also a relatively slow process, impeded by the heavy fixed investment in specialized facilities and lack of alternative farm opportunities and off-farm employment for dairy farmers in some major dairy areas such as Wisconsin and the Northeast. Changes in feeding and culling rates can alter milk production to only a limited extent. These production lags make milk supply relatively unresponsive to price changes over periods of less than a year.

The milk supply is more responsive to price changes in the long run. Most of the inputs—feed concentrates, labor, and equipment—can be acquired in greater volume for dairy production at modestly higher prices. High-quality forage appears to be an exception and a limiting factor for expanding milk production in some areas.

FIGURE 1

DAIRY SECTOR BIOLOGICAL LAGS

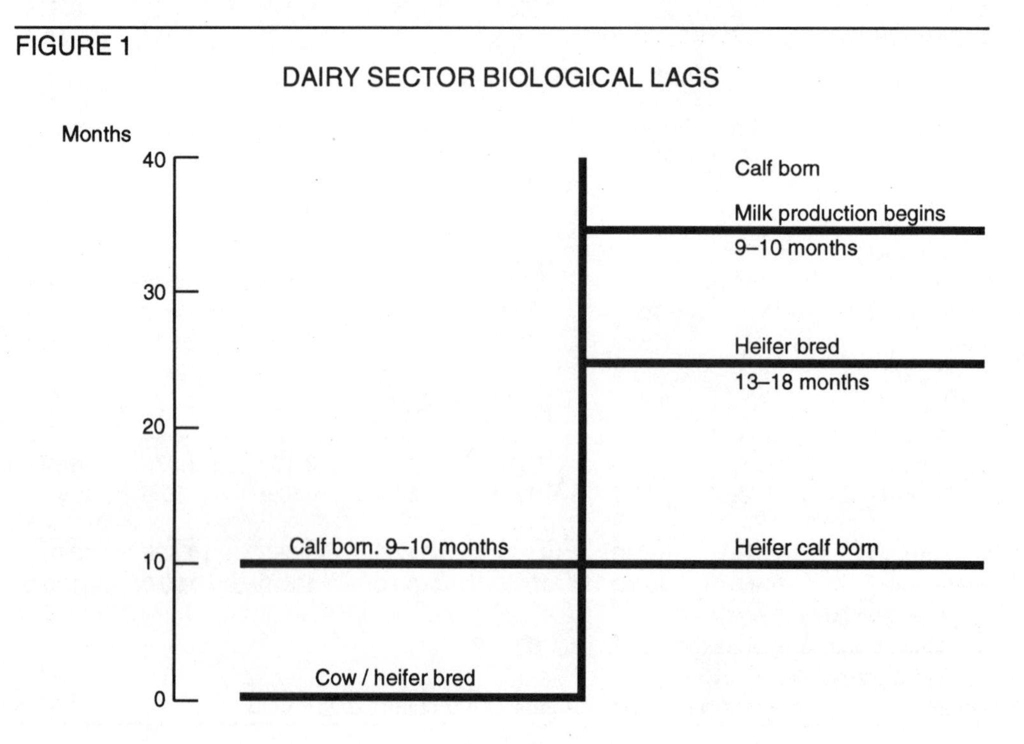

In the long run, a 10 percent change in farm milk price would change milk production about 5 percent in the same direction. The adjustment seems to be spread over a 4-year period, with very little change occurring during the year of the price change. During periods of milk price increases, U.S. milk production can be expected to increase 6 percent for every 10 percent increase in farm-level milk prices. However, when milk prices are decreasing, U.S. milk production can be expected to decrease only 4 percent for every 10 percent decrease in the farm milk price. Considerable regional variation exists, ranging from a change of 3 percent in the Southwest to 7 percent in the Southeast assuming a 10 percent change in price. The traditional dairy regions of the Northeast and Upper Midwest are close to the national average.

Two major implications for U. S . dairy price support policy are that: (1) most of the supply adjustment occurs in the first and second year after a price change, not in the year of the change; and (2) with decreasing prices, it takes more time to achieve a supply/demand balance.

Revenues, Costs, and Returns

Dairy enterprise returns above cash expenses and replacement costs are estimated to be $0.98 per cwt. of milk in 1989 compared with $0.87 in 1988 and $1.56 in 1986. Total cash expenses are estimated at $11.84 per cwt. in 1989, compared with $10.92 in 1988 and $10.29 in 1986. Feed costs, normally about 50 percent of cash expenses, increased to $6.55 per cwt. in 1989 (55 percent of cash expenses), compared with $5.89 in 1988 and $5.06 in 1986. The relatively high feed costs in 1988 and 1989 are primarily the result of the 1988 drought.

Returns consist of all current cash receipts generated from producing and marketing both milk and secondary products. Gains or losses occurring from asset appreciation or reduction are not included. Cash receipts are a function of both price (which may be heavily influenced by government programs) and production per cow. Receipts from secondary products typically include items such as the sale of breeding or culled livestock.

Cash expenses consist of both variable expenditures (those incurred only when production takes place in a given year) and fixed expenditures (items including taxes, insurance, overhead, interest, rent, and leasing costs for which the operator or landlord would be responsible whether or not production occurs). Replacement costs represent an imputed charge sufficient to maintain average machinery, equipment, and purchased breeding livestock investment and production capacity through time. The replacement charges are based on current prices of these capital assets.

Cash expenses are influenced by government programs and policies. For example, the feed grain program affects the cost of dairy feeds. The availability of water at an affordable price affects the cost of forage in some regions, especially in irrigated western regions. Conservation and disaster relief programs also affect the dairy farmer. Agricultural credit policy can affect interest rates as well as availability of credit for entry into dairying or expansion of an existing operation. Federal and state tax policy can also affect entry, expansion, or renovation decisions. Decisions of nonfarm investors are especially influenced by tax policy. Also, macroeconomic policy decisions, as they affect interest rates and agricultural trade, are becoming increasingly important to the well-being of dairy farmers and to agriculture in general.

A recurring problem of dairy programs is that benefits are often capitalized into asset values, especially cattle. An example of relatively high net returns, associated with the capitalization of program benefits into asset values, is the rise in replacement dairy cow prices to over $1,000 in 1979 from under $700 in 1978 and even lower prices in prior years (Table 3). The difference between dairy cow prices and slaughter cow prices increased from $233 in 1978 to $443 in 1979.

This difference reached a peak of nearly $700 in 1981, and then declined in the mid-1980s as the supply of replacements expanded coupled with dairy farmers' facing lower immediate and anticipated returns. Both dairy cow prices and slaughter cow prices increased again in the late 1980s.

Some entering dairy farmers, and those who expanded their dairy operations substantially during the late 1970s period of relatively high dairy cow prices, probably faced financial difficulty as the industry came closer to a workable supply-demand balance. This capitalization phenomenon also causes problems in costs and returns analyses and in attempts to assess industry well-being.

Another factor in the persistence of excess milk supplies in the 1980s was the apparent increase in dairy productivity and the willingness of U.S. dairy farmers to produce more milk in spite of lower real (adjusted for inflation) prices (Fig. 2). One effect of this phenomenon is that it has made the U.S. dairy industry more competitive in world markets. A key question for the 1990s is whether these milk supply shifts will continue. The drought of 1988 and wet weather conditions in parts of the country in early 1989 both adversely affected forage quality and milk production per cow. However, given normal weather conditions and the likely emergence of new technology such as bovine somatotropin (bST), the trends initiated in the 1980s are likely to continue.

Emerging Production Issues

A major emerging issue related to milk production is the use of bovine somatotropin (bST). bST is a naturally occurring protein in dairy cattle which has been linked to milk production. Recombinant DNA technology has made the production of a synthetic bST possible at reasonable cost. Herd trials have shown that injections of bST increase milk production. The increased milk production is not without costs; additional nutrients and more management expertise are required to obtain the most benefits from bST. Studies suggest that even with additional feed and management costs, dairy farmers can obtain more milk at less cost per cwt. using the product.

bST is the latest in a series of output-enhancing and cost-reducing technologies in the dairy industry. The bulk tank, improved parlor designs, automatic feeding systems, artificial insemination, DHIA (dairy herd improvement associations), embryo transfers, and 3X (three times per day) milking have all contributed to increased production and reduced milk production costs. The major difference in the case of bST is its biotechnological origin, and its appearance at a time when the industry had just come out of a lengthy debate concerning surplus milk production.

As of January 1990, bST was not yet commercially available. The Food and Drug Administration (FDA) must certify the safety and efficacy of the product prior to commercial release. Safety of milk and meat consumption by humans as well as safety of bST use on cows, in the environment, and in bST manufacturing must be assured.

TABLE 3

DAIRY COW AND SLAUGHTER COW PRICES, 1970–89

Year	Dairy cow prices[1]	Slaughter cow prices[2]	Difference
	Dollars per head		
1970	332	256	76
1971	358	259	99
1972	397	303	94
1973	496	394	102
1974	500	307	193
1975	412	253	159
1976	476	304	172
1977	504	304	200
1978	674	441	233
1979	1,044	601	443
1980	1,195	549	646
1981	1,201	503	698
1982	1,100	480	620
1983	1,020	472	548
1984	895	478	417
1985	861	460	401
1986	821	446	375
1987	917	538	379
1988	986	567	419
1989	1,027	589	438

1. Price per head received by farmers. *Agricultural Prices*, U.S. Department of Agriculture, National Agricultural Statistics Service, various issues.
2. Hundredweight price of utility cows at Omaha times 12 cwt. per cow. *Livestock, Meat and Wool Market News: Weekly Summary and Statistics*, U.S. Department of Agriculture Marketing Service, various issues.

Source: U.S. Department of Agriculture.

As of this time, only the safety of meat and milk consumption from bST-treated cows has been determined. Completed review of the product is likely in the early 1990s.

There is much debate as to the need for a product such as bST. Some farmers, consumer groups, dairy cooperatives, and state legislators have taken stands opposing bST. National policymakers have also expressed some concerns. Economic analyses of the effects of the product are, by necessity, speculative. Analyses over time suggest that the impact on the industry will be more modest than early studies suggested. As the latest in a long series of technological advances in the industry, bST reinforces, but does not fundamentally change, long-term trends in the dairy industry.

Demand for Dairy Products

Milk demand is composed of the purchases of many products, primarily fluid milk, cheese, butter, and nonfat dry milk. In periods when the industry is near a supply-demand balance, about half the milk supply is used in fluid milk products and the

FIGURE 2

DURING THE 1980S, FARMERS CONTINUED PRODUCING MORE MILK EVEN AS REAL PRICES DECLINED

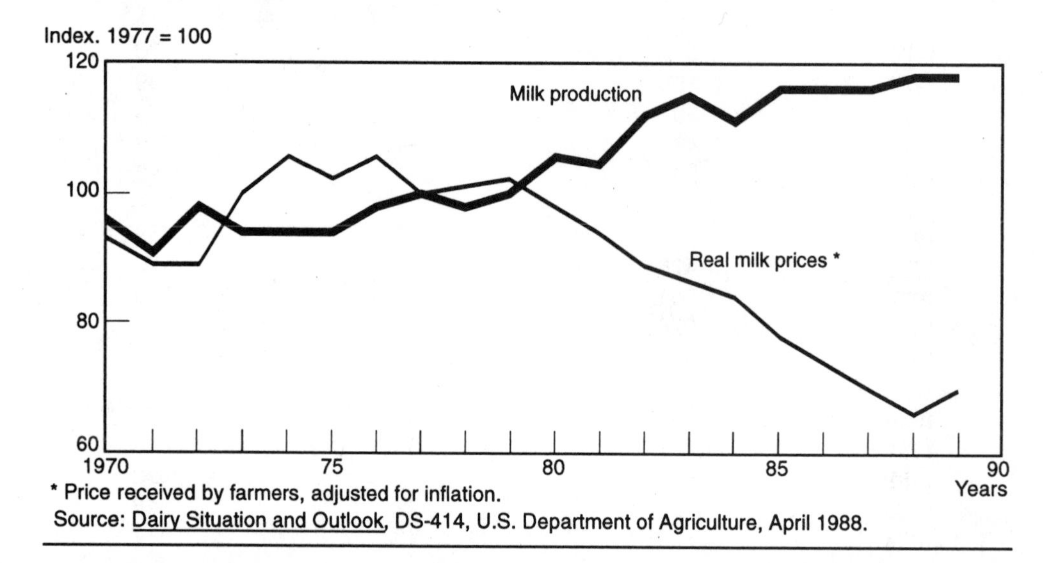

* Price received by farmers, adjusted for inflation.

Source: Dairy Situation and Outlook, DS-414, U.S. Department of Agriculture, April 1988.

remainder in manufactured dairy products. The roles of the various dairy products in the diet differ as to trends in their use. Thus, the demand for raw milk depends on both the product mix at a given time and the demand characteristics of individual products.

Changes in Commercial Use

Per capita commercial use of all dairy products has shown little change since 1970. This is in contrast to the downward trend of more than 1 percent per year during the 1960s. Total commercial disappearance in 1988 was 26 percent greater than in 1970, primarily due to increased population. Per capita fluid milk sales have decreased by an average of 2.7 pounds (about 1 percent) per year; this reflects an annual 6.5-pound average drop in whole milk use partially offset by a 3.8-pound increase in sales of lowfat milk.

The downtrend in fluid milk sales was accelerated during the 1970s and 1980s by changes in the age distribution of the population. The population bulge resulting from the post-World War II baby boom has moved beyond the peak milk-consuming ages to the lowest consuming-age bracket. Consumers began shifting from whole milk to lowfat milk in the early 1960s. In 1970, lowfat milk accounted for 19 percent of fluid milk sales and by 1988 it reached 53 percent. The past erosion of the fluid milk market has been fairly steady despite declining relative milk prices. However, demographic shifts indicate that the rate of decline in use per person might slacken.

Growth in commercial cheese use has been very important to the dairy industry over the past 25 years. Cheese production has used more than a fourth of the market supply of milk in recent years, compared with less than an eighth in 1960. Per capita consumption of American cheese grew about a fourth of a pound each year during 1970 to 1988, while consumption of other varieties rose about half a pound per year.

Over half the growth in sales of other cheese varieties (about a third of the total growth) came from mozzarella, used in making pizza. Most of the expansion in cheese sales has been in natural forms; per capita sales of processed cheese items have risen only slightly.

Among the attributes of cheese that appear to be aligned with changing life-styles are convenience in use, the range of flavors and textures, storability, and affordability. Increased consumption of food away from home, such as pizza and salads, has also increased cheese sales. Acceptable alternative cheeses have been developed for consumers concerned about the high fat or salt content of some traditional varieties.

Demand for butter appears to have stabilized since the early 1970s after declining for decades. Changes in use over the pact two decades seem to have been related mostly to changes in butter prices relative to margarine. Per capita sales will most likely be fairly stable at slowly declining relative prices. Both butter and cream demands are potentially vulnerable to concerns raised by recent dietary studies.

Progressive replacement of butter by margarine has apparently ended. Since 1974, market shares of butter and margarine have fluctuated but have shown little trend. Civilian per capita butter consumption appears to have stabilized at about 4.5 pounds and margarine at about 10.5 pounds. Butter sales, however, are still sensitive to relative prices of the two products.

Commercial use of nonfat dry milk has decreased. Per capita sales in 1988 were less than half those of 1970. Sharp declines were registered for almost every significant end use.

While nonfat dry milk sales have declined, production and use of whey products, particularly whey protein concentrates, have expanded. Increased cheese production and environmental regulations that limit whey disposal have combined to enable whey protein concentrates to fill the role (formerly held by nonfat dry milk) of an inexpensive source of very high quality protein. Increased demand for whey products has only a minor impact on overall milk demand since it primarily involves recovery of milk components not currently used.

Since 1970, per capita use of ice cream has remained unchanged, while per capita sales of ice milk and sherbet have slipped slightly. Use of mellorine (frozen dessert made with vegetable oil) has decreased substantially.

The Dairy and Tobacco Adjustment Act of 1983 authorized a dairy product promotion and research program. It is designed to strengthen the dairy industry's position in the marketplace and to maintain and expand domestic and foreign markets and uses for U.S. fluid milk and dairy products. The program is funded by an assessment of 15 cents per cwt. on all commercially marketed milk. Collections under the program for 1988 totaled approximately $215 million.

Consumption Response to Changes in Prices and Incomes

Dairy product sales respond relatively little to price changes, at least in the short run. A 10 percent decline in retail prices will increase sales of fluid milk by only about 2 percent. Butter and cheese sales would increase the most, perhaps 7 to 8 percent; other products would fall in between. Total commercial use would be expected to rise about 3 percent if retail prices fell 10 percent.

This low level of short-run demand response to price changes (price inelasticity) has several ramifications. First, small variations in milk output will result in substantial price movements as long as prices are determined by the market. Second, total consumer expenditures for dairy products vary directly and almost proportionately with price level. For example, a 10 percent increase in retail prices will result in a decrease in consumption of 3 percent and an increase in consumer expenditures of about 7 percent. Third, the small consumer responses to price are difficult to observe because they can easily be veiled by demographic changes, changes in consumer preferences, and other factors. According to economic theory, consumers are more responsive to prices in the long run than in the short run.

Some dairy products are affected more by incomes and general economic conditions than others, although the effects are relatively small in all cases. In general, fluid milk sales are not changed significantly by income changes. Butter consumption and cheese consumption are both positively related to income, but the effect is small. Sales of both of these products in recent years have varied with the state of the economy.

Substitute Products

Substitute dairy products have significantly affected demand for butter and, to a lesser extent, cheese. Margarine had taken most of the table-spread market by the early 1970s. More recently, imitation cheese (cheese made with vegetable fat and casein) had absorbed part of the growth in the cheese market. Census of Manufactures data indicate that cheese substitutes (products substituting for natural and processed cheese) totaled 449.4 million pounds in 1987, compared with 227.1 million pounds in 1982. This was about 8 percent of total cheese production in 1987 and 5 percent in 1982.

Other substitute products have had only slight effects on dairy product demand. Whipped toppings and coffee whiteners have significant markets, but cream sales have grown slightly since the introduction of these substitutes. Sales of products such as filled and imitation milk, vegetable fat frozen desserts, and filled evaporated milk have fallen after some initial successes.

Processing

The dairy processing industry has undergone marked change in recent decades, with substantial gains in efficiency and reductions in real costs. Changes in the industry in recent decades include: fewer but larger plants, increased importance of producer cooperatives, and regional shifts precipitated by population shifts and shifts in milk production in excess of fluid sales. The number of plants producing cottage cheese and butter dropped over 90 percent from 1950 to 1988. Hard cheese and ice cream plants declined by approximately three-fourths and nonfat dry milk plants by over 80 percent (Table 4). In 1988, average output per plant was over 15 times the 1950 level for butter and cheese, about 7 times for nonfat dry milk and ice cream, and 18 times for cottage cheese. Automation and technological advances, such as continuous churns, have increased economies of size in processing. However, changes in assembly and distribution costs were probably of equal importance.

Dairy producer cooperatives have become an important part of the manufactured dairy products sector. Between 1973 and 1987, cooperatives' share of total production rose from 35 to 45 percent for cheese and from 85 to 91 percent for dry products, while increasing from 60 to 83 percent for butter. Cooperatives' share of fluid products and cottage cheese stabilized at about 14 and 13 percent, respectively, while their share of ice cream was about 8 percent in 1987.

Important factors underlying the increased role of cooperatives include the transfer of the fluid milk procurement and daily and seasonal balancing functions to cooperatives, a perceived need for cooperatives to assure members of an outlet for all their milk, a desire to control more of the value added to milk, and the tendency of large traditional dairy corporations to specialize in dairy merchandising and to diversify into other products. As indicated above, cooperative integration into fluid and soft manufactured products has been considerably more modest.

Supermarket chains have also increased their manufacturing capacity, with fluid milk processing increasing from 3 percent of total sales in 1964 to almost 18 percent in 1980. Their share of relatively modern capacity is considerably higher, but food chain involvement in fluid milk processing seems to have declined somewhat since 1980 as a result of a few chains selling off bottling plants.

TABLE 4

NUMBER OF DAIRY PRODUCT MANUFACTURING PLANTS
SELECTED YEARS

Product	1950	1970	1980	1983	1988
			Number		
Hard cheese	2,158	963	737	696	573
Butter	3,060	622	258	222	165
Nonfat dry milk (human food)	459	219	113	101	76
Hard ice cream	3,269	1,628	949	862	765
Cottage cheese curd	1,900	593	269	240	185

Source: *Dairy Products, Annual Summary,* U.S. Department of Agriculture, National Agricultural Statistics, various issues.

The dairy industry moved from a relatively well-balanced supply-demand situation in 1978, when dairy product removals by the Commodity Credit Corporation (CCC) were 2.3 percent of total milk marketings, to a point where over 12 percent of U.S. milk marketings were purchased by CCC in 1983. From 1984 through 1989, the average was 6.8 percent. In the process of generating excess milk supplies, both the milk production and processing sectors attracted additional fixed resources, which increased capacity. In the mid-1980s, a financial and structural adjustment was necessary, especially in the milk production and manufactured dairy products industries, to achieve an overall supply-demand balance.

Structural adjustment especially affects the manufactured dairy products industry. Excess capacity developed because fluid milk consumption remained relatively stable while total milk marketings increased from 119 billion pounds in 1978 to nearly 143 billion pounds in 1989. The effect has varied across regions since the buildup of milk supplies was not geographically uniform. The Southeast, Corn Belt, and Plains regions increased production relatively little compared with the major milk production states of New York, Pennsylvania, Wisconsin, and California. Milk production generally shifted toward the West and Southwest, especially California.

Limited plant capacity curtailed the expansion of milk production in California during the early 1980s. However, even though large quantities of California butter, nonfat dry milk powder, and cheese were sold to the CCC, a high proportion of cheese consumed in California was imported from out of state, especially from Wisconsin. The California dairy industry has since moved to increase its cheese manufacturing capacity. In turn, California milk production increased 43 percent from 1980 to 1989, while U.S. milk production increased 12 percent over this same period.

Milk production shifts and aggregate levels cause some adjustment problems for the fluid milk processing industry in some regions, but the manufactured dairy products industry is generally affected the most. This is because fluid milk product sales are fairly stable; they account for about half of overall milk supplies. Therefore, reductions/increases in milk production will result in drops/jumps of twice that proportion in milk supplies available for processing into manufactured dairy products.

Trends in World Dairy Trade

International trade of agricultural commodities is under continual debate. But the current round of multilateral trade negotiations under the auspices of the GATT (General Agreement on Tariffs and Trade) has made agricultural trade a high-priority issue .

Every major developed dairy-producing nation operates government programs regulating its domestic dairy industry. Many subsidize part or all of domestic production, imports are commonly restricted, and exports are frequently subsidized. There have been significant strides taken in some major producing countries to address dairy industry problems in the last several years, mostly to reduce the burdens of excess milk supplies and the associated costs to government of handling the excess. The implementation of production quotas in the European Community (EC-12) in 1984 and legislation authorizing the milk diversion program and the dairy termination program in the United States are examples of alternative approaches for attacking the excess supply issue. In addition to the voluntary supply management programs, the United States implemented a flexible dairy price support mechanism.

Dairy trade is small relative to total world milk production. World milk production in 1988 was approximately 430 million metric tons, an estimate that covers about 90 percent of world production. From 1985 to 1988, world production grew by just over 3 percent. If intra-EC trade is excluded, about 5 percent of world production (milk equivalent) is traded—a world market slightly greater than 40 percent of 1988 U.S. milk production.

High dairy price supports in many countries tended to stimulate production to the extent that subsidized exports were required to maintain domestic dairy programs. The subsidized sale of butter by the EC to the Soviet Union is one example. The implementation of production quotas in the EC in 1984, which did not lower price supports, dramatically reduced the world's largest dairy product surpluses.

From 1985 to 1988, exports of the three major manufactured dairy products—butter, cheeses, and nonfat dry milk (NFDM)—were made primarily by countries with high dairy price support, the EC, other Western European nations, Canada, and the United States (Table 5). An interesting feature of the data is that although surpluses of dairy products have been reduced in the European Community, it is still the major

TABLE 5

AVERAGE EXPORTS AND MARKET SHARES FOR BUTTER, CHEESE, AND NONFAT DRY MILK, 1984–88

	BUTTER			CHEESE			NONFAT DRY MILK		
ITEM	84–85	86–87	1988	84–85	86–87	1988	84–85	86–87	1988
Average annual				1,000 metric tons					
Exports[1]	853	902	1,050	858	811	830	1,186	1,144	1,206
Shares:[1]				Percent					
EC[2]	45	51	57	48	45	46	26	29	51
Other Western Europe	5	4	3	20	19	18	5	4	2
United States	8	4	2	3	3	2	31	35	16
Canada	*	*	0	1	1	1	6	5	5
New Zealand	25	25	23	7	8	13	20	16	15
Australia	2	9	5	11	12	9	6	7	6
Total	88	93	90	90	88	88	94	96	96

* = Less than 0.5 percent.

1. Excluding intra-EC trade.

2. EC-10 in 1984–85, expanded to EC-12 in 1986 with inclusion of Portugal and Spain.

Source: *World Dairy Situation*, Circular Series FFD 2-89, U.S. Department of Agriculture, Foreign Agricultural Service, November 1989.

exporting area for the three major manufactured dairy products. U.S. participation in international markets, based on export shares, has fallen as butter and NFDM exports have declined.

As a result of export subsidies, international prices for manufactured dairy products were below what they would have been in the absence of such subsidies. As surplus products available for exports have declined, international prices have strengthened considerably (Table 6). The announced government purchase prices by the CCC in 1988 for butter, $2,900 per metric ton, and cheese, $2,540 per metric ton, were closer to international prices than in previous years. The U.S. price of $1,600 per metric ton for NFDM was actually below the international price; this resulted in the commercial export of NFDM without government assistance.

Restrictive import quotas have been used by the United States to prevent lower-cost and subsidized dairy products from undercutting U.S. dairy price supports. The import quotas on manufactured dairy products, which have essentially been fixed since the Tokyo round of GATT, limit imports to about 2.5 billion pounds milk equivalent, just under 2 percent of U.S. milk production in 1989. Under restrictive import quotas, consumers pay more for all dairy products than they would under lesser restrictions. The dairy product quotas, authorized by Section 22 of the Agricultural Adjustment Act of 1933, as amended, may be implemented, adjusted, or eliminated only by the President, usually based on the findings and recommendations of the International Trade Commission (ITC).

Imports of butter, NFDM, and American-type and processing cheeses compete directly with domestically produced products and displace them roughly pound for

TABLE 6

INTERNATIONAL PRICES FOR BUTTER, CHEESE, AND NONFAT DRY MILK, F.O.B. NORTHERN EUROPE AND SELECTED WORLD PORTS

PERIOD	BUTTER	CHEESE	NFDM
		U.S. dollars per metric ton	
1985:			
Spring	950–1,050	1,100–1,250	600–680
Fall	1,000–1,050	1,150–1,275	600–650
1986:			
Spring	1,050–1,150	1,100–1,200	680–720
Fall	800–1,100	1,000–1,100	680–720
1987:			
Spring	750–1,100	900–1,200	760–840
Fall	900–1,150	1,000–1,300	890–1,150
1988:			
Spring	1,150–1,350	1,250–1,500	1,150–1,550
Fall	1,350–1,500	1,800–2,050	1,750–2,050
1989:			
Spring	1,650–1,900	1,750–1,950	1,750–2,000
Fall	1,800–2,000	2,000–2,150	1,750–1,900

Source: *World Dairy Situation*, Circular Series FD2-89, U.S. Department of Agriculture, Agricultural Service, November 1989.

pound. Specialty cheese, the bulk of U.S. dairy product imports, compete less directly with domestically produced cheeses. It is unlikely that restricting imports of some specialty cheeses would result in increased sales of similar domestically produced cheeses of the same magnitudes.

Imports of casein are problematic. For some food products, there is direct substitution of imported casein for domestically produced dairy products such as nonfat dry milk. Restricting casein imports which enter nonfood uses—for example, glue and paint production—would not contribute to an increase in demand for U.S. domestic dairy products because there is no casein production in the United States and other dairy products are not good substitutes for casein in industrial uses.

Policy actions by major developed dairy-producing nations affect the international dairy trade more than any "market" determinations. The small size of international trade relative to the domestic dairy industries of these countries contributes to the dependence. The environment generated by the current multilateral trade negotiations has in turn led to a situation where the debate on U.S. domestic dairy policy and programs will include both domestic and international issues more than ever before.

International Trade Outlook

The current situation in international dairy markets owes much to the policy actions of two of the major developed dairy-producing areas: the European Community (EC-12) and the United States. The implementation of production quotas in the EC-12 and the implementation of voluntary supply management programs and a flexible dairy support mechanism in the United States led to reduced stocks in both areas. As stockpiles decreased, international prices strengthened to the extent that the United States was able to export dairy products, particularly nonfat dry milk, on a commercial basis (with no government subsidy).

The rather sudden availability of an international market for U.S. dairy products added a certain amount of volatility to the domestic industry. With a continuation of program provisions implemented under the 1985 Act, the United States would periodically have commercial export opportunities. Those opportunities would depend to a large extent on the maintenance of export "discipline" on the part of the EC-12. Even if domestic supply shifts in the United States were to ease, the international prices for dairy products would provide a realistic floor under domestic U.S. prices.

HISTORY OF DAIRY PROGRAMS

The U.S. dairy industry, while subjected to more government participation or regulation than most other domestic agricultural industries, is less regulated than the dairy industries in many other developed countries. The price support program authorized by the Agricultural Act of 1949 and the federal milk marketing order program authorized by the Agricultural Marketing Agreement Act of 1937 are the principal domestic dairy programs. With relatively high support prices compared with world prices, and because exports are subsidized by many countries, import quotas are imposed to keep imports of dairy products from overwhelming the dairy price support program. Federal policy has also fostered the growth of dairy cooperatives to promote the balance of market power between dairy farmers and those who buy from them.

The Dairy Price Support Program

The dairy price support program supports the milk price received by farmers through purchases of butter, nonfat dry milk, and American cheese. Purchase prices for the products are set at levels designed to enable manufacturers to pay farmers the announced support price for milk in surplus production periods.

In the Agricultural Act of 1949 and subsequent amendments to that act, Congress specified three major guidelines for the operation of the price support program. First, it provided for minimum and maximum levels at which farm milk prices were to be supported based on parity price guidelines. For many years, the minimum support price was 75 percent and the maximum was 90 percent of parity. (Legislation in 1981 departed from the parity concept for the first time and parity has not been used as a basis for establishing dairy price supports since then.)

Second, the program authorizes the Secretary of Agriculture to determine the specific price support level within the minimum and maximum prices specified in the legislation. The objective of the support program is to support the price of milk at a level that will assure an adequate supply of ". . .milk to meet current needs, reflect changes in the cost of production, and assure a level of farm income adequate to maintain productive capacity sufficient to meet anticipated future needs."

Third, the legislation specified that the price of milk would be supported through purchases of milk and milk products. Since milk is a bulky, perishable product, the government cannot reasonably buy raw milk. Therefore, the U.S. Department of Agriculture, through the CCC, purchases all the butter, nonfat dry milk, and cheese offered by processors at announced prices. These products are widely produced and take about two-thirds of the milk used in manufactured dairy products. The prices received by individual dairy farmers depend upon many factors other than the support level, including plant location, product manufactured, quantity of milk delivered, local competition, and plant operating efficiency.

The purchase prices announced by the CCC for butter, nonfat dry milk, and cheese include "manufacturing (make) allowances" or margins to cover the costs of processing milk into these products. These margins are administratively set at a level which should allow processors to pay, on average, dairy farmers at least the announced support price for Grade B milk. Prices to farmers for manufacturing grade milk are free to move above the support level if supply and demand conditions warrant. This occurred in the short-supply portion of the marketing season of most years until 1980 and, at times, even during the flush season.

In 1989, manufacturing grade milk prices ran substantially above the support level. They were below the support level, however, during much of the early and mid-1980s. The short-supply season usually occurs in October and November when milk production reaches a seasonal low point and fluid product demand is seasonally highest. The flush season normally occurs in May and June when milk production reaches its seasonal peak and fluid milk product sales are declining seasonally.

The Food and Agriculture Act of 1977 provided that, for the two marketing years beginning October 1977, the Secretary would adjust the support price of milk semiannually after the beginning of the marketing year to reflect any estimated change in the parity index during the semiannual period. These provisions were extended in 1979 for two more years.

Before 1977, support prices were set annually for the upcoming marketing year. However, support prices during the mid-1970s generally were also raised during the

FIGURE 3

MILK COW NUMBERS AFFECTED BY PRICES AND POLICY ACTIONS

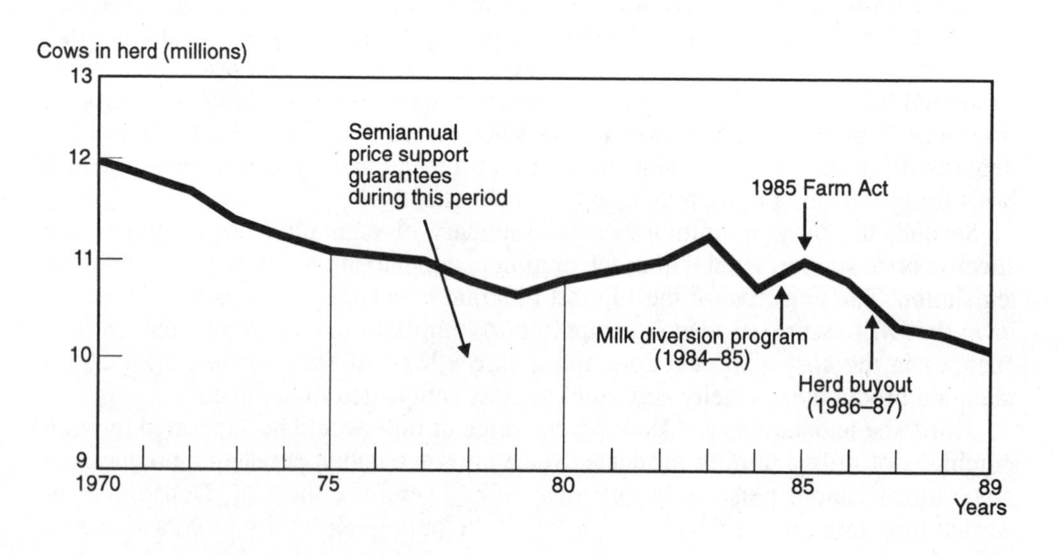

year to account for rapid inflation. The Food and Agriculture Act of 1977 required a midyear adjustment in the support price to reflect changes in the parity index during the first 6 months of each marketing year. This had the effect of raising the support prices in the middle of the marketing year to reflect increases in the index of prices paid by farmers. At the administration's request, the first step toward bringing supplies back into line with consumption was taken when legislation was enacted on March 31, 1981, which rescinded the scheduled April 1, 1981 increase in the support price. Figure 3 shows the effects on cow numbers of the income-enhancing features of the Food and Agriculture Act of 1977 and the various steps required in later years to bring the industry closer to market equilibrium.

The Agriculture and Food Act of 1981, passed at a time of large surpluses, used a set of triggers relating the minimum support level to the size of CCC purchases. This was a major departure from traditional price support policy under which price changes were tied directly to parity. As long as large CCC purchases continued, the support prices were specified in dollar terms with the 1981–82 price at the 1980–81 level of $13.10 per cwt., which was 72.9 percent of parity in September 1981, and modest increases thereafter. Only if surpluses declined to stated levels would supports at 70 to 75 percent of parity be required.

With continued surpluses, legislation was enacted in 1982 which froze support prices for 2 years and provided for deductions totaling $1 per cwt. from milk producers' marketing receipts to partially offset rising government costs. The 1983

FEDERAL PRICE SUPPORT PURCHASES

The federal government supports milk prices through purchases of butter, nonfat dry milk, and cheddar cheese. The following example illustrates the connection between the prices USDA pays for these dairy products and the price support rate for milk, which was $10.10 per hundredweight (cwt.) effective January 1, 1990.

Smith and Jones are average dairy farmers living near Plainville, U.S.A. Smith sells milk to the local processing plant that makes butter and nonfat dry milk. For each hundredweight (100 pounds) of milk Smith sells, the plant makes 4.48 pounds of butter and 8.13 pounds of nonfat dry milk. With the CCC prices of butter and nonfat dry milk set at $1.0925 and 79 cents per pound, respectively, the products made from Smith's 100 pounds of milk are worth $11.32. However, the plant's allowance for the cost of manufacturing these products is $1.22 per cwt.; this leaves $10.10 to Smith for the milk.

Jones sells milk to the cheese plant on the other side of town. For every hundredweight of milk purchased, the plant manufactures 10.1 pounds of cheese with some whey solids left over. The CCC pays about $1.11 per pound for the cheese. The fat in the whey solids is worth 27 cents; this makes the market value of the products made from Jones's milk equal to $11.47. Since the plant's allowance for manufacturing the cheese is $1.37 per cwt., Jones receives $10.10 per cwt. for the milk.

Dairy and Tobacco Adjustment Act lowered the minimum price support level from $13.10 to $12.60 effective December 1, 1983. It allowed for a further reduction in support of 50 cents per cwt. on April 1, 1985 if net government purchases in the succeeding 12 months were projected to be above 6 billion pounds milk equivalent. It further allowed the Secretary to reduce the support price another 50 cents on July 1, 1985 if net government purchases in the succeeding 12 months were projected to be above 5 billion pounds. Alternatively, the Secretary had authority to increase the support levels by not less than 50 cents per cwt. on July 1, 1985, if net government

MAJOR PRICE SUPPORT ACTIONS, 1970–90

1970–72 Support prices set at levels above the minimum of 75 percent of parity.

1970 The Agricultural Act of 1970 suspended the obligation to support prices of farm-separated cream.

1973 The Agricultural and Consumer Protection Act of 1973 set a minimum support level of 80 percent of parity through March 1974.

1974–77 Support prices adjusted frequently because of rapid inflation. No support price lasted more than 9 months. Support prices set at 80 percent or current parity.

1977 The Food and Agriculture Act of 1977 set a minimum of 80 percent of parity. It also required that the support price be adjusted semiannually to reflect changes in prices paid by farmers. These provisions were to be in effect for 2 years.

1979 The support price provisions of the 1977 Act were extended for 2 additional years.

1981–82 The support price was frozen at $13.10 per cwt. in effect since October 1, 1980.

1981–83 The Omnibus Budget Reconciliation Act of 1982 authorized a 50-cent deduction on all milk marketed that was first collected in April 1983. An additional 50-cent deduction, implemented on September 1, 1983, was refundable to producers who reduced marketings by a specified amount.

1984–85 The Dairy and Tobacco Adjustment Act of 1983 lowered the support price to $12.60 effective December 1, 1983. A 50-cent deduction was continued through March 1985. A dairy diversion program, operated between January 1984 and March 1985, paid contracting producers $10 per cwt. for reductions from base milk marketings. The support purchases in the next succeeding 12 months were projected to be 5 billion pounds or less milk equivalent.

The 1983 Act also amended the 1949 Act to provide for a milk diversion program. For the period December 1, 1983, through March 31, 1985, a mandated assessment of 50 cents per cwt. was made on all milk marketed for commercial use by U.S. producers in the 48 contiguous states. The funds collected were used to partially offset the cost of the program. Producers who elected to participate in the program and reduce their milk marketings between 5 and 30 percent below their base period production were paid $10 per cwt. for these reductions. The 1983 Act also authorized

	price was reduced 50 cents on both April 1 and July 1, 1985, because purchases were projected to exceed trigger levels.
1986–90	The Food Security Act of 1985 set the support price at $11.60 for calendar 1986, $11.35 for January–September 1987, and $11.10 thereafter. On January 1 of 1988, 1989, and 1990, the support price had to be adjusted by 50 cents if projected removals exceeded 5 billion pounds or were less than 2.5 billion pounds. The first such reduction was implemented on January 1, 1988. Deductions were set at 40 cents during April–December 1986 and at 25 cents during January-September 1987. Additional deductions, authorized to help reduce budget deficits, were 12 cents during April-September 1986 and 2.5 cents during calendar 1988. A January 1, 1989, support price reduction was prohibited by drought relief legislation passed in mid-1988. It also required a 50-cent increase on April 1, 1989, followed by a 50-cent reduction on July 1, 1989. On January 1, 1990, the support price was reduced another 50 cents to 10.10 per cwt. The Food Security Act also authorized the dairy termination program. Producers whose bids were accepted agreed to slaughter or export all female dairy cattle, have no interest in milk production or dairy cattle for 5 years, and ensure that their facilities were not used for these purposes during that time. Those producers, who had marketed more than 12 billion pounds of milk during 1985, left the industry during April 1986–August 1987. The act increased Class I differentials in most federal milk marketing orders, effective May 1, 1986. These differentials were not to be altered for a period of 2 years. As of February 1990, the congressionally mandated differentials remained in place.

a nonrefundable 15-cent-per-cwt. assessment on milk marketed by producers to finance a dairy product research and promotion program.

The 1949 Act was again amended by the Food Security Act of 1985 to authorize a voluntary dairy termination program, also known as the whole-herd buyout, in which producers could submit competitive bids during the period of April 1, 1986, through September 31, 1987, to remove milk production for at least 5 years.

The 1985 Act continued the dairy support price of $11.60 per cwt. for milk containing 3.67 percent milkfat (originally established at this level on July 1, 1985) for calendar year 1986 and established the support price at $11.35 per cwt. for January 1 through September 30, 1987, and $11.10 per cwt. for October 1, 1987 through December 31, 1990.

Changes in dairy price supports on January 1, 1988, 1989, and 1990 were linked to projected annual government purchases. The Secretary of Agriculture is to reduce the support price 50 cents per cwt. if net price support purchases in any of these respective calendar years are projected to exceed 5 billion pounds milk equivalent or increase the support price 50 cents per cwt. if net purchases are projected at not more than 2.5 billion pounds milk equivalent. Because it was estimated that net purchases would exceed 5 billion pounds in calendar year 1988, the support level was reduced to $10.60 per cwt. on January 1, 1988.

Other provisions of the 1985 Act included a 40-cent-per-cwt. assessment on all milk marketed within the 48 contiguous states during April 1 through December 31, 1986, and 25-cent-per-cwt. during January 1 through September 30, 1987. However, to reduce outlays required by the Balanced Budget and Emergency Deficit Control Act of 1985 (Gramm-Rudman-Hollings Act), the Food Security Improvement Act of 1986 further amended the 1949 Act to provide an additional 12 cents-per cwt. deduction for the period April 1, 1986 through September 30, 1986. The Omnibus Budget Reconciliation Act of 1987 required a 2.5-cent-per-cwt. assessment for calendar year 1988.

The 1985 Act required the Secretary to offer at least 1 million pounds of surplus nonfat dry milk on a bid basis for manufacture into casein[1] and to establish a program to encourage additional exports of dairy products. To avoid burdensome supplies, the Secretary was also provided the option to establish a milk diversion or milk production termination program for calendar years 1988, 1989, or 1990.

The 1985 Act also legislated higher minimum Class I differentials in 35 of the 44 federal milk orders that were operating in May 1986 (Table 7). Most of these increases were in milk-deficit southern markets.

Drought relief legislation passed in mid-1988 prohibited any January 1, 1989 reduction in the support price. It also required a 50-cent increase on April 1, 1989, to be followed by a 50-cent reduction on July 1, 1989. The support price was reduced to $10.10 per cwt. on January 1, 1990. See the box for highlights of price support actions over the past two decades.

Most of the legislative changes made during the early and mid-1980s were attempts to reduce the supply of excess milk and cut government purchases and costs. In 1983,

1. Due to the lack of interest on the part of the dairy industry, this program was discontinued in marketing year 1987–88. CCC accepted only one offer in 1986–87, totaling 79,926 pounds of nonfat dry milk

dairy farmers produced over 10 percent more milk than consumers were willing to buy at the supported prices. However, with strengthening international dairy product prices, the United States has become a significant participant in international markets and the persistent excess milk supply problem has been reduced. However, commercial export sales of manufactured dairy products by the United States will likely continue to be dependent on policy actions taken by both the United States and other countries, particularly the EC.

Dairy Program Costs

Nominal costs for price supports ranged from $69 million to $612 million between 1952–53 and 1972–73 averaging $325 million for the period. Over the 1970s, outlays fluctuated, with greater variability in milk production. Since 1979–80, program costs have exceeded $1 billion in each year. In the 1982–83 marketing year, costs reached a record $2.6 billion, about 13 percent of total cash receipts from farm marketings of milk and cream, or an average of about $13,000 per commercial dairy farmer. Program costs for the 1988–89 marketing year were down to $698 million, or an average of around $5,000 per commercial dairy farmer. [Note: As will be explained in detail in Chapter 10 of the text, these are not the full costs to society due to dairy price supports.][2]

2. This case is derived from R. Fallert, D. Blayney, and J. Miller, *Dairy: Background for 1990 Farm Legislation* (Washington, D. C.: Economic Research Service, Department of Agriculture, 1990).

TABLE 7
CLASS I DIFFERENTIALS UNDER THE 1985 FOOD SECURITY ACT[1]

FEDERAL ORDER	DIFFERENTIAL PRE-FOOD SECURITY ACT	DIFFERENTIAL FOOD SECURITY ACT	INCREASE
		Dollars per cwt.	
New England	3.00	3.24	0.24
New York-New Jersey	2.84	3.14	.30
Middle Atlantic	2.78	3.03	.25
Georgia	2.30	3.08	.78
Alabama-West Florida	2.30	3.08	.78
Upper Florida	2.85	3.58	.73
Tampa Bay	2.95	3.88	.93
Southeastern Florida	3.15	4.18	1.03
Upper Michigan	1.35	1.35	0
Southern Michigan	1.60	1.75	.15
East Ohio-Western Pennsylvania	1.85	2.00	.15
Ohio Valley	1.70	2.04	.34
Indiana	1.53	2.00	.47
Chicago	1.26	1.40	.14
Central Illinois	1.39	1.61	.22
Southern Illinios	1.53	1.92	.39
Louisville-Lexington-Evans	1.70	2.11	.41
Upper Midwest	1.12	1.20	.08
Easter South Dakota	1.40	1.50	.10
Black Hillls	1.95	2.05	.10
Iowa	1.40	1.55	.15
Nebraska-Western Iowa	1.60	1.75	.15
Kansas City	1.74	1.92	.18
Tennessee Valley	2.10	2.77	.67
Nashville	1.85	2.52	.67
Paducah	1.70	2.39	.69
Memphis	1.94	2.77	.83
Central Arkansas	1.94	2.77	.83
South West Plains	1.98	2.77	.79
Texas Panhandle	2.25	2.49	.24
Lubbock	2.42	2.49	.07
Texas	2.32	3.28	.96
Louisiana	2.47	3.28	.81
New Orleans-Mississippi	2.85	3.85	1.00
Eastern Colorado	2.30	2.73	.43
Western Colorado	2.00	2.00	0
South West Idaho-Eastern Oregon	1.50	1.50	0
Great Basin	1.90	1.90	0
Lake Mead	1.60	1.60	0
Central Arizona	2.52	2.52	0
Rio Grande Valley	2.35	2.35	0
Puget Sound-Inland	1.85	1.85	0
Oregon-Washington	1.95	1.95	0

1. Increased differentials effective May 1, 1986. May be changed by normal procedures after May 1, 1988.

QUESTIONS

1. Table 1 shows that there has been a remarkable increase in the average number of cows per farm (from 8 in 1955 to 49 in 1989). Why has this occurred? Has it been healthy for the dairy industry? Has it promoted the interests of consumers?
2. Table 1 also shows that there has been a substantial increase in the amount of milk produced per cow (from 5,842 pounds per year in 1955 to 14,244 pounds per year in 1989). Why has this occurred? Is it related to the increase in the number of cows per farm? If so, how? Has it been healthy for the dairy industry? Has it promoted the interests of consumers?
3. In a four-year period, what is the price elasticity of supply for milk? Is this price elasticity the same for price increases as for price decreases? Why or why not? Is it the same in the Southwest as in the Southeast?
4. Based on the estimates provided in this case, what is the price elasticity of demand for fluid milk? For butter and cheese?
5. According to this report, "small variations in milk output will result in substantial price movements as long as prices are determined by the market." Does this follow from your answer to Question 4? If so, why?
6. According to this report, "total consumer expenditures for dairy products vary directly and almost proportionately with the price level." Does this follow from your answer to Question 4? If so, why?
7. The federal government supports milk prices through purchases of butter, nonfat dry milk, and cheddar cheese. Explain how this is done.
8. Why did the Food Security Act of 1985 call for the Secretary of Agriculture "to reduce the support price 50 cents per cwt. if net price support purchases in [1988, 1989, or 1990] are projected to exceed 5 billion pounds milk equivalent or increase the support price 50 cents per cwt. if net purchases are projected at not more than 2.5 billion pound milk equivalent"? What was the objective of this provision?

Part 4 Market Structure, Price, and Output

Chapter 10 Perfect Competition

Case Study: The Widget Industry

In 1994, the widget industry is perfectly competitive. The lowest point on the long-run average cost curve of each of the identical widget producers is $4, and this minimum point occurs at an output of 1,000 widgets per month. When the optimal scale of a firm's plant is operated to produce 1,150 widgets per month, the short-run average cost of each firm is $5. The market demand curve for widgets is

$$Q_D = 140{,}000 - 10{,}000P$$

where P is the price of a widget and Q_D is the quantity of widgets demanded per month. The market supply curve for widgets is

$$Q_S = 80{,}000 + 5{,}000P$$

where Q_S is the quantity of widgets supplied per month. (P is expressed in dollars per widget.)

a. What is the equilibrium price of a widget? Is this the long-run equilibrium price?

b. How many firms are in this industry when it is in long-run equilibrium?

c. If the market demand curve shifts to

$$Q_D = 150{,}000 - 5{,}000P,$$

what is the new equilibrium price and industry output in the short run?

d. In the situation described in part (*c*), are firms making profits or losses? (Assume that the number of firms in the industry equals the number that would exist in long-run equilibrium in part (*b*).)

Completion Questions

1. If a firm's revenues cover its variable costs, but not its total costs, its economic profit is (zero, positive, negative) ________________, which means that its resources could yield (more, less, the same) ______________ elsewhere.
2. In the long run, a perfectly competitive firm's equilibrium position is at the place where its long-run average cost equals ________________.
3. A decreasing-cost industry has a ________________ long-run supply curve.
4. An increasing-cost industry has an ________________ long-run supply curve.
5. A constant-cost industry has a ________________ long-run supply curve.
6. At the optimal level of production, profit maximization requires that long-run marginal cost equals short-run ______________ equals price.
7. Short-run supply elasticities tend to be ______________ than long-run supply elasticities in American agriculture.
8. External economies may result in ________________ that occur when the industry expands.
9. Increasing-cost and constant-cost industries are more ______________ than

decreasing-cost industries.

10. ______________ is the amount producers receive above and beyond the minimum price that would be required to get them to produce and sell these units.

11. Suppose that a cotton textile firm's total costs and total revenue are as follows:

Total output	*Total cost (dollars)*	*Total revenue (dollars)*
0	30	0
1	33	10
2	37	20
3	42	30
4	50	40
5	60	50
6	90	60

This firm's marginal cost between 2 and 3 units of output is ______________.

12. On the basis of the figures in Question 11, this firm will produce ____________ units of output.

13. This firm will make a loss of ______________.

14. If the firm shut down completely, it would make a loss of ______________.

15. The price of this firm's product is ______________.

16. At an output of between 4 and 5 units, the firm's marginal cost is ______________.

17. If price is less than marginal cost, decreases in output will ______________ profit.

18. For a firm in a perfectly competitive market, if price is more than marginal cost, increases in output will ______________ profit.

19. ______________ is consumer surplus plus producer surplus.

True or False

____ 1. If a firm's price is fixed, then increases in output will have little effect on the firm's profits.

____ 2. If all firms are identical with respect to technology and managerial ability, the industry supply curve under perfect competition is horizontal.

____ 3. An industry with increasing returns to scale must be a decreasing-cost industry.

____ 4. In the short run, equilibrium price under perfect competition may be above or below average total cost.

____ 5. In the long run, equilibrium price under perfect competition may be above or below average total cost.

____ 6. Under perfect competition, one producer can produce a somewhat different good from other producers in his or her industry.

____ 7. Under perfect competition, each firm must be careful not to produce too much and spoil the market.

____ 8. One can derive the firm's supply curve in the short run by simply tracing out its average cost curve.

____ 9. One can always derive the industry's supply curve by summing up the firm's marginal cost curves.

___ 10. At the equilibrium price, price will equal marginal cost (for all firms that choose to produce) under perfect competition.

___ 11. Under perfect competition, marginal cost is the same for all producers of a particular product in equilibrium.

___ 12. Firms that appear to have lower average costs than others often have superior resources or managements.

___ 13. Total surplus is maximized when output is at the perfectly competitive level.

Multiple Choice

1. The market demand curve for a particular kind of desk chair is as follows:

Price (dollars)	*Quantity (demanded)*
30	200
20	300
10	400
5	600
3	800

The industry producing this kind of chair is a constant-cost industry, and each firm has the following long-run total cost curve

Output	*Total cost (dollars)*
1	10
2	12
3	15
4	30

(Each firm can produce only integer numbers of units of output.)
In the long run, the total number of firms in this industry will be about
a. 100.
b. 200.
c. 300.
d. 400.
e. 500.

2. The long-run average cost curve of a perfectly competitive firm is given on the next page. Given that this curve does not shift, the long-run equilibrium output of the firm will be
a. 4 units.

b. 5 units.
c. 6 units.
d. 7 units.
e. 8 units.

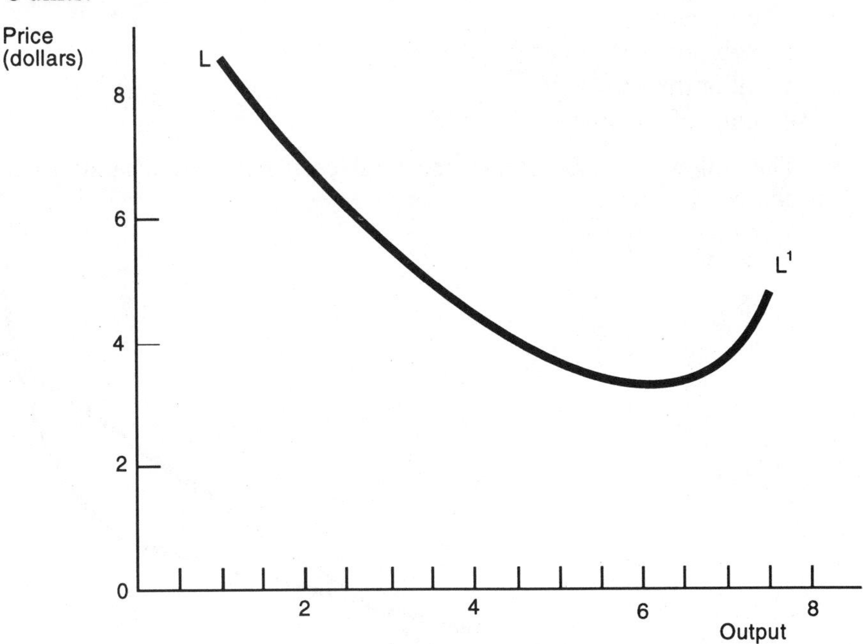

3. If the demand curve shifts to the right (and if the supply curve is upward sloping), equilibrium price will
   a. decrease.
   b. increase.
   c. stay the same.
   d. all of the above.
   e. none of the above.

4. Recognizing that the assumptions of perfect competition never hold at all precisely, the perfectly competitive model is
   a. interesting mainly for academic studies.
   b. outmoded and seldom used even by academic economists.
   c. of considerable use to industrial economists, as well as academic economists.
   d. all of the above.
   e. none of the above.

5. Under perfect competition, rivalry is
   a. impersonal.
   b. very personal and direct, advertising being important.
   c. nonexistent since the firms cooperate and collude.
   d. all of the above.
   e. none of the above.

6. If average total cost is less than marginal cost at its profit-maximizing output, a perfectly competitive firm
   a. will make positive profits.
   b. will operate at a point to the right of the minimum point on the average total cost curve.
   c. will not discontinue production.
   d. all of the above.
   e. none of the above.

7. The following graph shows the total cost and total revenue of a perfectly competitive firm:

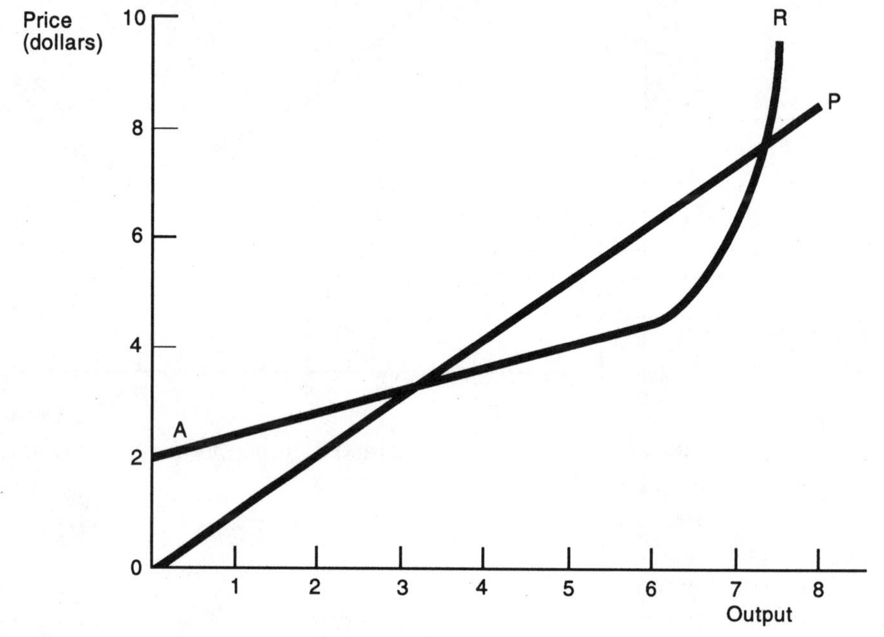

   The line *OP* is
   a. the total cost curve.
   b. the total revenue curve.
   c. the relationship between price and output.
   d. all of the above.
   e. none of the above.

8. The curve *AR* is
   a. the total cost curve.
   b. the total revenue curve.
   c. the relationship between price and output.
   d. all of the above.
   e. none of the above.

9. The optimal output of the firm
   a. is less than 5.

b. is more than 7.
c. is between 5 and 7.
d. all of the above.
e. none of the above.

Review Questions

1. Describe the four basic conditions that define perfect competition. How often are they encountered in the real world?
2. Describe how equilibrium price is determined in a perfectly competitive market if the supply curve is a vertical line.
3. Suppose that the supply of a certain kind of paintings is fixed at 8 units and that the demand curve for these paintings is as shown below. What is the equilibrium price for these paintings?

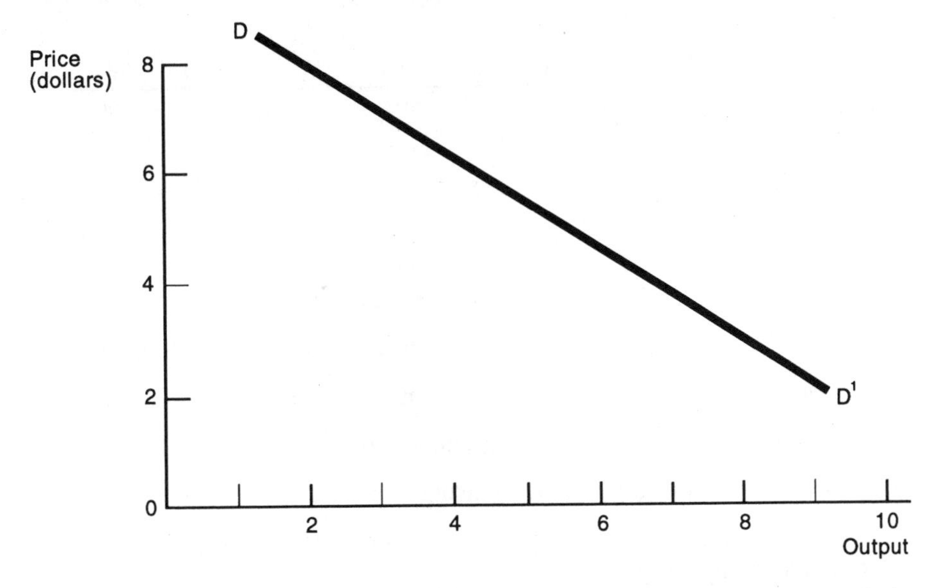

4. Show that the perfectly competitive firm's supply curve in the short run is its marginal cost curve.
5. If the supplies of inputs to the industry as a whole are perfectly elastic, show that the industry supply curve in the short run is the horizontal summation of the firm supply curves.

6. *a.* If the coal industry's demand and supply curves are as shown below, what are the equilibrium price and output?

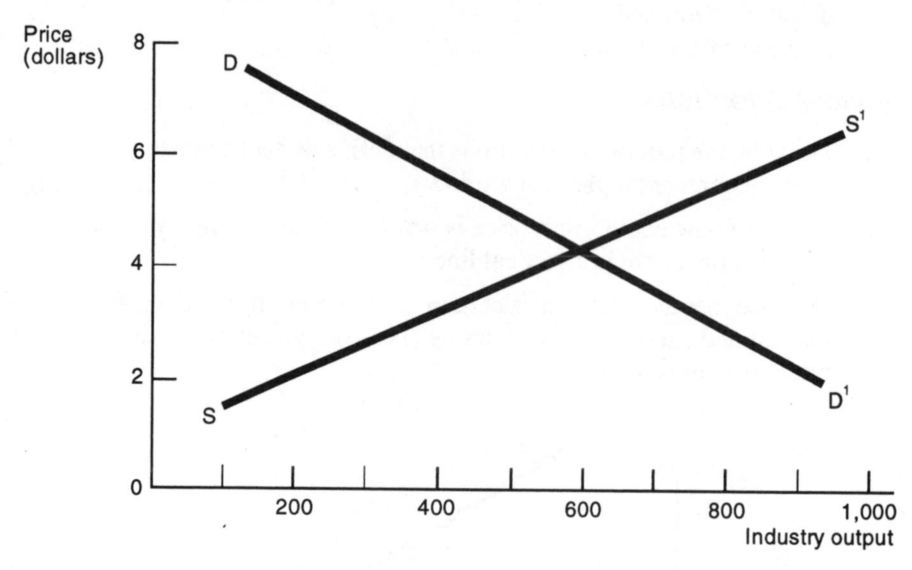

b. Why can't a price of $6 prevail in this market? Why can't a price of $2 prevail?

7. Describe the forces that determine whether there will be a net in-migration or out-migration of firms from a particular industry.

8. Describe the equilibrium conditions for the perfectly competitive firm in the long run.

9. Describe what is meant by constant-cost industries.

10. Describe what is meant by increasing-cost industries.

11. Describe what is meant by decreasing-cost industries.

12. To what extent does the cotton textile industry conform to the assumptions of perfect competition?

13. Explain why, if we correct for changes in the price level resulting from over-all inflation, there was a declining trend in farm prices up to 1973.

14. Explain why farm prices vary between good times and bad to a much greater extent than nonfarm prices.

15. According to the U.S. Census Bureau, the largest four producers of rubber products accounted for about 50 percent of the industry's value added. Do you think that the perfectly competitive model will work as well in rubber products as if the largest four producers accounted for 5 percent of the industry's value added? Why or why not?

16. According to Marc Nerlove and W. Addison, the long-run elasticity of supply of green peas is about 4.40. Of what use might this fact be to the Department of Agriculture?

17. What assumptions underlie the use of total surplus as a measure of how a market is functioning?

Problems

1. In the cotton textile industry, which is perfectly competitive, each firm's marginal cost curve is

$$MC = 5 + 3q$$

where MC is its marginal cost (in dollars per ton) and q is its output per day (in tons).

   a. If there are 1,000 firms in the cotton textile industry, derive the industry's short-run supply curve.

   b. If the price is $8, how much will the cotton textile industry produce per day?

   c. Can you derive each firm's total cost function? Why or why not?

2. According to a study by Henry Steele, if crude oil had been produced in the United States under perfectly competitive conditions, its short-run supply curve in 1965 would have been as shown below.

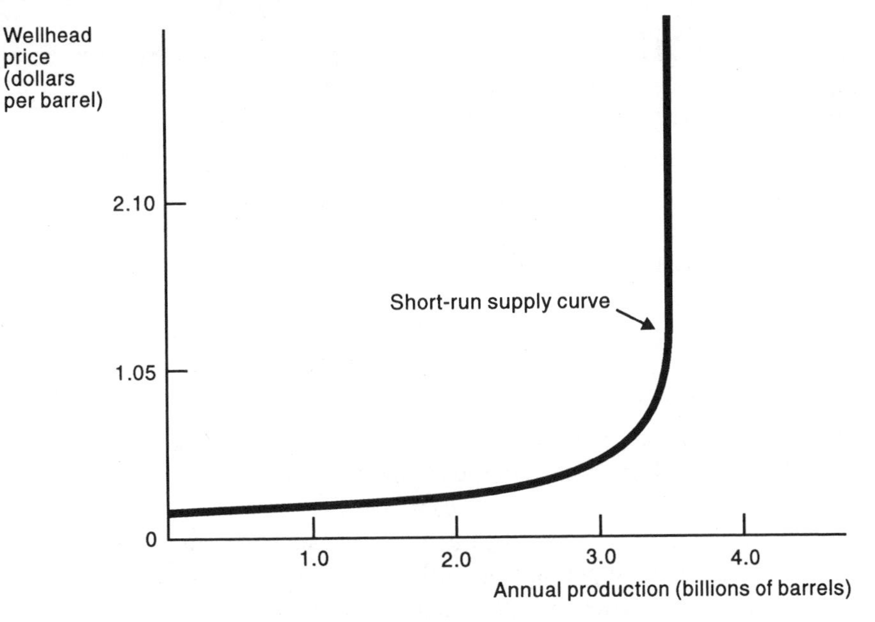

   a. If annual production had been 2.0 billion barrels in 1965, what would have been the marginal cost of a barrel of oil, according to Steele?

   b. If crude oil were produced under perfectly competitive conditions, and its supply curve was as shown below, what would have been its short-run equilibrium price if the demand for crude oil was 3 billion barrels per year, regardless of price?

   c. In fact, in 1965, there was prorationing in the oil industry, an arrangement which limited the amount that could be produced by low-cost wells. Given the existence of prorationing was the short-run supply curve to the left or right of the one in the graph below? Why?

3. Suppose that a perfectly competitive firm has the short-run total cost function

shown below.

Output	*Total cost (dollars)*
0	10
1	12
2	16
3	22
4	30
5	40

There are 1,000 firms in this industry, and the market demand curve is as follows:

Price (dollars)	*Quantity demanded*
3	3,000
5	2,000
7	1,500
9	1,000

a. What is the equilibrium price of the product? (Assume that the price must be $3, $5, $7, or $9.)
b. What will be the output of each firm?
c. In the long run, will firms tend to enter or leave this industry?

4. From time immemorial moralists have been fond of counseling that "anything worth doing is worth doing well."
   a. Indicate how the theory of the firm can be adapted to shed light on whether this counsel is correct.
   b. Consider cleanliness, which some say is second only to Godliness. Use your answer to part (*a*) to explain why people settle for a lower level of cleanliness in garages than in operating rooms.

5. The long-run supply curve for a good is a horizontal line at a price of $3 per unit of the good. The demand curve for the good is
$$Q_D = 50 - 2P$$
where Q_D is the quantity of the good demanded (in millions of units per year) and P is the price per unit (in dollars).
   a. What is the equilibrium output of the good?
   b. If a tax of $1.00 is imposed on the good, what is the equilibrium output of the good?
   c. After the tax is imposed, a friend of yours buys a unit of the good for $3.75. Is this the long-run equilibrium price?

6. (*Advanced*[1]) A firm's total cost function (where C is total cost in dollars and Q is quantity) is
$$C = 200 + 4Q + 2Q^2$$
   a. If the firm is perfectly competitive and if the price of its product is $24, what is its optimal output rate?

1. This problem involves the use of some elementary calculus.

b. At this output rate, what are its profits?
c. Derive the firm's short-run supply curve.

7. If the textile industry is a constant-cost industry and the demand curve for textiles shifts upward, describe the process by which a perfectly competitive market generates an increased quantity of textiles. What happens if the government will not allow the price of textiles to rise?

8. A perfectly competitive firm has the following total cost function:

Total output	*Total cost (dollars)*
0	20
1	30
2	42
3	55
4	69
5	84
6	100
7	117

How much will the firm produce, if the price is
a. \$13?
b. \$14?
c. \$15?
d. \$16?
e. \$17?

9. Suppose that there are 100 firms producing the good in Question 8 and that each firm has the total cost function shown there. If input prices remain constant, draw the industry supply curve in the graph at the top of the next page (when the price is between \$13 and \$17).

10. An economist estimates that, in the short run, the quantity of widgets supplied at each price is as follows:

Price (dollars)	*Quantity supplied per year (millions)*
1	5
2	6
3	7
4	8

Calculate the arc elasticity of supply when the price is between \$3 and \$4 per widget. (Review Chapter 2 if you do not recall the definition of the arc elasticity of supply. Note that this supply curve is quite different from that which was given in the case study at the beginning of this chapter, perhaps because they pertain to different time periods or markets.)

11. The Besser Manufacturing Company had only 465 employees in 1951 and sales of less than \$15 million. Since it is a small firm, can we be sure that it will have little or no control over the price of its product (concrete block machinery)?

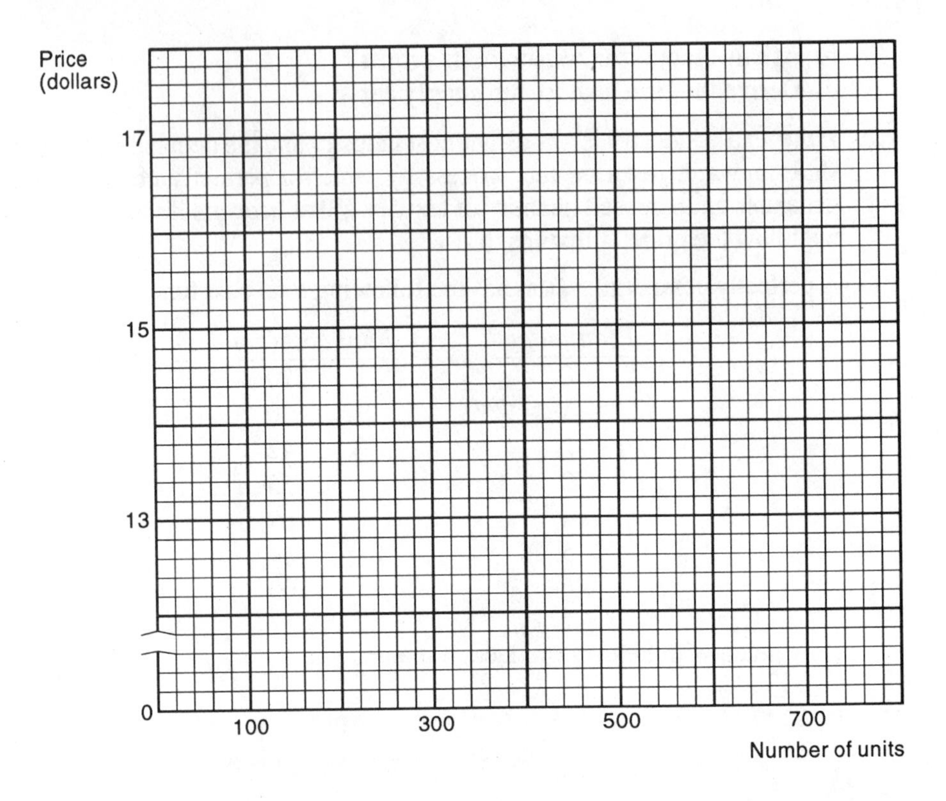

12. In late 1962, Karl Fox estimated that, based on a fairly typical farm situation in northeast Iowa, the most profitable number of dairy cows a farm operator should keep (at various expected prices of farm-separated cream) was as follows:

Expected price of cream (cents per pound)	*Number of cows*
60	2
80	4
100	6
120	11

If each cow yields the same amount of cream, what is the price elasticity of supply of cream for this farm when the expected price of cream is between 80 and 100 cents per pound?

13. According to D. Suits and S. Koizumi, the supply function for onions in the United States is $\log q = 0.134 + 0.0123t + 0.324 \log P - 0.512 \log C$, where q is the quantity supplied in a particular year, t is the year (less 1924), P is the price last season, and C is the cost index last season. Suppose that price is estimated by one forecaster to be 10 cents this season, whereas another says that it will be 11 cents. How much difference will this make in the quantity supplied next season?

14. The Alliance Delivery Company operates under perfectly competitive conditions, its only variable input being labor. In the short run, its production function is as follows:

Quantity of labor (people hired per day)	*Quantity of output (pounds per day)*
0	0
1	50
2	75
3	95
4	110

In 1994, the daily wage of a worker is $40.

a. Plot the Alliance Delivery Company's total variable cost function in the graph below.

b. Plot the Alliance Delivery Company's marginal cost function in the graph at the top of the next page.

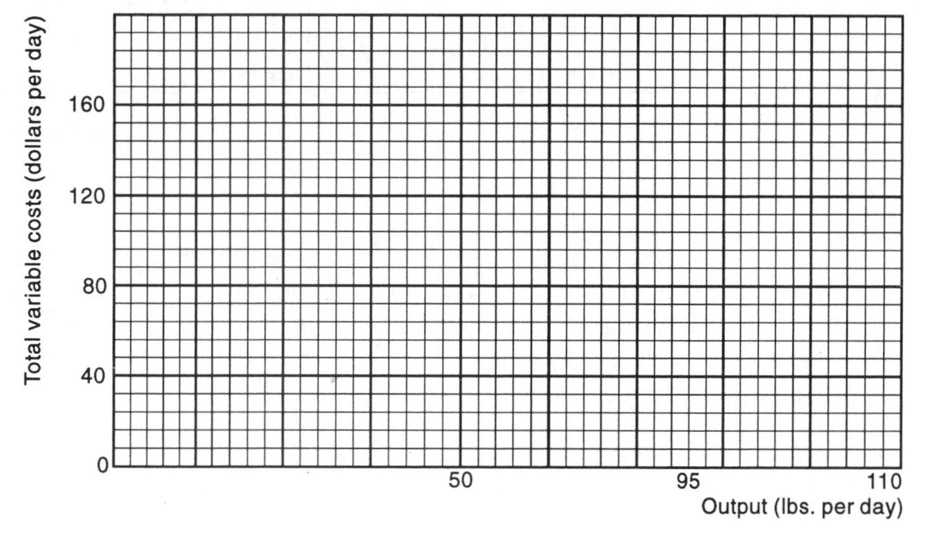

c. What is the optimal output of the firm if the price of a pound of output is $2?

d. Would the answers to the previous questions be the same if the firm could hire fractional numbers of workers per day?

e. In fact, is it possible for a firm to hire a fractional number of workers per day? Explain.

15. The Miller Company is a perfectly competitive firm. Its output is 50 units per month, its total revenue is $35,000 per month, its average total cost is $700; and its average variable cost is $500. It is operating at the output where average total cost is a minimum. What is each of the following?

    a. Price.

    b. Total cost.

    c. Total fixed cost.

    d. Total profit.

    e. Marginal cost.

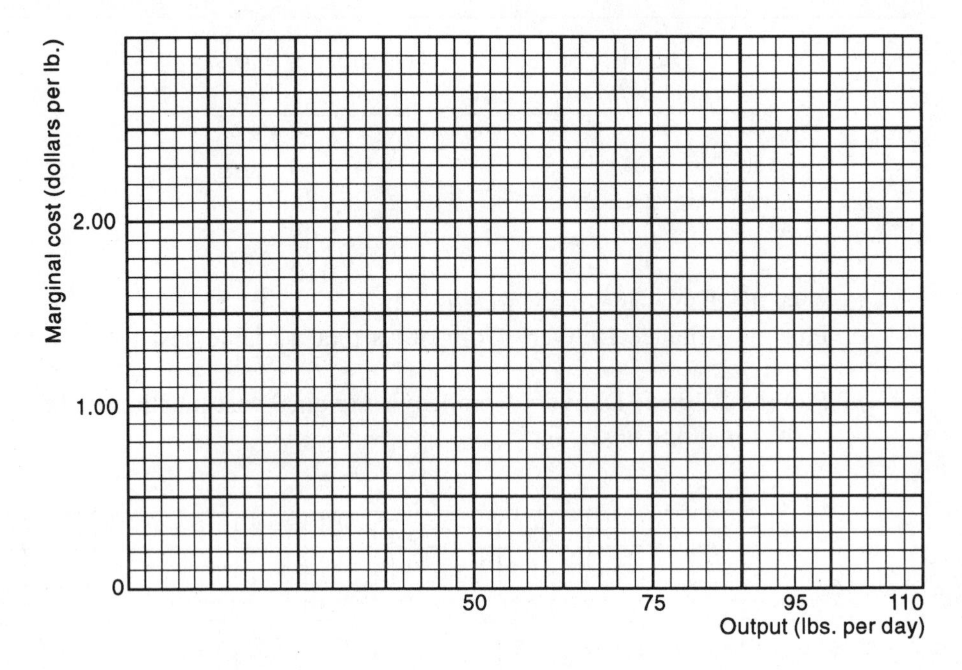

16. The Milton Corporation's marginal cost curve is
$$MC = 4 + 3Q$$
where MC is the cost (in dollars) of producing the Qth unit of its product and Q is the number of units of its product produced per day. The price of a unit of its product is \$3. A student at a local business school who works for the firm during the summer argues that, based on this evidence, the firm would make more money by shutting down than by continuing to operate. Do you agree? Explain.

17. The Zoraster Company's average cost curve is
$$AC = \frac{400}{Q} + 3Q$$
where AC is its average cost (in dollars) and Q is the number of units of its product that is produced per day. The price of a unit of its product is \$3.
    a. If the firm produces more than one unit per day, is it making profits or losses? Explain.
    b. Is the firm better off to shut down or operate? Explain.

18. Suppose that a firm's marginal cost curve is a horizontal line at \$3 per unit of output, if output is less than or equal to 100 units per month. If the price of the product is \$4 per unit, should the firm produce at least 100 units per month? Why or why not?

19. Suppose that you own a car wash and that its total cost function is
$$C = 20 + 2Q + 0.3Q^2$$
where C is total cost (in dollars) per hour and Q is number of cars washed per hour. Suppose that you receive \$5 for each car that is washed. What is the optimal number of cars to wash per hour? What is the maximum profit that you can

obtain?

20. Suppose once again that you are the owner of the car wash described in the previous problem. If the price you receive for each car washed falls from \$5 to \$2, will you continue to stay in operation? Why or why not?

21. *a.* Suppose that a perfectly competitive firm's total costs are as follows:

Output rate	*Total cost (dollars)*
0	10
1	12
2	15
3	19
4	24
5	30

If the price of the product is \$5, how many units of output should the firm produce?

b. Draw the firm's total costs and total revenues in the graph below.

c. Draw the firm's marginal cost curve in the graph at the top of the next page, and find the output where price equals marginal cost.

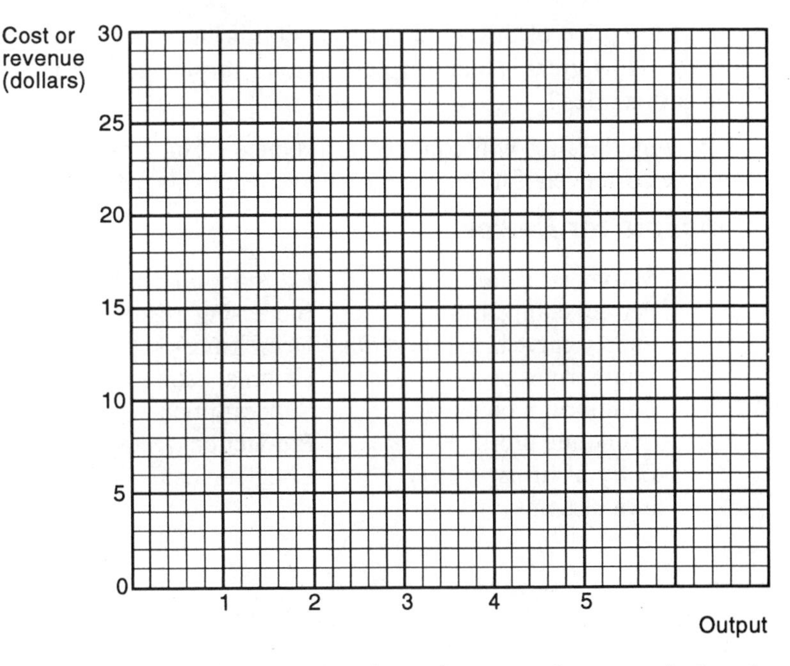

22. A researcher estimates the demand and supply curves for an agricultural good. For this good,

$$Q_D = 10 - 2P$$
$$Q_S = 3P$$

where Q_D is the quantity demanded per year (in millions of tons), Q_S is the quantity supplied per year (in millions of tons), and P is the price (in dollars per

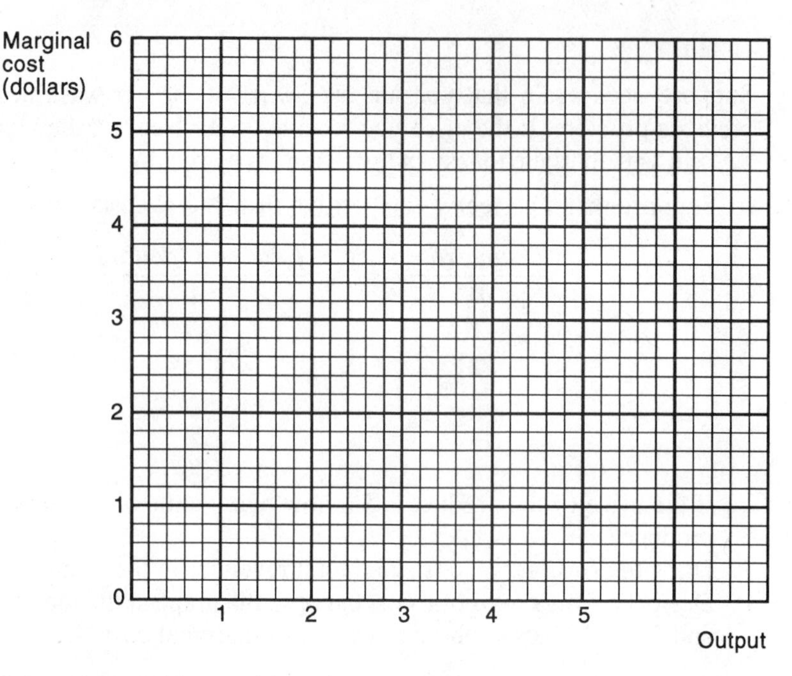

ton). If the price of this good is at its equilibrium level, how great is producer surplus? How great is total surplus? Is total surplus greater at some other price than at this price?

23. Show why the firm will continue to produce so long as price exceeds average variable cost, even if it is smaller than average total cost.

Key Concepts for Review

Market structure
Perfect competition
Supply curve in the short run
Equilibrium price
Equilibrium output
Economic profit
Entry of firms
Exit of firms
Constant-cost industries
Increasing-cost industries
Decreasing-cost industries
External economies
Support price
Production quota
Long-run equilibrium of the firm

ANSWERS

Case Study: The Widget Industry

a. If we set $Q_D = Q_S$,

$$140{,}000 - 10{,}000P = 80{,}000 + 5{,}000P$$
$$140{,}000 - 80{,}000 = (5{,}000 + 10{,}000)P$$
$$P = \frac{60{,}000}{15{,}000} = 4$$

Thus, the equilibrium price equals \$4. Since the long-run equilibrium price equals the minimum point on the firms' long-run average cost curves (that is, it equals \$4), this is the long-run equilibrium price.

b. When $P = 4$, $Q = 140{,}000 - (10{,}000)(4) = 100{,}000$. Thus, since industry output is 100,000 and the minimum point on each firm's long-run average cost curve occurs at an output of 1,000, there are 100 firms in the industry when each firm operates at the minimum point on its long-run average cost curve.

c. Setting $Q_D = Q_S$, we have

$$150{,}000 - 5{,}000P = 80{,}000 + 5{,}000P$$
$$150{,}000 - 80{,}000 = (5{,}000 + 5{,}000)P$$
$$P = \frac{70{,}000}{10{,}000} = 7$$
$$Q = 150{,}000 - 5{,}000\,(7) = 115{,}000$$

Thus, price equals \$7 and output equals 115,000.

d. If the industry output is 115,000, if there are 100 firms, and if each firm produces the same amount, the output of each firm is 1,150 widgets per month. It is stated at the outset that a firm producing this amount has a short-run average cost of \$5. Since price equals \$7, each firm is making profits.

Completion Questions

1. negative; more
2. price
3. downward-sloping
4. upward-sloping
5. horizontal
6. marginal cost
7. less
8. cost reductions
9. common
10. producer surplus
11. \$5
12. 4 or 5
13. \$10
14. \$30
15. \$10
16. \$10
17. increase

18. increase
19. total surplus

True or False

1. False	2. False	3. False	4. True	5. False	6. False
7. False	8. False	9. False	10. True	11. True	12. True
13. True					

Multiple Choice

1. *b*	2. *c*	3. *b*	4. *c*	5. *a*
6. *d*	7. *b*	8. *a*	9. *c*	

Review Questions

1. First, perfect competition requires that the product of any one seller be the same as the product of any other seller.

   Second, perfect competition requires each participant in the market, whether buyer or seller, to be so small, in relation to the entire market, that he or she cannot affect the product's price.

   Third, perfect competition requires that all resources be completely mobile. Fourth, perfect competition requires that consumers, firms, and resource owners have perfect knowledge of all relevant economic and technological data.

   No industry meets all of these characteristics, but some, like particular agricultural markets, may be reasonably close.

2. Equilibrium price is determined by the intersection of the demand curve (*DD'*) and the supply curve (*SS'*); that is, it is *OP*.

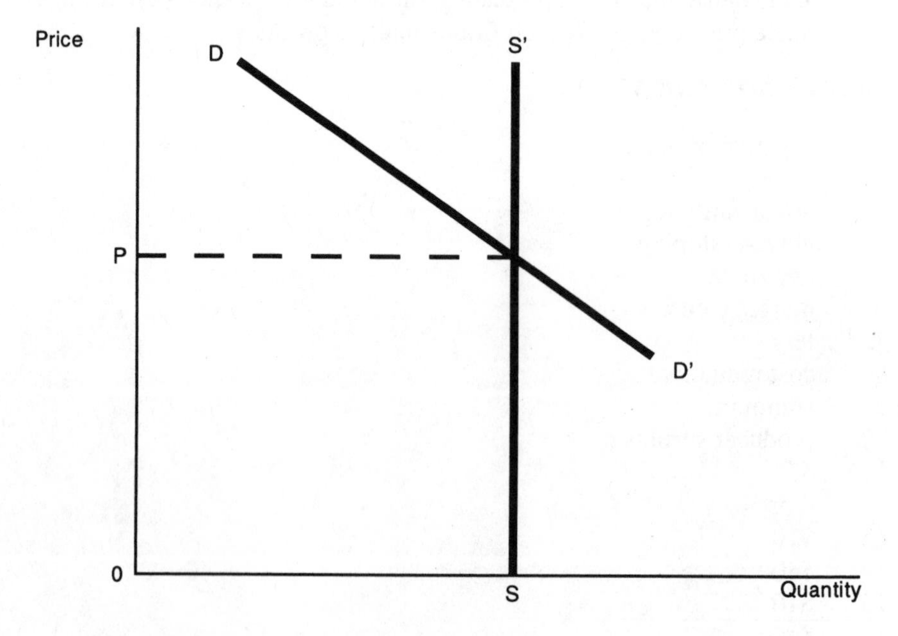

3. $3, as shown below

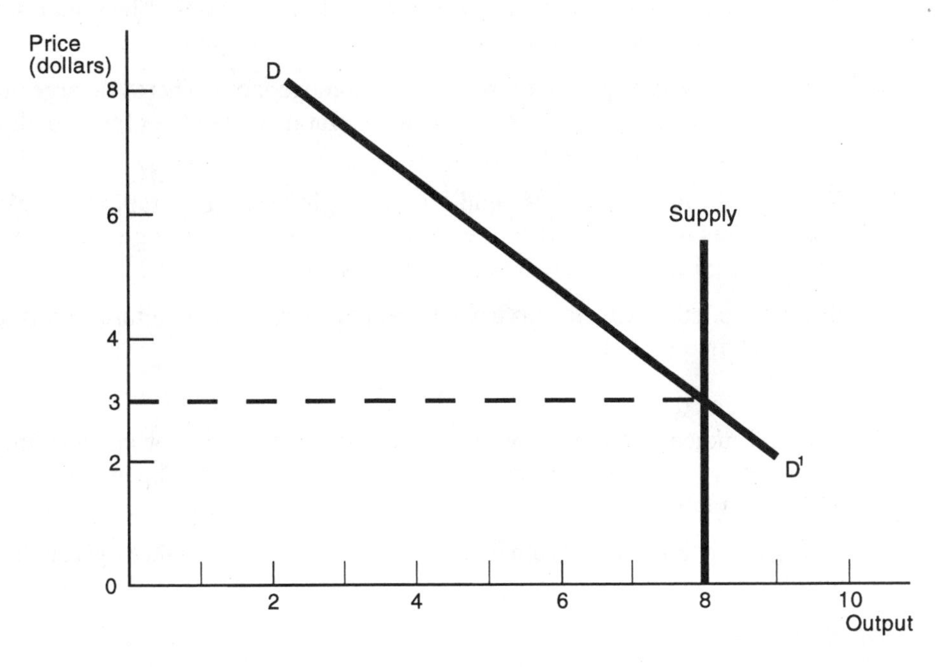

4. Given a certain price, the firm will produce an amount such that price equals marginal cost. Thus, if we vary the price the amount that the firm will produce at each price will be given by the marginal cost curve (as long as price exceeds average variable cost).

5. If the level of the industry output does not affect the cost curves of the individual firms, the industry supply at a given price can be determined by summing up the amount supplied by each of the firms, as indicated by their marginal cost curves.

6. *a.* Equilibrium output is 600 units. Equilibrium price is $4⅓.
   b. If price were $6, the amount demanded would be less than the amount supplied. If price were $2, the amount supplied would be less than the amount demanded.

7. If there are economic profits, new firms will tend to enter an industry. If there are economic losses, there will tend to be an out-migration of firms from the industry.

8. In the long run, firms operate where long-run marginal cost equals short-run marginal cost equals long-run average cost equals short-run average cost equals price. This means that firms operate at the minimum point of the long-run average cost curve.

9. A constant cost industry is an industry where expansion of the industry does not result in a change in input prices. The long-run supply curve of the industry is horizontal.

10. An increasing cost industry is an industry where expansion of the industry results in an increase in input prices. The long-run supply curve of the industry is upward-sloping.

11. A decreasing cost industry is an industry where expansion of the industry results in a decrease in input prices and costs. External economies may be important. The long-run supply curve of the industry is downward-sloping.
12. It is fairly close. The product is reasonably homogeneous. There is a large number of buyers and sellers. Entry is not difficult. But it does not conform to all of the assumptions of perfect competition.
13. The supply function shifted rapidly to the right because of rapid technological change. This more than offset the slower movements to the right of the demand curve stemming from increases in population and income.
14. Because both the demand curve and the supply curve of agricultural commodities are quite price-inelastic.
15. No.
16. It would indicate the long-run effect on quantity supplied of changes in price. Apparently, a 1 percent increase in price would result in a 4.4 percent increase in quantity supplied.
17. It is assumed that a dollar gained or lost by every person should be given the same weight.

Problems

1. *a.* If Q is the quantity supplied by the industry (in tons per day) and P is the price (in dollars per ton),

$$Q = \frac{-5{,}000}{3} + \frac{1{,}000P}{3}$$

   b. 1,000 tons.

   c. No.

2. *a.* Based on the graph, it appears that the marginal cost would have been about 30 cents, since this is the price corresponding on the supply curve to an annual production of 2 billion barrels.

   b. The equilibrium price would have been about 42 cents per barrel, since, if the demand curve were a vertical line at an annual production of 3 billion barrels, it would intersect the supply curve at 42 cents.

   c. It was to the left of the one in the graph, because less could be supplied at each price.

3. *a.* $5.

   b. 2 units.

   c. They will tend to leave this industry.

4. *a.* A person should increase the quality of the job he or she is carrying out so long as the marginal benefit from an extra unit of quality exceeds the marginal cost from this extra unit. Thus, to maximize his or her satisfaction, the quality of the job should be set at Ou units of quality in the case shown at the top of the next page.
   Obviously, the optimal quality level may be quite low if the marginal benefit

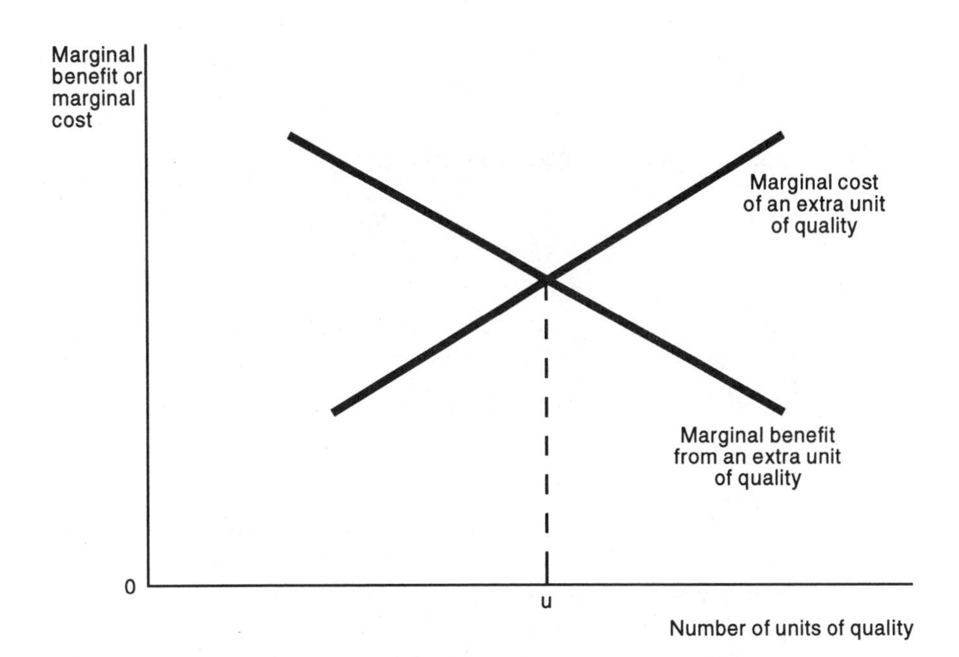

from an extra unit of quality is small and/or the marginal cost of an extra unit of quality is high. This graph is analogous to Figure 10.2 in the text, which shows the marginal cost and marginal revenue of a competitive firm. Just as a firm should produce no more output when it reaches the point where the marginal cost exceeds the marginal revenue, so a person should add no more quality to a job when the marginal cost of an extra unit of quality exceeds the marginal benefit from an extra unit of quality.

b. All other things equal the marginal benefit from an extra unit of cleanliness (that is, one less unit of dirt) tends to be higher in an operating room than in a garage. An extra unit of cleanliness may save a life in an operating room whereas it may have little effect in a garage. Thus the marginal-benefit curve in the graph above will tend to be higher in an operating room, and the optimal quality level—that is, the optimal level of cleanliness—will tend to be higher there than in a garage.

5. *a.* Since the equation for the supply curve is $P = 3$, the long-run equilibrium output can be determined by finding the value of Q_D if $P = 3$. (This is the value where the supply and demand curves intersect.) Thus, since $Q_D = 50 - 2P$, the equilibrium output is 50 – (2)(3), or 44 million units per year.

b. As we know from Chapter 2 of the text, a tax of $1 will raise the supply curve by $1, which means that its equation will be $P = 4$. Thus the long-run equilibrium output can be determined by finding the value of Q_D if $P = 4$. (This is the value where the demand curve intersects the new supply curve.) Thus, since $Q_D = 50 - 2P$, the equilibrium output is 50 – (2)(4), or 42 million units per year.

c. No. The long-run equilibrium price after the imposition of the tax is $4.

6. *a.* Marginal cost equals the following:

$$MC = \frac{dC}{dQ} = 4 + 4Q$$

Setting marginal cost equal to price, we have

$$4 + 4Q = 24$$
$$4Q = 20$$
$$Q = 5$$

Thus the optimal output rate is 5.

b. Profit equals total revenue minus total cost. Since total revenue equals $24Q$, profit equals

$$\pi = 24Q - 200 - 4Q - 2Q^2 = -200 + 20Q - 2Q^2$$

Because $Q = 5$,

$$\pi = -200 + (20)(5) - (2)(5)^2 = -200 + 100 - 50 = -150$$

Thus the firm loses $150.

c. The firm's short-run supply curve is its marginal cost curve, $4 + 4Q$, so long as price exceeds average variable cost. Total variable cost equals $4Q + 2Q^2$, so average variable cost equals

$$4 + 2Q$$

and its minimum value is $4. Thus, so long as price (P) exceeds $4, the firm's short-run supply curve is

$$P = 4 + 4Q$$

7. The shift in the demand curve will cause the price of textiles to rise, with the result that textile firms will increase production and will earn economic profits. The profits will attract entrants, with the result that the supply curve for textiles will shift to the right. Eventually the price of textiles will go back to its original level, but the output of textiles will be higher because of the entrants. If the government does not allow the price of textiles to rise, this process is nipped in the bud. There is no incentive for existing firms to increase production or for firms to enter the industry.

8. The firm's marginal cost curve is:

Output	*Marginal cost (dollars)*
0 to 1	10
1 to 2	12
2 to 3	13
3 to 4	14
4 to 5	15
5 to 6	16
6 to 7	17

a. If the price is $13, the firm will produce 2 or 3 units.
b. 3 or 4 units.
c. 4 or 5 units.
d. 5 or 6 units.
e. 6 or 7 units.

9. The industry supply curve is shown below.

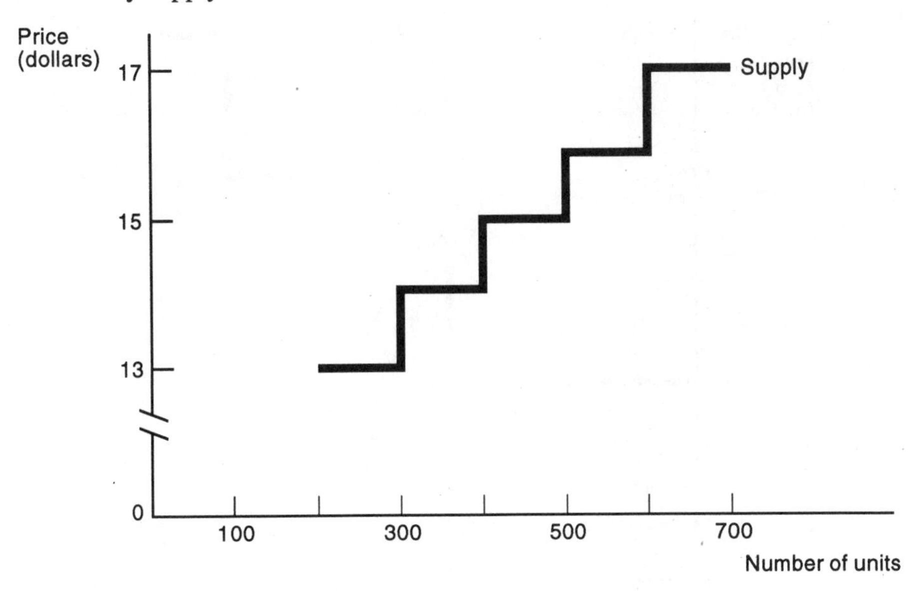

10. $$\eta_S = \frac{8-7}{(8+7)/2} \div \frac{4-3}{(4+3)/2} = \frac{1}{7.5} \div \frac{1}{3.5} = 0.47$$

11. No. In fact. Besser was found guilty of illegally monopolizing the industry in 1951.

12. The arc elasticity of supply is (6 – 4)/5 ÷ (100 – 80)/90 = 1.8.

13. A difference of about 3.24 percent.

14. *a.* The firm's total variable cost function is:

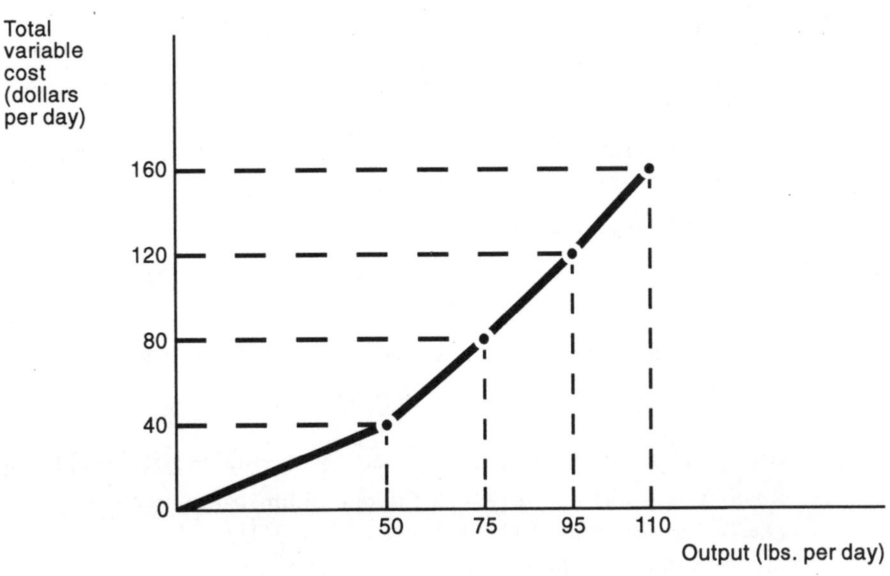

b. The firm's marginal cost function is:

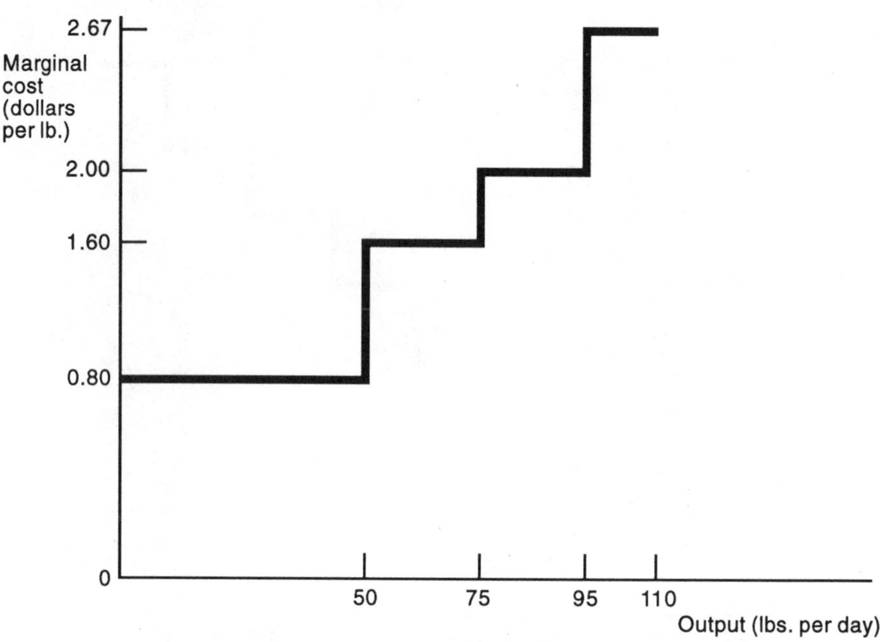

c. 75 or 95 pounds per day.

d. No.

e. Yes. For example, 1½ workers per day can be hired by having one worker work every day and one worker work every other day.

15. *a.* \$700.

b. \$35,000.

c. \$10,000.

d. Zero.

e. \$700.

16. According to this evidence, the marginal cost of the first unit of output is 4 + (3)(1) = \$7. Clearly, if the price of a unit of the product is \$3, the firm loses money on each unit it sells. Indeed it does not even cover its variable costs. Thus, the student is right; the firm should produce nothing.

17. *a.* The firm's total cost function is

$$C = AC \times Q = \left(\frac{400}{Q} + 3Q\right)Q = 400 + 3Q^2$$

The firm's total revenue equals $R = 3Q$. Thus, its profit equals

$$\pi = R - C = 3Q - 400 - 3Q^2 = -400 - 3(Q^2 - Q) = -400 - 3Q(Q-1)$$

If $Q > 1$, the firm incurs losses because $\pi = -400 - 3Q(Q-1)$ is negative.

b. Since $\pi = -400 - 3Q(Q-1)$, the maximum value of π is not achieved at $Q = 0$. For example, if $Q = 1/2$, $\pi = -400 - 3(1/2)(-1/2) = -399\frac{1}{4}$. This is a higher value of π than -400, which is the value of π when $Q = 0$. Thus, assuming that the firm can produce fractional units of output per day, it is better off to operate than to shut down.

18. Yes, because each additional unit produced (up to and including 100) brings in $1 more than it costs.

19. Profit $= 5Q - 20 - 2Q - 0.3Q^2$. If you plot profit against Q, you will find that it is a maximum at $Q = 5$. At $Q = 5$, profit equals –$12.50 per hour. This is the best you can do under these circumstances.

20. If the price is $2, profit $= 2Q - 20 - 2Q - 0.3Q^2$. If you plot profit against Q, you will find that it is a maximum at Q = 0. Thus, you will shut down.

21. *a.* 3 or 4 units.

b.

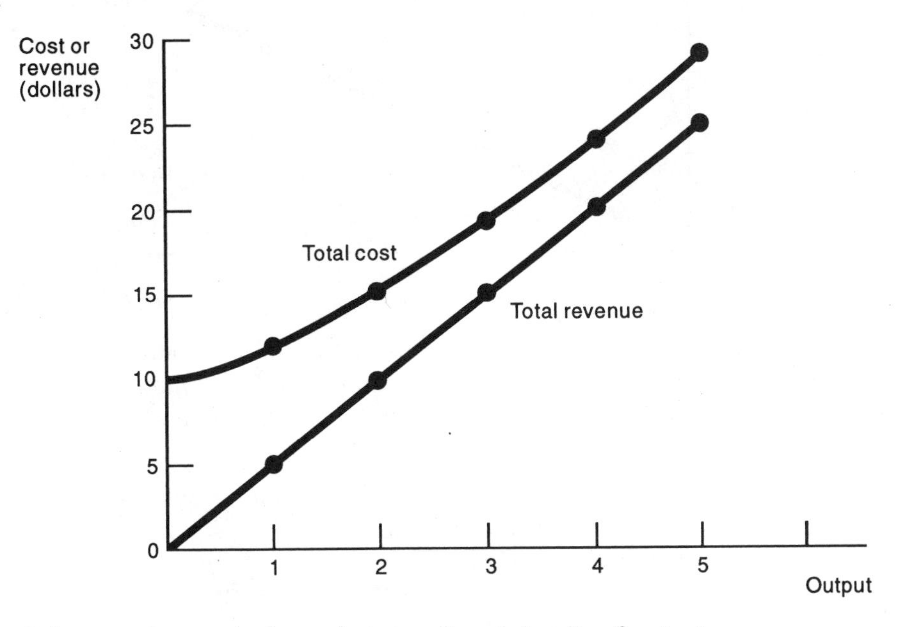

c. Price equals marginal cost between 3 and 4 units of output.

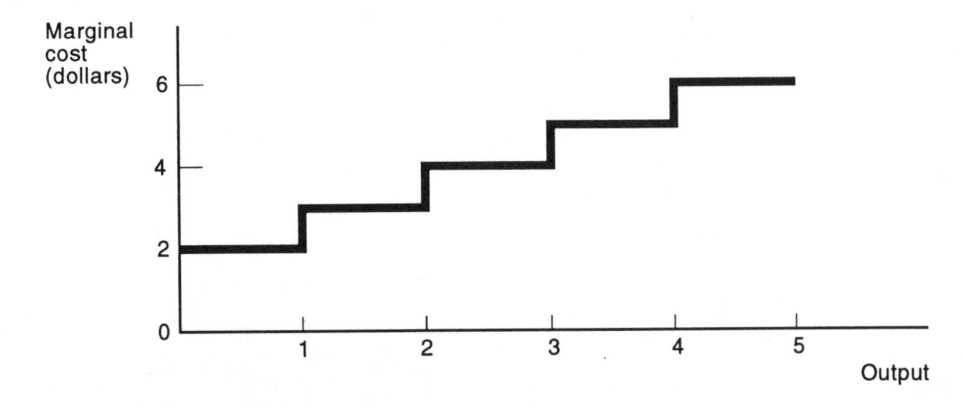

22. The demand and supply curves are

$$P = 5 - 0.5Q_D$$
$$P = \frac{1}{3}Q_S$$

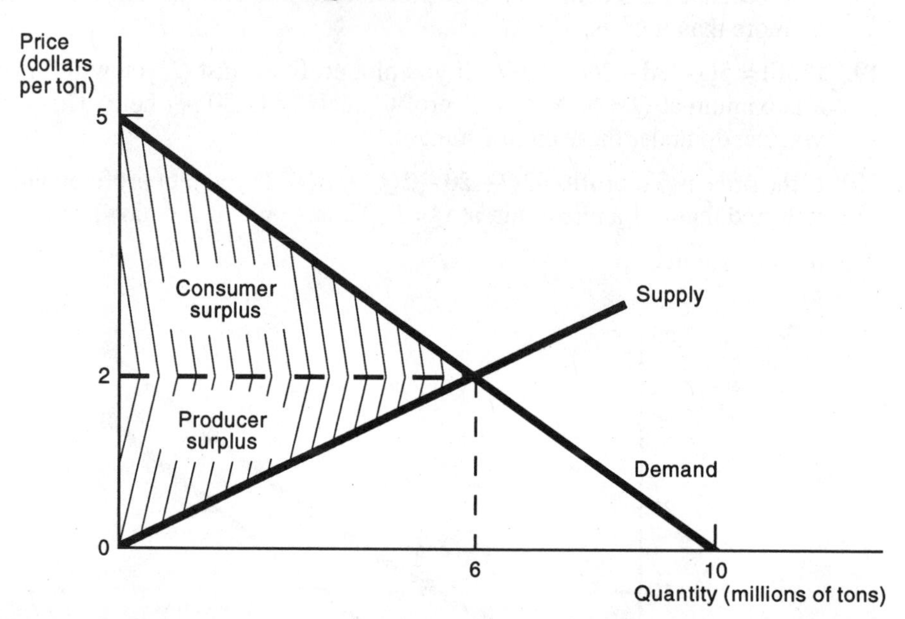

Producer surplus equals (½) ($2) (6 million) = $6 million.
Consumer surplus equals (½) ($3) (6 million) = $9 million.
Total surplus equals $6 million + $9 million = $15 million.
No.

23. It will lose less money than if it shuts down completely. As long as price (P) exceeds average variable costs (AVC),

$$ATC - AFC < P$$

where ATC is average total cost and AFC is average fixed cost. The loss is $Q(ATC - P)$, which must be less than $Q \times AFC$, since

$$ATC - P < AFC$$

Chapter 11 Monopoly

Case Study: Subway Fares

The problems of the New York subway are by no means new. In the early 1950s, it was felt by many responsible citizens that the burden on the city budget due to the subway deficit was perhaps the single most important factor in New York City's financial problem. The increased operating cost resulting from the adoption of the forty-hour week, together with higher prices of materials, was expected to cause an operating deficit for 1951–52 of almost $30 million for the city's transit system.

In response to this problem, the question was raised of whether or not subway fares should be increased. William Vickrey of Columbia University was asked by the Mayor's Committee on Management Survey of the City of New York to make a study of this question. Vickrey made four estimates of the demand curve, each based on a different assumption regarding its form. In Case *A*, he assumed that equal absolute changes in fare result in equal absolute changes in traffic. In Case *B*, he assumed that equal absolute changes in fare result in equal percentage changes in traffic. In Case *C*, he assumed that equal percentage changes in fare result in equal absolute changes in traffic. And in Case *D*, he assumed that equal percentage changes in fare result in equal percentage changes in traffic.[1] His results are shown on the next page.

Suppose that you were a member of the Mayor's Committee on Management Survey and that you were presented the data in the table below. How would these data be useful in deciding whether or not to raise the subway fare from the then-prevailing level of 10 cents? What additional data would you need in order to come to a conclusion on this score? Do you think that data of this sort would be as useful now as in the early 1950s? Do you think that the demand schedule for subway travel in New York is the same in 1994 as in 1952?

1. This very brief sketch cannot do justice to Vickrey's study, which was concerned with many other important aspects of the problem. See W. Vickrey, *The Revision of the Rapid Transit Fare Structure of the City of New York*, Technical Monograph No. 3, Finance Project, Mayor's Committee on Management Survey of the City of New York.

Alternative estimates of demand schedule for subway travel, New York, 1952

Fare (cents)	Case A Passengers	Case A Total revenue (dollars)	Case B Passengers	Case B Total revenue (dollars)	Case C Passengers	Case C Total revenue (dollars)	Case D Passengers	Case D Total revenue (dollars)
	(passengers and revenues in millions per year)							
5	1945	97.2	1945	97.2	1945	97.2	1945	97.2
10	1683	168.3	1683	168.3	1683	168.3	1683	168.3
15	1421	213.2	1458	218.7	1530	229.5	1547	232.0
20	1159	231.8	1262	252.4	1421	284.2	1457	291.5
25	897	224.2	1092	273.0	1347	336.8	1390	348.2
30	635	190.5	945	283.5	1278	383.4	1340	402.0

Source: W. S. Vickrey

Completion Questions

1. A monopolist can sell 12 units of output when it charges $8 a unit, 11 units of output when it charges $9 a unit, and 10 units of output when it charges $10 unit. The marginal revenue from the eleventh unit of output equals ________. The marginal revenue from the twelfth unit of output equals ____________.
2. The threat of potential competition often acts as a ______________ on the policies of a monopolist.
3. A monopolist's demand curve is the same as the industry ____________.
4. For a monopolist, marginal revenue is ______________ than price.
5. Under monopoly, there is not always a ______________ relationship between price and output.
6. In the long run, a monopolistic firm will ______________ incur losses, but it may make ____________.
7. Under multiplant monopoly, the monopolist will equalize the ______________ of the output of each plant if it minimizes cost.
8. If the demand curve for the product shifts to the right, the monopolist ______ increase price.
9. Price will be ______________ under monopoly than under perfect competition.
10. Output will be ____________ under monopoly than under perfect competition.
11. Average cost will be __________ under monopoly than under perfect competition.
12. The Lerner index is a measure of the amount of ________ possessed by a firm.

True or False

____ 1. A profit-maximizing monopolist will always choose an output in the short run where average total cost is less than average revenue.

____ 2. If the government sets a price ceiling below the equilibrium price, a perfectly competitive firm will reduce output, but a monopolist may increase output.

____ 3. If a monopolist sells in two markets, each of which has a linear demand curve, and the demand curves have equal slopes, the monopolist will charge the same price in both markets.

____ 4. At the point where total revenue is maximized, the price elasticity of demand equals 1.

____ 5. A monopolist always produces an output such that the price elasticity of demand equals 1.

____ 6. Price discrimination cannot occur unless consumers can be segregated into classes and the commodity cannot be transferred from one class to another.

____ 7. Price discrimination always occurs when differences in price exist among roughly similar products, even when their costs are not the same.

____ 8. Price discrimination is profitable even when the price elasticity of demand is the same among each class of consumer in the total market.

____ 9. If a monopolist practices price discrimination, he or she does not set marginal revenue equal to marginal cost.

____ 10. First-degree price discrimination occurs frequently, particularly in the auto industry.

____ 11. Sometimes a commodity cannot be produced without price discrimination, for the industry would have to produce at a loss otherwise.

____ 12. A two-part tariff requires the consumer to pay an initial fee for the right to buy the product as well as a usage fee for each unit of the product that he or she buys.

____ 13. Tying and bundling are the same thing.

____ 14. One reason why tying is used by firms is that it allows them to charge higher prices to customers that use their products intensively than to those who make little use of them.

____ 15. If a firm's price is $10 and its marginal cost is $5, the Lerner index is 2.

____ 16. If a firms's price is $10 and its marginal cost is $5, the price elasticity of demand for the firm's product is 2 if it maximizes profit.

Multiple Choice

1. A monopolist's total cost equals $100 + 3Q$, where Q is the number of units of output it produces per month. Its demand curve is $P = 200 - Q$, where P is the price of the product. The marginal revenue from the twentieth unit of output per month equals
   a. $3,600.
   b. $3,439.
   c. $180
   d. $140.
   e. none of the above.

2. Monopolies can arise as a consequence of

a. patents.
b. control over the supply of a basic input.
c. franchises.
d. the shape of the long-run average cost curve.
e. all of the above.

3. A monopolistic firm will expand its output when
a. marginal revenue exceeds marginal cost.
b. marginal cost exceeds marginal revenue.
c. marginal cost equals marginal revenue.
d. marginal revenue is negative.
e. none of the above.

4. A profit-maximizing monopolist will never produce at a point where
a. demand is price inelastic.
b. demand is price elastic.
c. marginal cost is positive.
d. marginal cost is increasing.
e. none of the above.

5. A profit-maximizing monopolist, if he or she owns a number of plants, will always:
a. produce some output at each plant.
b. transfer output from plants with high marginal cost to those with low marginal cost.
c. transfer output from plants with low marginal cost to those with high marginal cost.
d. produce all of the output at a single plant and shut down the rest.
e. none of the above.

6. Suppose that the monopolist's marginal cost is constant in the relevant range and that his or her demand curve is downward-sloping and linear. Suppose that an excise tax is imposed on the monopolist's product. If he or she maximizes profit, price will increase by
a. an amount less than the tax.
b. an amount equal to the tax.
c. an amount more than the tax.
d. no amount at all.
e. none of the above.

7. To determine how a firm that practices discrimination will allocate output between two classes of consumers, one must
a. compare the marginal revenues in the classes.
b. compare the prices in the classes.
c. compare the slopes of the demand curves in the classes.
d. compare the heights of the demand curves in the classes.
e. none of the above.

Review Questions

1. Define monopoly. Is a monopolist subject to any kind of competition?
2. What are some of the most important factors that give rise to monopoly?
3. How does the monopolist's demand curve differ from that of a perfectly competitive firm? How do its costs differ from those of a competitive firm?
4. *a.* Suppose that a monopolist's demand curve and total cost curve are as follows:

Output	*Price (dollars)*	*Total cost (dollars)*
1	10	20
2	9	21
3	8	22
4	7	23
5	6	24
6	5	26
7	4	29
8	3	32

Plot the monopolist's total revenue and total cost curves in the graph below.

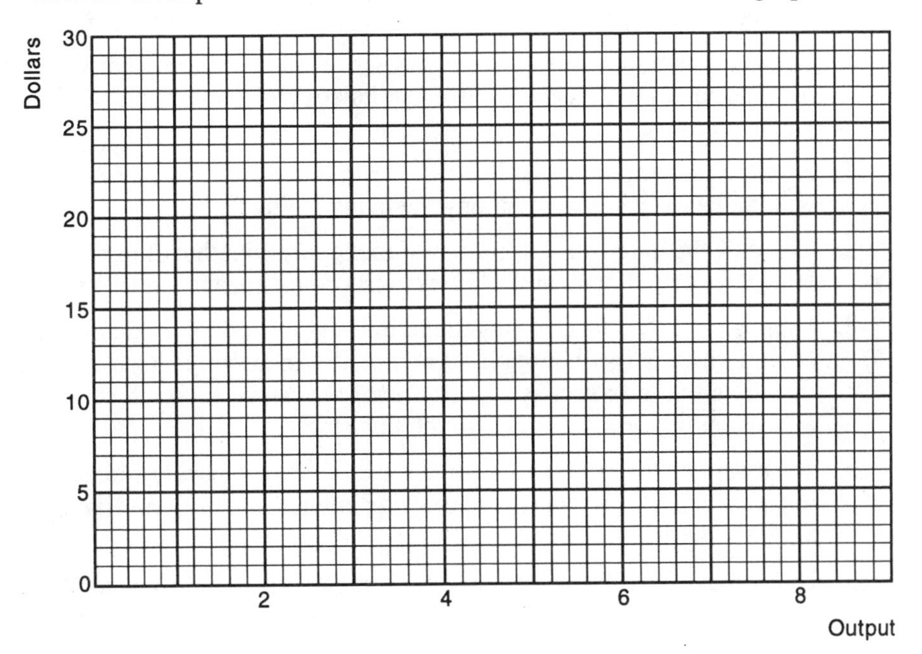

b. Plot the monopolist's marginal revenue and marginal cost curves in the graph below.

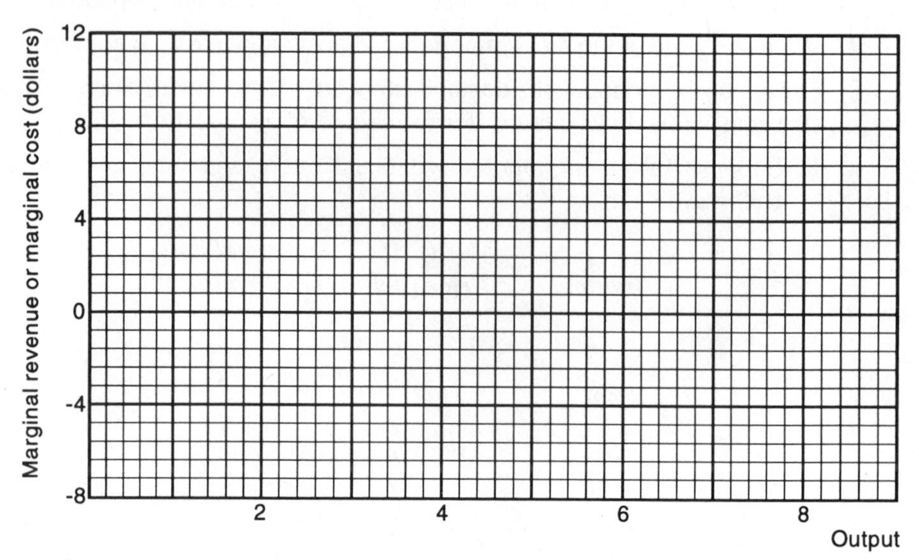

c. What output will maximize the monopolist's profit? Is this output the one where marginal cost equals marginal revenue? What price should the monopolist charge?

5. Show why a monopolist will always set its output at the point where marginal revenue equals marginal cost, if it maximizes profit and if price exceeds average variable cost.
6. Suppose that a monopolist has more than one plant. If it minimizes costs, how will it allocate production among the plants? Why?
7. Compare the long-run equilibrium of monopoly with that of perfect competition. How does output compare? How do costs compare? How does price compare?
8. Suppose that a perfectly competitive industry is monopolized and that its supply curve (which is the monopolist's marginal cost curve) is horizontal. Show how the welfare loss to society due to monopoly can be represented by a triangular area under the demand curve for the industry's product.
9. What is price discrimination? Under what circumstances is it feasible for a firm to practice discrimination? Under what circumstances is it profitable for the firm to practice price discrimination?
10. Under price discrimination, how does a monopolist decide how much to produce? How does it decide how to allocate this output among markets?
11. According to C. Emery Troxel, a general rate case is the common sort of regulatory "contest" in the Michigan telephone industry. What sorts of concepts and methods are used by these commissions in such cases?
12. Many industries—for example, the steel industry—claim that the demand for their product is price inelastic. Suppose that you were able to bring together all of the

firms in an industry and form a monopoly. Would you operate at a point on the demand curve where demand is price inelastic?

13. According to Milton Friedman of the Hoover Institution at Stanford University, it is preferable in cases of natural monopoly for society to tolerate private unregulated monopoly rather than to attempt to regulate natural monopolies. Do you agree? Why or why not?
14. The Aluminum Company of America (Alcoa) was the sole manufacturer of virgin aluminum ingot in the United States from its inception in the late nineteenth century until World War II. Suppose that in the 1920s you had had the power to break up Alcoa into several smaller firms. Would you have done so? Why?
15. According to Sir John Hicks, "The best of all monopoly profits is a quiet life." If so, how would this affect the theory of monopoly?
16. According to Henry C. Simons, "The great enemy of democracy is monopoly in all of its forms." Do you agree? Why or why not?
17. Describe tying and the reasons for its use.

Problems

1. The Morton Company's marginal cost of production is zero if its output is less than (or equal to) 50 units per day.
   a. If its output rate is 20 units per day, does its average total cost exceed its average fixed cost?
   b. If the Morton Company is a perfectly competitive firm, will it produce less than 50 units per day? Why or why not?
   c. If the Morton Company is a monopolist, under what conditions (if ever) will it produce less than 50 units per day? Explain.
2. *a.* Under what conditions will a monopolist be able and willing to engage in price discrimination?
   b. Under price discrimination, will price tend to be relatively high in markets where the price elasticity of demand is relatively high? Why or why not?
3. A firm sells in two distinct markets, which are sealed off from one another. The demand curve for the firm's output in the one market is

$$P_1 = 200 - 10Q_1$$

where P_1 is the price of the product and Q_1 is the amount sold in the first market. In the second market, the demand curve for the firm's product is

$$P_2 = 100 - 5Q_2$$

where P_2 is the price of the product and Q_2 is the amount sold in the second market. The firm's marginal cost curve is

$$MC = 10Q$$

where Q is the firm's entire output.
   a. How many units should the firm sell in the first market?
   b. What price should it charge in the first market?
   c. How many units should the firm sell in the second market?
   d. What price should it charge in the second market?

4. Firm X has a complete monopoly over the production of nutmeg. The following information is given;

$$\text{Marginal revenue} = 1{,}000 - 20Q$$
$$\text{Total revenue} = 1{,}000Q - 10Q^2$$
$$\text{Marginal cost} = 100 + 10Q$$

where Q equals the output of nutmeg per unit of time. How much nutmeg would be sold and at what price if

a. the firm sets price as a monopoly?

b. the industry (firm) behaves perfectly competitively?

5. A garbage collector in an Ohio community is a monopolist. The community imposes a lump sum tax of $20,000 per year on the garbage collector. Can the garbage collector shift the burden of the tax to consumers?

6. A firm has two plants with the following marginal cost functions:

$$MC_1 = 20 + 2Q_1$$
$$MC_2 = 10 + 5Q_2$$

where MC_1 is marginal cost in the first plant, MC_2 is marginal cost in the second plant, Q_1 is output in the first plant, and Q_2 is output in the second plant. If the firm is minimizing its costs and if it is producing 5 units of output in the first plant, how many units of output is it producing in the second plant? Explain.

7. The QXR Corporation has bought exclusive rights to sell hot dogs in a local sports arena. The fee it paid for this concession was $200. The variable cost of obtaining and marketing each hot dog is 10 cents. The demand schedule for hot dogs in this arena is as follows:

Price per hot dog (cents)	*Thousands of hot dogs sold per game*
20	10
25	9
30	8
35	7
40	6
45	5
50	4

It is assumed that prices must be multiples of a nickel.

a. What price should the QXR Corporation charge for a hot dog?

b. What is the maximum amount that the QXR Corporation should pay for this concession for a single game?

8. A monopolist has the following total cost function and demand curve:

Price (dollars)	*Output*	*Total cost (dollars)*
8	5	20
7	6	21
6	7	22
5	8	23
4	9	24
3	10	30

   What price should it charge?

9. Authors receive a royalty that is a fixed percentage of the price of the book. For this reason, economists have pointed out that an author has an interest in a book's price being lower than the price which maximizes the publisher's profits. Prove that this is true.

10. According to Spencer, Seo, and Simkin, the makers of methyl methacrylate used to sell it at 85 cents per pound for commercial purposes.[2] However, for denture purposes, it was sold to the dental profession for $45 per pound. Assuming that there was no difference in quality, why would the producers of methyl methacrylate, Du Pont and Rohm and Haas, find it profitable to charge different prices? In which of these markets (the commercial market or the dental market) do you think that the price elasticity of demand was lower?

11. According to Meyer, Peck, Stenason, and Zwick, trucks "have a comparative long-run marginal cost advantage over rails only in hauls of less than roughly 100 miles. Yet 97 percent of all manufactured goods ton-miles transported by common motor carriers in the early 1950s were on hauls of more than 100 miles."[3] What are the implications for resource allocation in transportation? What policies of the Interstate Commerce Commission help to account for this fact?

12. *a.* Suppose that you are the owner of a metals-producing firm that is an unregulated monopoly. After considerable experimentation and research you find that your marginal cost curve can be approximated by a straight line, $MC = 60 + 2Q$, where MC is marginal cost (in dollars) and Q is your output. Moreover, suppose that the demand curve for your product is $P = 100 - Q$, where P is the product price (in dollars) and Q is your output. If you want to maximize profit, what output should you choose?
    b. What price should you charge?

2. M. Spencer, K. Seo, and M. Simkin, *Managerial Economics.*
3. F. M. Scherer, *Industrial Market Structure and Economic Performance.*

13. For a particular good that is monopolized,

$$P = 12 - 2Q_D$$
$$MR = 12 - 4Q_D$$
$$MC = 2Q$$

where P is its price (in dollars per ton), Q_D is the quantity demanded per year (in millions of tons), MR is its marginal revenue (in dollars per ton), MC is its marginal cost (in dollars per ton), and Q is the quantity produced per year (in millions of tons). How great is the deadweight loss due to monopoly?

14. *a.* Suppose that you are hired as a consultant to a firm producing ball bearings. This firm is a monopolist which sells in two distinct markets, one of which is completely sealed off from the other. The demand curve for the firm's output in the one market is $P_1 = 160 - 8Q_1$, where P_1 is the price of the product and Q_1 is the amount sold in the first market. The demand curve for the firm's output in the second market is $P_2 = 80 - 2Q_2$, where P_2 is the price of the product and Q_2 is the amount sold in the second market. The firm's marginal cost curve is $5 + Q$, where Q is the firm's entire output (destined for either market). The firm asks you to suggest what its pricing policy should be. How many units of output should it sell in the second market?
    b. How many units of output should it sell in the first market?

Key Concepts for Review

Monopoly
Natural monopoly
Patents
Market franchise
Multiplant monopoly
Deadweight loss from monopoly
Price discrimination
First-degree price discrimination
Second-degree price discrimination
Third-degree price discrimination
Fair rate of return
Regulatory commissions
Value of plant
Two-part tariffs
Tying
Bundling
Lerner index

ANSWERS

Case Study: Subway Fares

Regardless of which case (A, B, C, or D) you choose, the demand for subway travel is shown by the table to be price inelastic at the then-prevailing fare of 10 cents. This means that increases in the fare would increase total revenues. Also, they would reduce the deficit because fewer passengers would mean lower costs (or at least no higher costs). But this is not the only consideration. You would probably want to investigate how such a fare increase would affect various parts of the population—the poor, the rich, rush-hour traffic, non-rushhour traffic, and so on. Data of this sort would be just as relevant now as in the early 1950s, but the demand schedule almost certainly is different in 1994 than in 1952.

Completion Questions

1. –$1; –$3
2. brake
3. demand curve
4. less
5. unique
6. not; profits
7. marginal cost
8. may or may not
9. higher
10. lower
11. higher
12. monopoly power

True or False

1. False	2. True	3. False	4. True	5. False	6. True
7. False	8. False	9. False	10. False	11. True	12. True
13. False	14. True	15. False	16. True		

Multiple Choice

1. *e*	2. *e*	3. *a*	4. *a*
5. *b*	6. *a*	7. *a*	

Review Questions

1. Monopoly occurs when there is one, and only one, seller in a market. The monopolist is not completely insulated from the effects of actions taken in the rest of the economy. He or she is affected by indirect and potential forms of competition.

2. First, a single firm may control the entire supply of a basic input that is required to make a product.

   Second, a firm may become monopolistic because the average cost of producing the product reaches a minimum at an output rate that is big enough to satisfy the entire market at a price that is profitable.

   Third, a firm may acquire a monopoly over the production of a good by having patents on the product or on certain basic processes that are used in its production.

   Fourth, a firm may become monopolistic because it is awarded a franchise by a government agency.

3. The monopolist's demand curve is the industry demand curve. Thus it is downward-sloping, whereas a perfectly competitive firm has a horizontal demand curve.

   If the monopolist is a perfect competitor in the market for inputs, the theory of cost is the same for a perfect competitor or a monopolist. Chapter 16 takes up the case where the firm is not a perfect competitor in the market for inputs.

4. *a.*

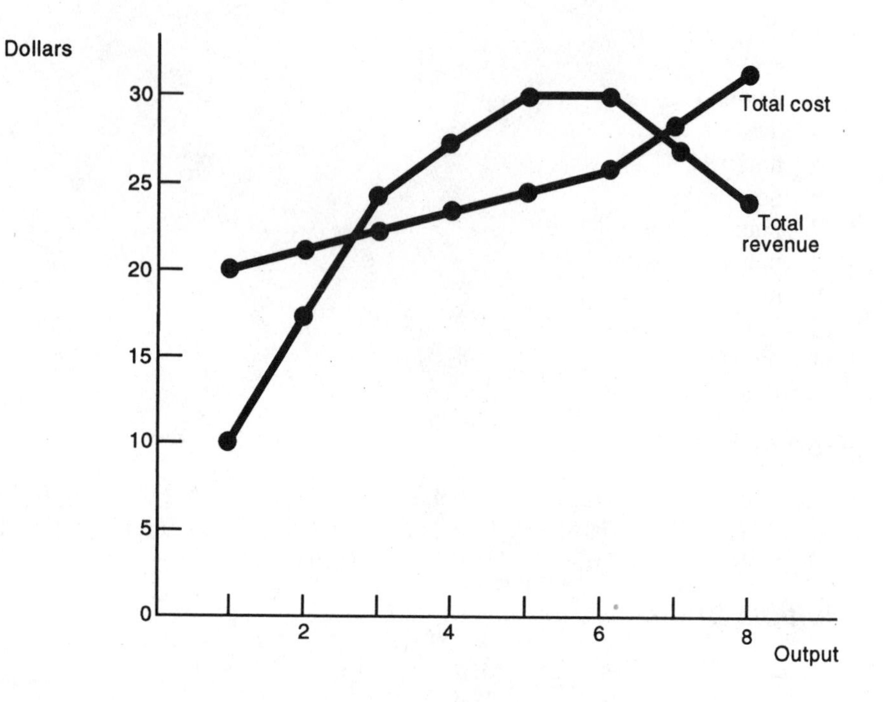
Dollars
30
25
20
15
10
5
0
2
4
6
8
Output
Total cost
Total revenue

b.

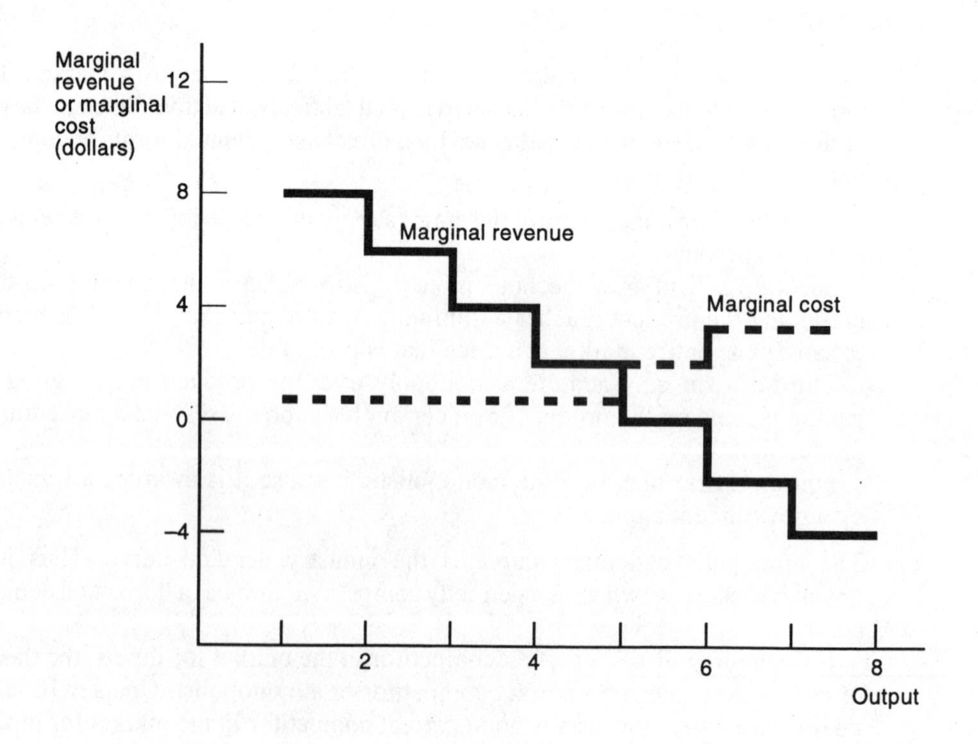
Marginal revenue or marginal cost (dollars)
12
8
4
0
-4
2
4
6
8
Output
Marginal revenue
Marginal cost

c. Five units of output will be produced, and the price will be $6. The output is the one where marginal cost equals marginal revenue.

5. If marginal revenue exceeds marginal cost, profit can be increased by increasing output. If marginal revenue is less than marginal cost, profit can be increased by decreasing output. Profit will be a maximum when marginal revenue equals marginal cost (as long as price exceeds average variable cost).
6. The monopolist will allocate output so that the marginal costs are the same at each plant. Why? Because this is a condition for cost minimization.
7. Output is less, average costs are higher, and price is higher under monopoly than under perfect competition.
8. Basically, the idea is that the value to society of the extra output resulting from perfect competition is equal to Q_1CAQ_0 (in the graph below), whereas the cost to society of the extra output is equal to Q_1BAQ_0. Thus the net loss due to the smaller output under monopoly is equal to the triangle *ABC*.

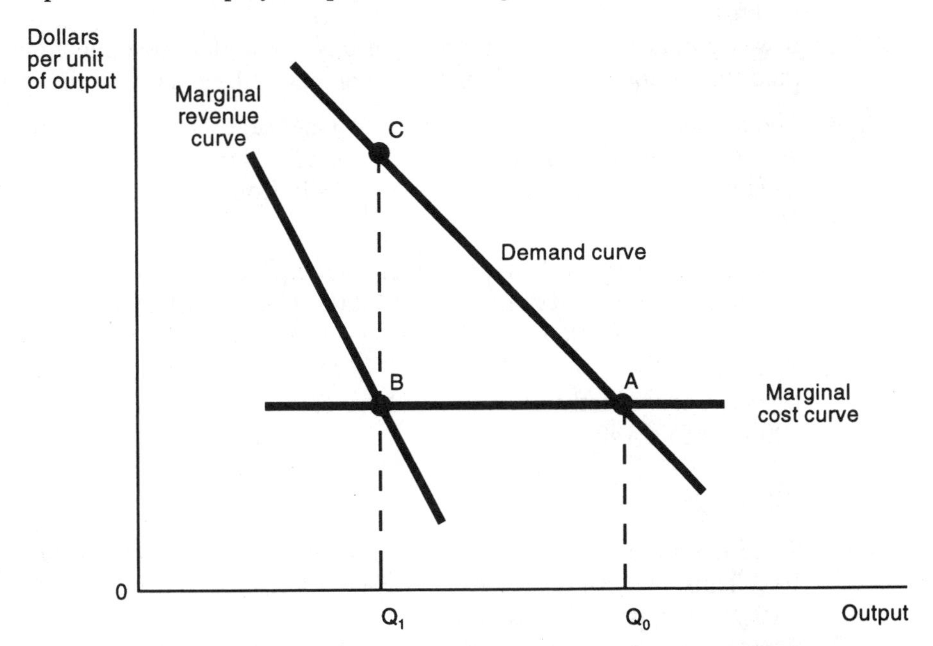

9. Price discrimination occurs when the same commodity is sold for more than one price. For price discrimination to be feasible, it must be possible for the seller to segregate buyers into classes and the buyers must be unable to transfer the good from low-price classes to high-price classes. For price discrimination to be profitable, the price elasticity of demand must vary among classes.
10. The monopolist will set marginal revenue in the various classes equal, and the common value of marginal revenue will be set equal to marginal cost.
11. The commissions try to establish a reasonable return on a firm's existing plant. There are many difficult questions involved in determining what is "reasonable" and what is the "value" of a firm's existing plant.

12. No. At such a point marginal revenue would be negative. Thus you could increase revenue—and consequently profit—by reducing output.
15. It would mean that monopolists would probably worry less than competitive firms about maximizing profits.
17. If a monopolist produces a product that will function properly only if used in conjunction with another product, it may require its customers to buy the other product from it, rather than from alternate suppliers. This is called tying. One reason why firms adopt this technique is that it enables them to charge higher prices to customers who use their products intensively than to those who make little use of them.

Problems

1. *a.* No. They are equal.
   b. No, since marginal cost (which is zero) must be less than price (unless it too is zero).
   c. It will produce less than 50 units per day if marginal revenue equals zero (and thus is equal to marginal cost) at an output less than 50 units per day.
2. *a.* The monopolist must be able to identify and segregate buyers into classes with different price elasticities of demand for the product, and the buyers must be unable to transfer the product easily from one class to another.
   b. No.
3. *a.* Since marginal revenue in the first market equals $200 - 20Q_1$ and marginal revenue in the second market equals $100 - 10Q_2$, it follows that
   $$200 - 20Q_1 = 100 - 10Q_2 = 10(Q_1 + Q_2)$$
   Thus, $Q_1 = 10 - 2Q_2$, and $200 - 20(10 - 2Q_2) = 100 - 10Q_2$, which means that $Q_2 = 2$. Consequently, $Q_1 = 10 - 2(2)$, or 6.
   b. $200 - 10(6) = 140$.
   c. 2.
   d. $100 - 5(2) = 90$.
4. *a.* Since marginal revenue would equal marginal cost, $1{,}000 - 20Q = 100 + 10Q$. Thus, $Q = 30$. Because $PQ = 1{,}000Q - 10Q^2$, it follows that $P = 1{,}000 - 10Q$. Thus, price equals 700.
   b. Since the demand curve is $P = 1{,}000 - 10Q$, and the supply curve is $P = 100 + 10Q$, they intersect at $Q = 45$ and $P = 550$.
5. No. The lump sum tax adds to the garbage collector's fixed costs but does not affect his marginal costs. Thus, since both the marginal revenue curve and marginal cost curve are unaffected by the tax, the profit-maximizing output is the same, with or without the tax, and the price too must be the same. Because the price is unaffected, none of the tax burden is shifted to consumers.
6. If the firm is producing 5 units in the first plant, the marginal cost in the first plant equals $20 + (2)(5)$, or 30. Thus, if the firm is minimizing costs, marginal cost in the second plant must also equal 30 which means that

$$10 + 5Q_2 = 30$$
$$5Q_2 = 20$$
$$Q_2 = 4$$

Consequently, the second plant must be producing 4 units of output.

7. *a.* Since total cost (excluding the fee) is 10 cents times the number of hot dogs sold, the QXR Corporation's total profit per game is shown below, under various assumptions concerning the price.

Thousands of hot dogs	*Price (cents)*	*Total revenue (dollars)*	*Total cost (dollars) (excluding fee)*	*Total profit (dollars)*
10	20	2,000	1,000	1,000
9	25	2,250	900	1,350
8	30	2,400	800	1,600
7	35	2,450	700	1,750
6	40	2,400	600	1,800
5	45	2,250	500	1,750
4	50	2,000	400	1,600

Thus the QXR Corporation will maximize profit if it sets a price of 40 cents.

b. \$1,800, since this is the maximum amount it could make per game.

8. The monopolist's total revenue, total cost, and total profit at each level of output are as follows:

Output	*Total revenue (dollars)*	*Total cost (dollars)*	*Total profit (dollars)*
5	40	20	20
6	42	21	21
7	42	22	20
8	40	23	17
9	36	24	12
10	30	30	0

Thus, the optimal output is 6, which means that price should be \$7.

9. Since the author's royalty is a fixed percentage of the price of the book, the total royalties per year earned by the author are proportional to the total revenue per year from the book. Consequently, the author would like to maximize total revenue, since this maximizes his or her royalties. To maximize total revenue, marginal revenue should be set equal to zero. On the other hand, the publisher would like to maximize profit, which means that marginal revenue should be set equal to marginal cost. Thus, the author would like to set marginal revenue equal to a lower value than would the publisher. Holding constant the price elasticity of demand, marginal revenue decreases as price is lowered. Thus, the author tends to prefer a lower price than does the publisher.

10. Because the price elasticity of demand was different in the two markets.
    The dental market.

11. Apparently, trucks were carrying a great deal of traffic that could be carried more cheaply by railroads. In the past, the ICC often has not allowed railroads to cut their rates in order to take some of this business away from the trucks.

12. *a.* Marginal revenue = $100 - 2Q$.
    Marginal cost = $60 + 2Q$
    Consequently,

$$\begin{aligned} 100 - 2Q &= 60 + 2Q \\ 40 &= 4Q \\ 10 &= Q \end{aligned}$$

    That is, you should choose an output of 10 units.

    b. Since $P = 100 - Q$, if Q equals 10, P must equal 90. So you should charge a price of $90.

13. The deadweight loss is the shaded area in the graph below.

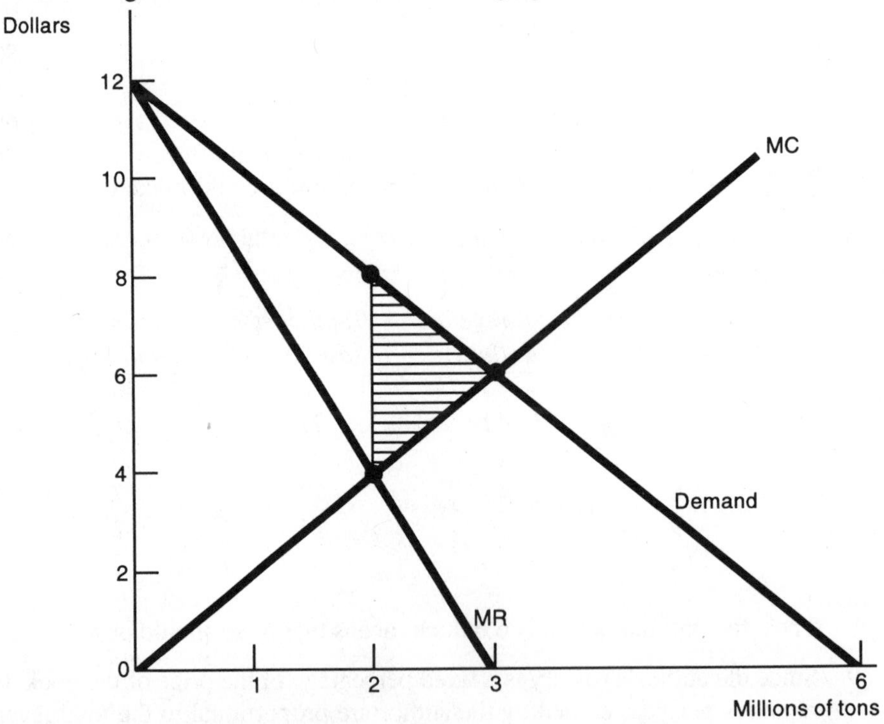

The deadweight loss equals ½ (1 million tons times $2 per ton) + ½(1 million tons times $2 per ton), or $2 million per year.

14. *a.*

$$MR_1 = 160 - 16Q_1$$
$$MR_2 = 80 - 4Q_2$$
$$MC = 5 + (Q_1 + Q_2)$$

Therefore,

$$160 - 16Q_1 = 5 + Q_1 + Q_2$$
$$80 - 4Q_2 = 5 + Q_1 + Q_2$$

Or,

$$155 - 17Q_1 = Q_2$$
$$75 - 5Q_2 = Q_1$$

Thus,

$$155 - 17(75 - 5Q_2) = Q_2$$
$$155 - 1275 + 85Q_2 = Q_2$$
$$84Q_2 = 1120$$
$$Q_2 = \frac{1120}{84}$$

It should sell 1120/84 units in the second market.

b.

$$Q_1 = 75 - 5Q_2$$
$$= 75 - 5(1120/84)$$
$$= 75 - 5600/84$$
$$= 75 - 66\tfrac{2}{3}$$
$$= 8\tfrac{1}{3}$$

It should sell $8\frac{1}{3}$ units in the first market.

Chapter 12 Monopolistic Competition and Oligopoly

Case Study: The Pricing of American Small Cars

The major U.S. automobile manufacturers were not very enthusiastic about producing small cars during much of the period after World War II, due in part to the fact that they made a larger profit on a large car than a small one. But the explosion of gasoline prices and the press of foreign competition during the 1970s led the American producers to introduce new lines of smaller, fuel-efficient cars in the fall of 1980. The prices at which they and their Japanese competitors were offered were as follows:

Chevrolet Citation	$6,337	Subaru wagon	$5,612
GM "J" Cars	6,300	Datsun 310	5,439
Ford Escort	6,009	Subaru hatchback	5,212
Dodge Omni	5,713	Mazda GLC	4,755

a. According to some observers, "Detroit . . . priced these cars relatively high to replenish its depleted coffers as soon as possible." Is it always true that this is the effect of charging relatively high prices?

b. Some auto executives stated that the U.S. demand for smaller cars would be so great in the early 1980s that Detroit would be able to sell all such cars that it could produce. What implicit assumptions were they making?

c. In 1979, the International Trade Commission decided not to recommend curbing the imports of Japanese cars into the United States. In terms of promoting the vitality and efficiency of the U.S. automobile industry, what arguments could be made in favor of this decision?

d. Were imports of Japanese cars into the United States subsequently curbed?

Completion Questions

1. A monopolistically competitive firm is in long-run equilibrium. Its marginal revenue equals $10. If its marginal revenue plus its economic profit equal one-half of its price, its price equals ______________, and its marginal cost equals ______________.

2. Chamberlin's theory assumes that there are a ______________ of firms and a ______________ product.

3. Under monopolistic competition, the theory claims that ______________ will be produced at a ______________ price than under monopoly.

4. If firms A and B are the only firms in an industry, firm A's ______________ curve shows how firm A will react, as a function of how much it thinks ______________ will produce and sell.

5. According to the ______________ model, an ______________ will occur at the point where the firms' reaction curves intersect.

6. The Cournot model provides ________________ satisfactory description or explanation of the way in which firms move toward equilibrium.
7. Under monopolistic competition, the theory suggests that ____________ will be produced at a ____________ price than under perfect competition.
8. It is difficult to compare perfect competition with monopolistic competition because the product is ___________ under perfect competition but ___________ under monopolistic competition.
9. In monopolistic competition each firm expects its actions to have __________ on its competitors.
10. Excess capacity under monopolistic competition may be fairly small if the demand curve facing the monopolistically competitive firm is _____________.
11. All firms in an industry have marginal cost curves that are horizontal lines at $10 per unit of output. If they combine to form a cartel, the cartel's marginal cost (will, will not) ______________ be $10 per unit of output. If the cartel maximizes profit, its marginal revenue will be (greater than, smaller than, equal to) _______________$10 per unit. Its price will be (greater than, smaller than, equal to) _______________ $10 per unit of output.
12. Under oligopoly, each firm is aware that its actions are likely to elicit ____________ in the policies of its competitors.
13. A good example of oligopoly in the United States is the ___________ industry.
14. In a contestable market, entry is absolutely ______________.
15. According to the kinked oligopoly demand curve, firms think that, if they raise their price, their rivals will _______________.
16. According to the kinked oligopoly demand curve, firms think that, if they lower their price, their rivals will _______________.
17. The Cournot model assumes that each firm thinks that the other will hold its output constant at _______________.

True or False

____ 1. A monopolistically competitive firm's short-run demand and cost curves are as follows:

Price (dollars)	*Quantity demanded*	*Output*	*Total cost (dollars)*
8	1	1	5
7	2	2	7
6	3	3	9
4	4	4	11
3	5	5	20

This firm, if it maximizes profit, will choose an output rate of 5.

____ 2. Because monopolistically competitive firms cannot earn monopoly profits, monopolistic competition does not result in economic inefficiency.

____ 3. Monopolistic competition is likely to result in more brands than perfect competition.

____ 4. Under monopolistic competition, firms can vary the characteristics of their products.

____ 5. Chamberlin assumes that firms in the same product group under monopolistic competition have the same cost and demand curves.

____ 6. Long-run equilibrium under monopolistic competition will be achieved when economic profits are zero.

____ 7. Entry and exit under monopolistic competition are as free as they are under perfect competition.

____ 8. Critics have charged that Chamberlin's definition of the group of firms in the product group is very ambiguous.

____ 9. It is obvious that product differentiation is a waste and that all industries should standardize their products.

____ 10. Collusion is often difficult to achieve and maintain because an oligopoly contains an unwieldy number of firms, or because the product is quite heterogeneous.

____ 11. If a perfectly competitive industry is operating along the elastic portion of its demand curve, there is no point in trying to cartelize the industry.

____ 12. Because a cartel maximizes profit, there is no incentive for a member to violate the rules of the cartel.

____ 13. Cartels have been common and legally acceptable in Europe.

____ 14. Cartels tend to be unstable because the demand curve facing a "cheater" is highly inelastic.

____ 15. According to the dominant-firm model, the dominant firm allows the smaller firms to sell all they want at the price it sets.

____ 16. Whether or not an industry remains oligopolistic in the face of relatively easy entry depends on the size of the market for the product relative to the optimal size of firm.

____ 17. The basic purpose of advertising is to make the demand curve for the product more elastic.

Multiple Choice

1. The Cournot model
   a. illustrates the theory of monopolistic competition.
   b. says that an equilibrium will occur at the point where the firm's reaction curves intersect.
   c. does not illustrate a Nash equilibrium.
   d. all of the above.
   e. none of the above.

2. One major assumption of Chamberlin's theory of monopolistic competition is that
   a. demand curves will be the same for each firm but cost curves will be different.
   b. each firm's product is a fairly close substitute for the others in the group.
   c. there are just a few firms in each group.
   d. each firm expects its actions to influence those of its rivals.
   e. all of the above.
3. Under monopolistic competition, a firm can use selling expenses to increase its profits. We know that such expenses are
   a. a social waste.
   b. entirely socially productive.
   c. large in many industries, but we are able to say little with confidence about their social productivity.
   d. all of the above.
   e. none of the above.
4. Firm A's demand curve, given below, is kinked.

Price (dollars)	*Quantity demanded*
3	100
4	80
5	60
6	55
7	50

   If the price must be an integer number of dollars, at what price is the kink?
   a. \$3.
   b. \$4.
   c. \$5.
   d. \$6.
   e. \$7.
5. If DD' is the industry demand curve and marginal costs are zero, the price that maximizes a cartel's profits is (see graph at the top of the next page.)
   a. OP_0.
   b. OP_1.
   c. OP_2.
   d. zero.
   e. none of the above.
6. In the previous question, the RR' curve is
   a. the marginal revenue curve.
   b. the marginal cost curve.
   c. the reaction curve.
   d. all of the above.
   e. none of the above.

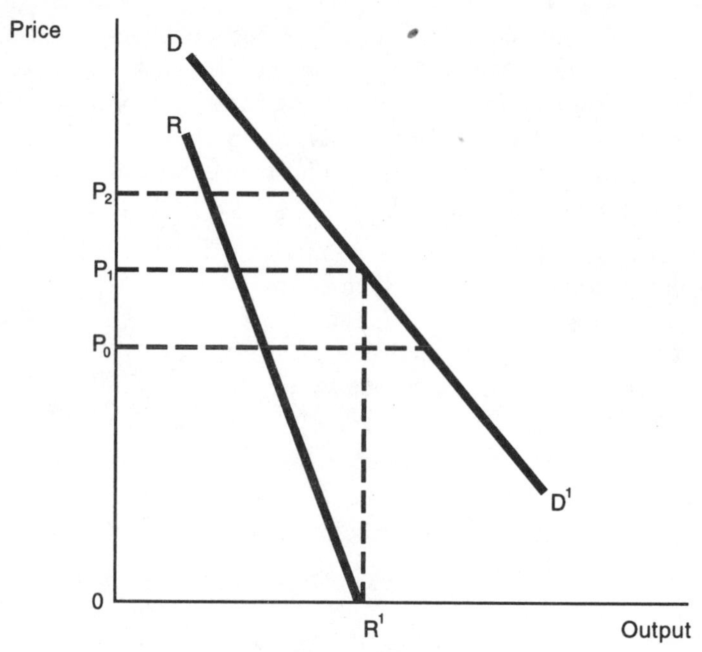

7. In the graph below, which of the curves is a kinked oligopoly demand curve?
   a. *AA′*.
   b. *BB′*.
   c. *CC′*.
   d. All of the above.
   e. None of the above.

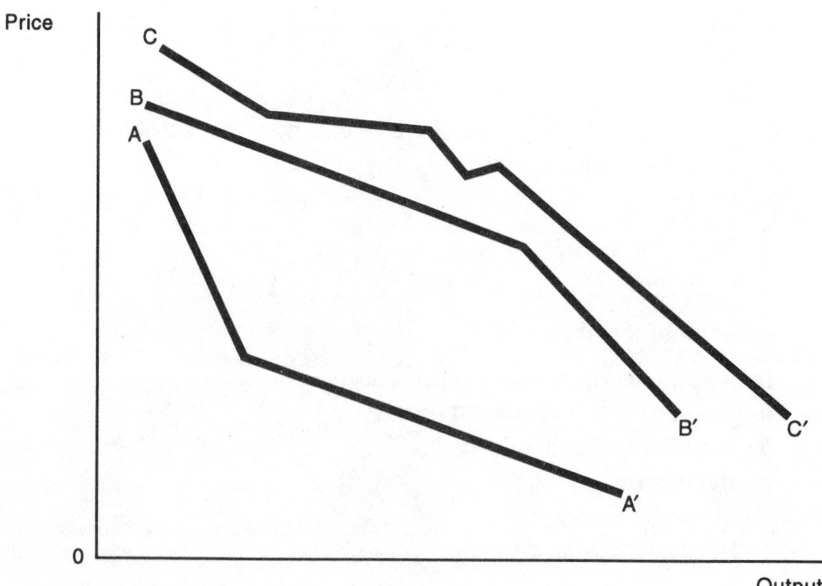

8. In the dominant-firm model, the dominant firm finds the demand curve for its output by
   a. using the unadjusted industry demand curve.
   b. adding up the small firms' demand curves.
   c. subtracting the small firms' supply from the industry demand curve.
   d. all of the above.
   e. none of the above.

Review Questions

1. Describe the Cournot model. Comment on the reasonableness of its assumptions.
2. What is a Nash equilibrium?
3. According to one estimate, perhaps a fourth of the sales of drug stores, liquor stores, and gasoline stations were made by firms whose costs were 10 percent higher than the optimum, the resulting annual cost being about 0.13 percent of GNP.[1] Why did such inefficiencies come about and persist?
4. What sorts of dissatisfaction with the theory of perfect competition and monopoly led to the theory of monopolistic competition? Who were the principal founders of the theory of monopolistic competition? When did their theories first appear?
5. Describe the basic assumptions underlying Chamberlin's theory of monopolistic competition.
6. Compare the long-run equilibrium of a monopolistically competitive industry with the long-run equilibria of perfectly competitive and monopolistic industries.
7. Describe and discuss some major criticisms of Chamberlin's theory of monopolistic competition. How important do you think these criticisms are?
8. Give five real-world examples of industries in which there is product differentiation.
9. According to some economists, there are too many gas stations and grocery stores. Is this in keeping with the theory of monopolistic competition?
10. A firm estimates its average total cost to be $10 per unit of output when it produces 10,000 units, which it regards as 80 percent of capacity. Its goal is to earn 20 percent on its total investment, which is $250,000. If it uses markup pricing, what price should it set? Can it be sure of selling 10,000 units if it sets this price?
11. In one of its annual reports, Aluminum, Ltd. of Canada noted that "World stocks of aluminum are not excessively large. They are in firm hands and do not weight unduly upon the world market."[2] What did it mean by "firm hands"? Do you think that this phenomenon was at all related to the firmness of world aluminum prices during this period? Under perfect competition, do you think that aluminum prices would have remained fairly constant in the face of sharp reductions in sales during the Great Depression?

1. F. M. Scherer, *Industrial Market Structure and Economic Performance*.
2. Ibid.

12. According to F. M. Scherer, Reynolds and American shared price leadership in cigarettes, U S. Steel (now USX) often was the price leader in steel, Alcoa has led most frequently in virgin aluminum, American Viscose in rayon, and Du Pont in nylon and polyester fibers.[3] What characteristics tend to distinguish price leaders from other firms? If you had to predict which of a number of firms in an industry was the price leader, what variables would you use to make the forecast?
13. What is meant by oligopoly? How does oligopoly arise?
14. What is meant by the kinked oligopoly demand curve? Does this theory explain the level of price? What does it help to explain?
15. What is a cartel? How will a perfect cartel determine its price and output? Is it likely that a cartel will allocate output among its members so as to minimize total cost?
16. Why are cartels and collusive agreements inherently unstable? Document your answer with evidence from the electrical equipment conspiracy in the United States.
17. Describe the dominant-firm model of price leadership.
18. What are some important barriers to entry?
19. According to Milton Friedman, "Few trends could so thoroughly undermine the very foundations of our free society as the acceptance by corporate officials of a social responsibility other than to make as much money for their stockholders as possible." Do you agree? Why or why not?
20. A major aluminum firm raised its price for sheet by a penny a pound. Almost a month later, the increase was rescinded because the other aluminum producers did not go along with the increase. Is this consistent with the kinked oligopoly demand curve?
21. Joe Bain estimated that the output of an efficient-sized fresh-meat-packing plant was at most 1 percent of industry sales. He also estimated that the output of an efficient tractor-manufacturing plant was equal to 10 to 15 percent of industry sales. Discuss the relevance of these facts for the market structure of these industries.
22. According to the Federal Trade Commission, there once was collusion among the leading bakers and food outlets in the state of Washington. Prior to the conspiracy, bread prices in Seattle were about equal to the U.S. average. During the period of the conspiracy, bread prices in Seattle were 15 to 20 percent above the U.S. average. Is this consistent with the theory of collusive behavior? Why do you think that bread prices in Seattle were not double or triple the U.S. average during the conspiracy?
23. According to John Stuart Mill, "Where competitors are so few, they always agree not to compete. They may run a race of cheapness to ruin a new candidate, but as soon as he has established his footing, they come to terms with him." Do you agree or not? Why?

3. Ibid., p. 167.

24. The following table shows, for food manufacturing, the relationship between the percentage of industry output accounted for by the top four firms and the net profit on stockholder equity of firms in the industry.

Percentage held by top four firms	*Net profit on stockholder equity (percentage)*
31–40	6.2
41–49	9.2
50–59	12.9
60–69	14.6
70–90	16.3

Are these results what you would expect from the theory of oligopoly? Why or why not? (The above data are from the Federal Trade Commission.)

Problems

1. The Jones Company believes that the price elasticity of demand for its product equals 2. It also believes that an extra $1 million in advertising would increase its sales by $1.5 million.
   a. Is the Jones Company spending the optimal amount on advertising? Why or why not?
   b. If not, should it spend more or less on advertising? Explain.
2. The Hamilton Company and the Jefferson Company are duopolists. Hamilton's reaction curve is
$$Q_H = 150 - 0.5Q_j$$
where Q_H is annual output (in units) of the Hamilton Company and Q_j is output (in units) of the Jefferson Company. Jefferson's reaction curve is
$$Q_j = 150 - 0.5Q_H$$
   a. What is the equilibrium output of the Hamilton Company?
   b. What is the equilibrium output of the Jefferson Company?
   c. Is this a Nash equilibrium?
   d. Does the Cournot model explain how these two firms move toward this equilibrium?
3. For the Cronin Company, the marginal revenue from an extra dollar of advertising is as shown on the next page.

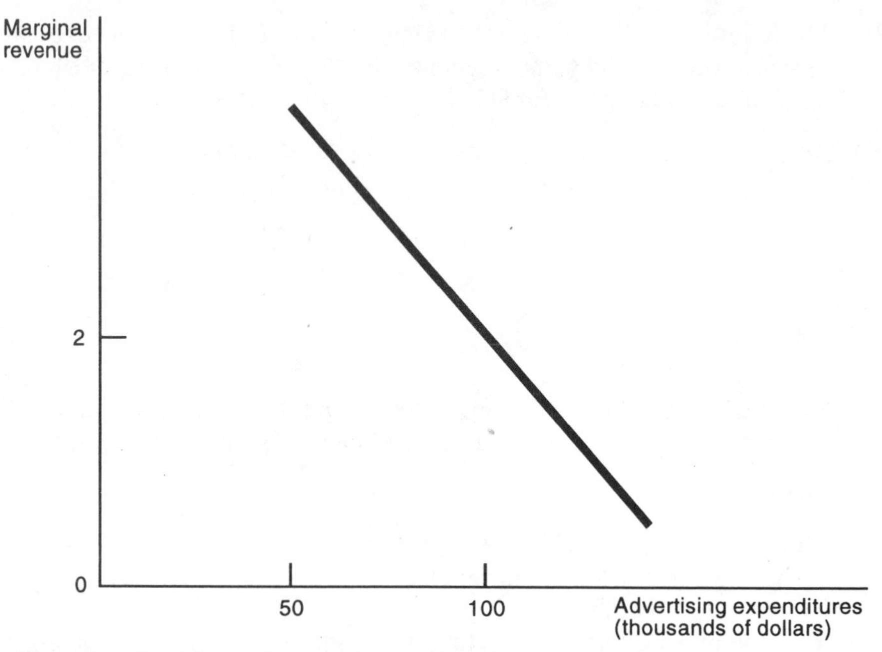

If the price elasticity of demand for the Cronin Company's product is 2, regardless of how much is spent on advertising, how much should the firm spend on advertising?

4. Under what circumstances would you expect a product's price elasticity of demand to decrease as it spends increasing amounts on advertising?

5. In Question 3 above, suppose that there is a leftward shift in the curve showing the marginal revenue from an extra dollar of advertising by the Cronin Company.
   a. What could cause such a leftward shift in this curve?
   b. What would be the effect of such a leftward shift in this curve on the amount spent by the Cronin Company on advertising?

6. The Chrome Corporation is a monopolistically competitive firm. In the short run, its marginal revenue equals \$4, its total revenue equals \$21, and its marginal revenue curve is $10 - 2Q$. Its average total cost is at its minimum value. If it is maximizing profit, what does each of the following equal?
   a. The firm's price.
   b. The firm's output.
   c. The firm's total cost.
   d. The firm's average total cost.
   e. The firm's marginal cost.

7. Explain why the extent of the excess capacity in a particular monopolistically competitive firm is inversely related to the price elasticity of demand of its product.

8. According to Mann, Henning, and Meehan,, there is a fairly substantial positive correlation between the ratio of advertising expenditure to sales in an industry and the percentage of the industry's sales accounted for by the four largest firms. What factors may account for such a correlation?

9. If the market for computer-aided design systems is a contestable market, what will economic profit in the industry be in the long run? Will price equal marginal cost? In answering these questions, does it matter whether there are two or twenty firms in the industry? Explain.

10. Suppose that there are ten identical producers of spring water, that each of these firms has zero costs of production, so long as it produces less than (or equal to) 10 gallons of water per hour, and that it is impossible for each to produce more than 10 gallons per hour. Suppose that the market demand curve for this water is as follows:

Price (dollars per gallon)	*Number of gallons demanded per hour*	*Price (dollars per gallon)*	*Number of gallons demanded per hour*
11	0	5	60
10	10	4	70
9	20	3	80
8	30	2	90
7	40	1	100
6	50	½	105

a. If the firms take the price of water as given (as in the case of perfect competition), what will be the price and output of each firm?

b. If the firms form a completely effective cartel, what will be the price and output of each firm?

c. How much money will each firm be willing to pay to achieve and enforce the collusive agreement described in part (*b*)?

d. Suppose that one of the firms secretly breaks the terms of the agreement and shades price. What effect will this have on its profits?

11. The can industry is composed of two firms. Suppose that the demand curve for cans is

$$P = 100 - Q$$

where P is the price of a can (in cents) and Q is the quantity demanded of cans (in millions per month). Suppose that the total cost function of each firm is

$$C = 1{,}000{,}000 + 15q$$

where C is total cost per month (in cents) and q is the quantity produced (in millions per month) by the firm. What are the profit-maximizing price and output if the firms collude and act like a monopolist? (Elementary calculus is useful for this problem.)

12. An oligopolistic industry selling a homogeneous product is composed of two firms. The two firms set the same price and share the total market equally. The demand curve confronting each firm (assuming that the other firm sets the same price as this firm) is shown on the next page. Also, each firm's total cost function is shown on the next page.

Price (dollars)	*Quantity demanded*	*Output*	*Total cost (dollars)*
10	5	5	45
9	6	6	47
8	7	7	50
7	8	8	55
6	9	9	65

a. Assuming that each firm is correct in believing that the other firm will charge the same price as it does, what is the price that each should charge?
b. Under the assumption in part (*a*), what output rate should each firm set?

13. The International Air Transport Association (IATA) has been composed of 108 American and European airlines that fly transatlantic routes. For many years, IATA acted as a cartel: it fixed and enforced uniform prices. If IATA wanted to maximize the total profit of all member airlines, what uniform price would it charge? How would the total amount of traffic be allocated among the member airlines?

14. Suppose that two firms are producers of spring water, which can be obtained at zero cost. The marginal revenue curve for their combined output is
$$MR = 10 - 2Q$$
where MR is marginal revenue and Q is the number of gallons of spring water sold by both together per hour. If the two producers decide to collude and share the monopoly profits, how much will be their combined output? Why?

15. Suppose that a cartel is formed by three firms. Their total cost functions are as follows:

Units of output	*Firm 1*	*Firm 2*	*Firm 3*
0	20	25	15
1	25	35	22
2	35	50	32
3	50	80	47
4	80	120	77
5	120	160	117

If the cartel decides to produce 11 units of output, how should the output be distributed among the three firms, if they want to minimize cost?

16. *a.* Suppose that you are on the board of directors of a firm which is the price leader in its industry; that is, it lets all of the other firms sell all they want at the existing price. In other words, the other firms act as perfect competitors. Your firm, on the other hand, sets the price, which the other firms accept. The demand curve for your industry's product is $P = 300 - Q$, where P is the price and Q is the total quantity demanded. The total amount supplied by the other firms is equal to Q_r, where $Q_r = 49P$. If your firm's marginal cost curve is $2.96Q_b$, where Q_b is the output of your firm, at what output level should you operate to maximize profit? (P is measured in dollars per

barrel; Q, Q_r and Q_b are measured in millions of barrels per week.)

b. What price should you charge?

c. How much will the industry as a whole produce at this price? Is your firm the price leader because it is the dominant firm in the industry?

Key Concepts for Review

Monopolistic competition
Product differentiation
Product group
Oligopoly
Collusion
Contestable market
Kinked oligopoly demand curve
Duopoly
Cournot model
Excess capacity
Nonprice competition
Cartel
Price leadership
Dominant firm
Markup pricing
Reaction curve
Nash equilibrium

ANSWERS

Case Study: The Pricing of American Small Cars

a. No. If a firm charges relatively high prices, it may sell relatively few units of its product, the result being that its profits are lower than if its price was somewhat lower.

b. They were assumed that U.S. consumers would not buy imported smaller cars (instead of U.S. smaller cars) in such numbers that Detroit would be unable to sell all it could produce.

c. American auto producers would have to reduce their costs and change the features of their cars to meet foreign competition rather than rely on the government to protect them from such competition.

d. In 1981, the Japanese agreed to "voluntary" restraint on exports of cars to the United States.

Completion Questions

1. $20; $10
2. large number; differentiated
3. more; lower
4. reaction; firm B
5. Cournot; equilibrium
6. no
7. less; higher
8. homogeneous; differentiated
9. no effect
10. highly elastic
11. will; equal to; greater than
12. changes
13. steel (among many others)

14. free
15. maintain their present price
16. lower their price too
17. its existing level

True or False

1. False	2. False	3. True	4. True	5. True	6. True
7. True	8. True	9. False	10. True	11. False	12. False
13. True	14. False	15. True	16. True	17. False	

Multiple Choice

1. *b*	2. *b*	3. *c*	4. *c*
5. *b*	6. *a*	7. *b*	8. *c*

Review Questions

1. The Cournot model assumes that each firm believes that, regardless of what output it produces, the other firm will hold its output constant at the existing level. This is not a very reasonable assumption under a wide set of circumstances. However, the model is frequently used to illustrate a Nash equilibrium.
2. A Nash equilibrium is a situation where each player's strategy is optimal, given the strategies chosen by the others.
3. Because of resale price maintenance and other techniques used by manufacturers to keep retail prices high, as well as because of the tendency for excess capacity to occur under monopolistic competition. (In 1975, Congress passed a bill nullifying resale price maintenance agreements.)
4. Perfect competition and pure monopoly are polar extremes. There was a feeling that more attention should be devoted to the important "middle ground" between them. Edward Chamberlin and Joan Robinson were the principal founders of the theory of monopolistic competition. Their books appeared in 1933.
5. First, he assumes that the product is differentiated and that it is produced by a large number of firms, each firm's product being a fairly close substitute for the products of the other firms in the product group.

   Second, he assumes that the number of firms in the product group is sufficiently large so that each firm expects its actions to go unheeded by its rivals and to be unimpeded by any retaliatory measures on their part.

   Third, he assumes that both demand and cost curves are the same for all of the firms in the group.
6. It is difficult to make this comparison because the product is homogeneous under perfect competition or pure monopoly but differentiated under monopolistic competition. But it seems likely that price will be higher under monopolistic competition than under perfect competition, but lower than under monopoly. Output is likely to be lower under monopolistic competition than under perfect competition, but higher than under monopoly. There is likely to be some excess capacity under monopolistic competition, but not under perfect competition.

7. George Stigler attacked the concept of the group and the basic structure of Chamberlin's argument. Dixit, Stiglitz, Lancaster, and Spence question the conclusions concerning excess capacity. The criticisms have a considerable amount of merit.

8. Automobiles, furniture, cigarettes, tires, razor blades, and many others.

9. Yes. It is quite in keeping with the excess capacity theorem.

10. To earn 20 percent on its total investment of $250,000, its profit must equal $50,000 per year. Thus, if it operates at 80 percent of capacity (and sells 10,000 units), it must set a price of $15 per unit. (Since average cost equals $10, profit per unit will be $5, so total profit per year will be $50,000.) However, based on the information that is given, there is no assurance that it can sell 10,000 units per year if it charges a price of $15 per unit.

11. It meant that the holders of these inventories were unlikely to sell them (and depress the price).

    Yes, by holding output off the market, producers could maintain prices.

    No.

12. The price leaders tend to be the largest firms. Also, historical factors play a role, and there is sometimes a tendency for low cost (or medium-cost) firms to be leaders.

13. Oligopoly is a market characterized by a small number of firms and a great deal of interdependence, actual and perceived, among firms.

    Oligopoly can arise from economies of scale, or barriers to entry of various kinds.

14. The kinked oligopoly demand curve is more elastic for price increases than for price decreases. It does not explain the level of price, but it does help to explain the rigidity of price.

15. When a collusive arrangement is made openly and formally, it is called a cartel.

    A perfect cartel wilt determine the marginal cost curve for the cartel as a whole. Then it will produce the output where marginal cost equals marginal revenue, and it will charge the price at which it can sell that output.

    No.

16. Because members have a great incentive to "cheat." This tendency existed among the electrical equipment manufacturers. One collusive agreement after another was drawn up in the 1950s, but after a while some firms began in each case to pursue an independent price policy.

17. The dominant-firm model assumes that the dominant firm sets the price for the industry, but that it lets the other firms sell all they want at that price.

    On the other hand, economists sometimes assume the existence of a barometric firm, a firm that is usually the first to make changes in price that are generally accepted by other firms in the industry. The barometric firm may not be the largest or most powerful but it is a reasonably accurate interpreter of changes in basic cost and demand conditions in the industry as a whole.

18. Smallness of the market relative to optimum size of firm, large capital requirements, unavailability of natural resources, patents, and franchises.

20. Yes.
21. Oligopoly is very likely in tractor manufacturing, but not (for this reason at least) in fresh-meat packing.
22 Yes. Because of possible entry, for one thing.
24. Yes. One might expect that the profit rate would be higher in more concentrated industries. However, a number of prominent economists challenge this conclusion.

Problems

1. *a.* No.
   b. Less.
2. *a.* 100 units.
   b. 100 units.
   c. Yes.
   d. No.
3. $100,000.
4. If the product would be regarded as similar to many other products if little advertising were carried out, its price elasticity of demand would be relatively high under these circumstances. However, if more advertising were carried out, it would induce consumers to attach importance to the product's distinguishing features and reduce the price elasticity of demand.
5. *a.* A decrease in the effectiveness of the firm's advertising staff could cause such a shift.
   b. It would tend to decrease the amount spent on advertising.
6. *a.* $7.
   b. 3.
   c. $12.
   d. $4.
   e. $4.
7. If the demand for the product is very price elastic, the demand curve will be tangent to the firm's long-run average cost curve at a point that is close to its minimum. Thus, the extent of the excess capacity will be relatively small. On the other hand, if the demand curve is not very price elastic, it will be tangent to the firm's long-run average cost curve at a point that is relatively far away from the minimum point, with the result that the extent of the excess capacity will be relatively great.
8. One would expect relatively tightly held oligopolies to spend more on advertising than more competitive industries. Also, the line of causation may run the other way. To some extent, the high level of concentration may be due to the fact that an industry is one where advertising is an important commercial weapon.
9. Economic profit in the long run will be zero. Price will tend to equal marginal cost. It does not matter whether there are two or twenty firms, according to the theory.

10. *a.* The industry supply curve would be as shown below. This is the horizontal sum of the firm's marginal cost curves. Since this supply curve intersects the demand curve at a price of $1, this is the equilibrium price. Since the supply and demand curves intersect at an output of 100 gallons, this is the industry output. Each firm produces one-tenth of this output, or 10 gallons.

The industry supply curve is:

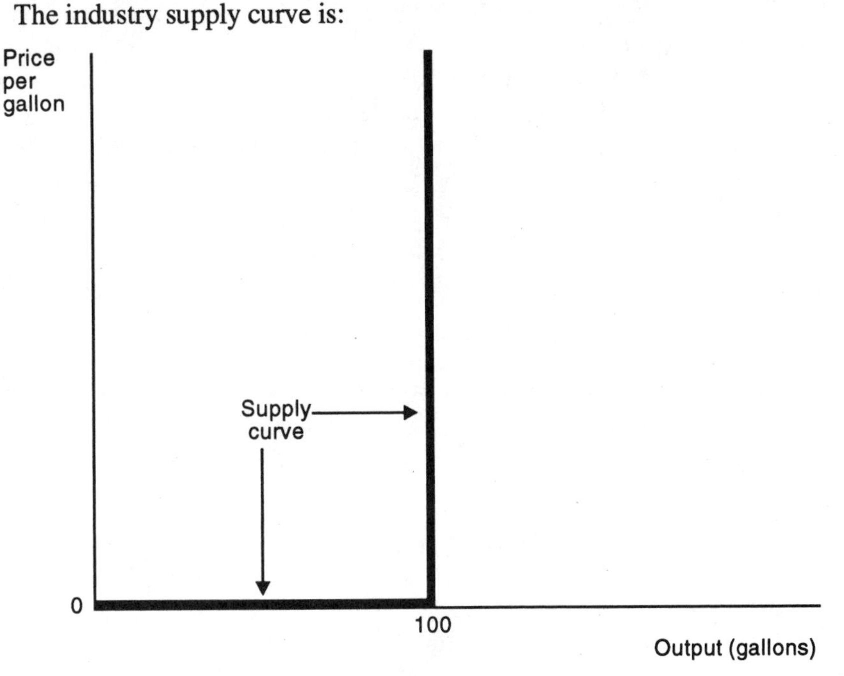

b. The cartel would expand output only so long as marginal revenue exceeds marginal cost. Since marginal cost is zero, this means that the cartel would expand output only so long as marginal revenue is positive. Marginal revenue is shown below.

Output	*Total revenue (dollars)*	*Marginal revenue (dollars)*
0	0	
		10
10	100	
		8
20	180	
		6
30	240	
		4
40	280	
		2
50	300	
		0
60	300	
		−2
70	280	
		−4
80	240	
		−6
90	180	
		−8
100	100	

Clearly, the cartel would restrict output to 50 gallons and charge $6 per gallon. (Alternatively, it could set output at 60 gallons, and charge $5 per gallon. The profit would be the same.)

c. If the alternative to the cartel is the arrangement described in part (*a*), each firm would be willing to pay up to $20 per hour to achieve and enforce the cartel. With the cartel, each firm makes $30 per hour, since each produces 5 gallons per hour, and charges $6 per gallon. Under the arrangement in part (*a*), each firm makes $10 per hour, since each produces 10 gallons per hour and charges $1 per gallon. Thus, each firm's profit per hour is $20 higher under the cartel.

d. If the cartel sets a price of $6 per gallon, this firm can increase its profits by selling an additional 5 gallons per hour at a price per gallon of $5. But if a substantial number of firms begin to act in this way, the cartel will break apart.

11. Let Q be the combined output of the two firms, and π be their combined profit.

$$\pi = PQ - 2{,}000{,}000 - 15Q$$
$$= (100 - Q)Q - 2{,}000{,}000 - 15Q$$
$$\frac{\partial \pi}{\partial Q} = 100 - 2Q - 15 = 0$$

Thus, Q equals 42.5, and price equals $100 - 42.5 = 57.5$. If they share the output equally, each will produce 21.25 million cans per month.

12. *a.* $9.
    b. 6.

13. To find the profit-maximizing price, the IATA should construct the marginal cost curve for the cartel as a whole. Then, as shown in Figure 12.6 of the text, it should determine the amount of traffic (which is the output of this industry) where marginal revenue equals marginal cost. The price that will elicit this amount of traffic is the profit-maximizing price. If IATA wants to maximize profit, it will allocate this traffic among the airlines in such a way that the marginal cost of all airlines is equal. (However, for reasons discussed in the text, it may not want to maximize profit.)

14. The two producers will choose the monopoly output, where marginal revenue equals marginal cost. Since marginal cost equals zero, this means that

$$MR = 10 - 2Q = 0$$

so

$$Q = 10/2 = 5$$

Thus, their combined output will be 5 gallons per hour.

15. They should set the marginal cost at one firm equal to the marginal cost at each other firm. If firm 1 produces 4 units, firm 2 produces 3 units, and firm 3 produces 4 units, the marginal cost at each firm equals $30. Thus, this seems to be the optimal distribution of output.

16. *a.* Since $Q = 300 - P$, and the demand for the firm's output is $Q - Q_r$, it follows that the firm's demand curve is

$$Q_b = Q - Q_r = (300 - P) - 49P$$
$$= 300 - 50P$$

or

$$P = 6 - 0.02Q_b$$

Thus, the firm's marginal revenue curve is $MR = 6 - 0.04Q_b$. Since its marginal cost curve is $2.96Q_b$

$$6 - 0.04Q_b = 2.96Q_b$$
$$Q_b = 2$$

That is, your output level should be 2 million barrels per week.

b. Since $P = 6 - 0.02Q_b$, and $Q_b = 2$, it follows that

$$P = 6 - 0.02(2) = 5.96$$

That is, the price should be $5.96 per barrel.

c. Since $Q = 300 - P$, and $P = 5.96$, it follows that

$$Q = 300 - 5.96 = 294.04$$

That is, the industry output is 294.04 million barrels per week. Your firm is not a dominant firm in terms of its output. It will produce less than 1 percent of the industry's output. There must be some other reason why it is the price leader.

Chapter 13 Game Theory and Strategic Behavior

Case Study: Du Pont and ICI

Suppose that Du Pont and ICI (a British firm) are the only two firms that can produce a particular type of chemical. Each has the choice of producing or not producing it. The payoff matrix is as follows:

Possible strategies for Du Pont	Possible strategies for ICI: Produce	Possible strategies for ICI: Do not produce
Produce	Du Pont's profit: \$10 million ICI's profit: –\$40 million	Du Pont's profit: \$250 million ICI's profit: 0
Do not produce	Du Pont's profit: 0 ICI's profit: \$200 million	Du Pont's profit: 0 ICI's profit: 0

Du Pont has a headstart over ICI, and can make the first move.

a. What will be the outcome?

b. Suppose the British government commits itself to pay ICI a subsidy of \$50 million if it produces the chemical. What now will be the outcome?

c. Will the subsidy discourage Du Pont from producing the chemical?

Completion Questions

1. If each player has a ______________ strategy, this strategy is its best choice regardless of what other players do.
2. In a ______________ equilibrium, each player adopts a strategy that is its best choice given what the other players do.
3. Duopoly is a situation where there are ______________ sellers.
4. In the game of the prisoners' dilemma, the dominant strategy for each prisoner is to ______________.
5. If the game of the prisoners' dilemma is repeated indefinitely, the best strategy for each player may not be to ______________.
6. According to Robert Axelrod of the University of Michigan, a good strategy is ______________.
7. A commitment tends to be more persuasive if it seems ______________ and ______________.
8. Not ______________ games have a dominant strategy for every player.
9. If one player can convince the other player that he or she is ______________ going to take the course of action that maximizes profit, this player can sometimes make a threat credible.
10. A firm may succeed in convincing a potential entrant not to enter its market by ______________ its own profits if it does not resist entry.

11. Firms often try to commit themselves ______________ to a particular move, even if they cannot be ______________ to actually carry out this move.

True or False

____ 1. To be credible, a firm's commitments must be backed up with the assets and expertise required to carry out the commitment.

____ 2. Firms frequently send signals to one another indicating their intentions, motives, and objectives.

____ 3. Firms never can alter the payoff matrix.

____ 4. A firm may succeed in convincing a potential entrant not to enter its market by reducing its own profits if it does not resist entry.

____ 5. A limit price is the highest price that the firm feels it can safely charge.

____ 6. A firm may set a low price to signal potential entrants that it is a very low-cost producer.

____ 7. In many situations, the firm that makes the first move has a substantial advantage.

____ 8. In many oligopolistic industries, firms tend to engage in nonprice competition, using advertising and variation in product characteristics, more than price, as competitive weapons.

____ 9. Most oligopoly models indicate that price under oligopoly will be lower than under perfect competition.

____ 10. A Nash equilibrium exists for any and every game.

Multiple Choice

1. A government, by subsidizing one of its firms, can
   a. change the payoff matrix in a particular market.
   b. induce foreign firms to withdraw from the market.
   c. get other governments to retaliate.
   d. all of the above.
   e. none of the above.

2. A dominant strategy is a strategy where
   a. the player loses.
   b. a second-best approach is used.
   c. pure strategies are mixed with impure strategies.
   d. all of the above.
   e. none of the above.

3. Games may have
   a. Nash equilibrium.
   b. no Nash equilibrium.
   c. more than one Nash equilibrium.
   d. a dominant strategy for each player.

e. all of the above.

4. A strategic move may be
   a. threatening to the firm's rivals.
   b. not threatening to the firm's rivals.
   c. met with substantial retaliation by the firm's rivals.
   d. met with little or no retaliation by the firm's rivals.
   e. all of the above.

5. A firm may be able to deter entry into its market by
   a. building excess production capacity.
   b. gaining a reputation for "irrational" resistance to entry.
   c. threatening to resist even though potential entrants believe that it is not in this firm's interest to resist.
   d. only *a* and *b*.
   e. all of the above.

6. If two rival firms both adopt a preemptive strategy with regard to capacity expansion,
   a. each tries to expand before the other does so.
   b. there may be two Nash equilibria.
   c. both firms are likely to profit greatly if the demand for the product grows more slowly than expected.
   d. both *a* and *b*.
   e. all of the above.

7. If a firm can convince its rivals that it is unequivocally committed to a particular move,
   a. they may be convinced that they would lose more than they would gain from a protracted struggle.
   b. they may back down without retaliating.
   c. they may engage in moves designed to counter this firm's move.
   d. both *a* and *b*.
   e. all of the above.

Review Questions

1. Suppose that Bonnie and Clyde are arrested for holding up a bank. The district attorney talks *separately* to each one, saying, "We've got lots of evidence indicating you robbed the bank, so you'd better confess. If you do, I'll give you a break. If you *alone* confess, I'll see to it you get only 6 months in jail. If you both confess, I'll see to it you get 10 years. But if you don't confess, and your partner *does*, I'll see to it you get 20 years." Both Bonnie and Clyde are sure that, if neither confesses, they will get 2 years.
   a. In this situation, what is the payoff matrix?
   b. What strategy will each choose, given that each tries to save his or her own skin?

2. What are the rules of the game? What is a strategy?

3. What is a player's payoff? What is a two-person game?
4. What factors tend to make a firm's commitment to a particular strategic move more persuasive?
5. Describe how a reputation for "irrational" resistance to entry may be very valuable to a firm.
6. Describe what is meant by a limit price and why such a price may be established.
7. Explain why the firm that makes the first move often has a substantial advantage.
8. Describe what is meant by a preemptive strategy (with regard to capacity expansion) and explain why it can be risky.

Problems

1. Firms E and F are duopolists. They each have two possible strategies for product development. The payoff matrix is as follows:

Possible strategies for Firm E	*Possible strategies for Firm F*	
	A	*B*
1	Firm E's profit: \$5 million Firm F's profit: \$6 million	Firm E's profit: \$4 million Firm F's profit: \$5 million
2	Firm E's profit: \$6 million Firm F's profit: \$5 million	Firm E's profit: \$5 million Firm F's profit: \$4 million

   a. What is firm E's optimal strategy?
   b. What is firm F's optimal strategy?
   c. Is this an example of the prisoner's dilemma?
   d. Is there a dominant strategy for firm E? Explain.
   e. Is there a dominant strategy for firm F? Explain.

2. If two players are engaged in a game with the characteristics of the prisoners' dilemma, can the equilibrium depend on whether the game is played repeatedly or just once? Explain.
3. Firms A and B agree to maintain their prices at the monopoly level. In the event that either cheats on this agreement, both firms will adopt a strategy of "tit for tat." Describe how each firm may attempt to determine whether the other firm is cheating.
4. Suppose the payoff matrix is as given below. What strategy will firm I choose? What strategy will Firm II choose? (All figures are in millions of dollars.)

Possible strategies for Firm I	*Possible strategies for Firm II*		
	1	*2*	*3*
	(Profits for Firm I, or losses for Firm II)		
A	\$10	\$9.0	\$11
B	8	8.5	10

5. Firms C and D are about to stage rival advertising campaigns, and each firm has a choice of strategies. The payoff matrix is given below.

Possible strategies for Firm C	*Possible strategies for Firm D* 1	2
A	Firm C's profit: $7 million Firm D's profit: $10 million	Firm C's profit: $6 million Firm D's profit: $7 million
B	Firm C's profit: $8 million Firm D's profit: $7 million	Firm C's profit: $7 million Firm D's profit: $8 million

a. Does firm C have a dominant strategy? If so, what is it?
b. Does firm D have a dominant strategy? If so, what is it?
c. Is there a Nash equilibrium? If so, what is it?

6. The industrial robot is one of the most important technological innovations of this century. In 1985, about 50 American firms produced robots, with the following six firms accounting for about 70 percent of total sales:

	Sales (millions of dollars)		*Sales (millions of dollars)*
GMF Robotics	180	ASEA	39
Cincinnati Melacron	59	GCA	35
Westinghouse	45	DeVilbiss	33

a. Robotics firms devoted about 17 percent of their sales to research and development. Is nonprice competition important in this industry?
b. During 1979–82, the robotics industry experienced substantial losses, although its sales increased at a relatively rapid rate. Why did firms stay in this industry in the face of these losses?
c. Does the fact that they stayed in the industry mean that they were irrational?

7. The Brooks Company's managers begin to sense that the Harris Corporation may attempt to enter their market.
a. What steps might they take to dissuade Harris from doing so?
b. What factors are likely to determine whether they will succeed?
c. What actions that they have taken (or not taken) in the past may play an important role in influencing whether or not Harris tries to enter?

8. The Miller Company must decide whether to advertise its product or not. If its rival, the Morgan Corporation, decides to advertise its product, Miller will make $4 million if it advertises and $2 million if it does not advertise. If Morgan does not advertise, Miller will make $5 million if it advertises and $3 million if it does not.
a. Is it possible to determine the payoff matrix? Why or why not?
b. Can you tell whether Miller has a dominant strategy? If so, what is it?

9. The Adam Company has two possible strategies and the Burr Company has two possible strategies. The payoff matrix is shown below.

Possible strategies for Adam Company	*Possible strategies for Burr Company*	
	1	*2*
A	Adam's profit: $-20 million Burr's profit: $-20 million	Adam's profit: $40 million Burr's profit: $20 million
B	Adam's profit: $20 million Burr's profit: $40 million	Adam's profit: $-20 million Burr's profit: $-20 million

a. If the Adam Company has the first move, what strategy is it likely to choose? What strategy is Burr likely to choose?

b. If the Burr Company has the first move, what strategy is it likely to choose? What strategy is Adam likely to choose?

Key Concepts for Review

Theory of games
Player
Strategy
Tit for tat
First-mover advantages
Nonprice competition
Payoff matrix
Dominant strategy
Prisoners' dilemma
Strategic move
Limit pricing
Preemption

ANSWERS

Case Study: Du Pont and ICI

a. Du Pont will produce the chemical, and ICI will not produce it.
b. Both firms will produce the chemical.
c. No.

Completion Questions

1. dominant
2. Nash
3. two
4. confess
5. confess
6. tit for tat
7. binding; irreversible
8. all
9. not
10. reducing
11. first; first

True or False

1. True	2. True	3. False	4. True	5. False
6. True	7. True	8. True	9. False	10. False

Multiple Choice

1. *d*	2. *e*	3. *e*	4. *e*
5. *d*	6. *d*	7. *e*	

Review Questions

1. *a.* The payoff matrix is as follows:

Possible strategies for Clyde	Bonnie's payoff Possible strategies for Bonnie Confess	 Not confess	Clyde's payoff Possible strategies for Bonnie Confess	 Not confess
	(Years of imprisonment)		(Years of imprisonment)	
Confess	10	20	10	½
Not confess	½	2	20	2

b. The payoff matrix shows that, if Clyde confesses, Bonnie is better off to confess than not to confess. If he does not confess, she is also better off to confess than not to confess. Thus, her dominant strategy is to confess. Similarly, the payoff matrix shows that Clyde is better off to confess if Bonnie confesses and that he is also better off to confess if she does not confess. Thus, his dominant strategy too is to confess. A game of this sort is called the *Prisoners' Dilemma.* Note that, in this situation, each will do better by being altruistic than by selfishly trying to minimize his or her own stay behind bars.

2. Each player has a certain amount of resources; the *rules of the game* describe how these resources can be used. For example, the rules of poker indicate how bets can be made and which hands are better than other hands. A *strategy* is a complete specification of what actions a player will take under each contingency in the playing of the game. For example, a corporation president might tell her subordinates how she wants an advertising campaign to start, and what should be done at subsequent points in time in response to various actions of competing firms.

3. A player's *payoff* varies from game to game. It is win, lose, or draw in checkers, and various sums of money in poker. A *two-person game* is a game with only two players. The relevant features of a two-person game can be shown by constructing a *payoff matrix.*

4. A commitment tends to be more persuasive if it seems binding and irreversible. For example, suppose that a firm commits itself to enter a particular market. If this firm buys a plant rather than leases it, or if it signs a long-term, rather than a short-term, contract for raw materials, this an indication that this firm is irreversibly committed to entering the market in question. Or suppose that a firm commits itself to meet price reductions by its rivals. If this firm makes written or verbal agreements with customers to meet price cuts, these agreements can make

such a commitment irreversible—and hence more persuasive. On the other hand, if a firm's rivals feel that it can easily renounce or ignore a particular commitment that it makes, they are not likely to pay as much attention to this commitment. To be credible, a firm's commitments must be backed up with the assets and expertise required to carry out the commitment. Thus, if a firm commits itself to invade another firm's market if the latter invades its market, it must have the financial and technological power and skills needed to carry out this commitment. Also, a firm's commitments are more credible if it has a long history of adherence to past commitments.

5. If a firm has imposed huge losses on every firm that has tried to enter its market in the past, and has a reputation for "irrational" resistance to entry, potential entrants may decide that this firm is too tough an opponent to challenge. Thus, it may be in this firm's interests to foster such a reputation by declaring total war on every entrant that appears, since the short-term losses from these wars may be more than offset by the longer-term gains from the prevention of entry.

6. A limit price is a price that discourages or prevents entry. A limit price may be established to signal potential entrants that the firm is a very low cost producer.

7. To illustrate why the firm that makes the first move often has a substantial advantage, suppose that two firms, X and Y, are about to introduce a new product and that each must choose whether to tailor its product for the civilian or military market. If both firms tailor their products for the same market (civilian or military), both will lose $10 million because these markets are too small to support two (profitable) producers. If one firm tailors its product for the civilian market while the other firm focuses on the military market, it will make $30 million, and its rival will make $15 million, since the civilian market for this product is more profitable than the military market.

   If one of these firms—say firm X—can introduce its product before the other firm does so, it enjoys a great advantage. It can tailor its product to the civilian market since this market is more profitable than the military market. Of course, if the other firm—firm Y—enters the civilian market too both will lose money, but given that firm X has already entered the civilian market, firm Y is unlikely to enter it too. Why? Because this would result in a substantial loss for firm Y, whereas if firm Y enters the military market it will earn a tidy profit—not as big a profit as if it had made the first move (and thus could have been the sole producer in the civilian market), but a tidy profit nonetheless.

8. If a firm adopts a preemptive strategy, it tries to expand before its rivals do, thus discouraging them from expanding. If the demand for the product grows less rapidly than expected, the firm could be stuck with an unprofitable addition to its plant. Also, if its rivals expand anyway, the industry could be plunged into a war.

Problems

1. *a.* Strategy 2.
   b. Strategy A.
   c. No.

   d. Yes, Strategy 2.
   e. Yes, Strategy A.

2. Yes. See the text.

3. In some cases, trade associations have been authorized to collect detailed information concerning each firm's transactions. Also, each firm may ask their customers to report whether the other firm is cutting price.

4. Firm I will choose strategy A.
   Firm II will choose strategy 2.

5. *a.* Yes. Strategy B.
   b. No.
   c. Yes. Firm C chooses strategy B, and firm D chooses strategy 2.

6. *a.* Yes.
   b. Because they felt that their losses were only temporary.
   c. No.

7. *a.* They might signal Harris that they will fight hard to resist its entry into their market. They might build extra productive capacity and do other things to convince Harris that the payoff matrix is such that, if Harris enters, it will be more profitable for Brooks to resist than not to resist.
   b. It will depend on how convinced Harris is that Brooks really will resist if Harris enters, and on how effective Harris believes such resistance would be.
   c. If Brooks has a reputation for fighting hard in the past to resist entry, this may be important.

8. *a.* No. We are not given Morgan's profit figures.
   b. Yes. Its dominant strategy is to advertise.

9. *a.* Strategy A.
   Strategy 2.
   b. Strategy 1.
   Strategy B.

Case 3 Cineplex Odeon Corporation

Joseph Wolfe

In mid-February 1989, Jack Valenti, head of the Motion Picture Association, reaffirmed the film industry's basic health by citing 1988 movie theater attendance figures surpassing 1 billion people for the seventh year in a row. While this magnitude translated into box office revenues of over $4.4 billion, there are indications the industry is in a state of both absolute and relative decline. It is also undergoing a restructuring that is fundamentally changing the nature of competitive practices for those in the film exhibition business. In the first instance, a lower proportion of America's aging population attends the movies each year, partially due to the use of VCRs for film viewing, the presence of television in both its broadcast and cable versions, and other uses of the consumer's leisure time dollars. In the second instance, a great degree of owner concentration is occurring, due to separate actions by both the Hollywood producers of films and their exhibitors.

Despite the apparent decline of the motion picture theater as the major supplier of America's needs for mass entertainment, the Toronto-based firm of Cineplex Odeon has quickly become North America's second largest and most profitable theater chain through a series of shrewd and adventuresome acquisitions while creating a large number of up-scaled, multiscreened theaters in key cities and market areas. With 482 theaters and 1,809 screens in 20 states, the Washington, D.C., area, six Canadian provinces, and the United Kingdom, the firm posted record sales of $695.8 million and profits of $40.4 million in 1988 while standing on the verge of developing and operating more than 110 screens in the United Kingdom by 1991. Central to Cineplex Odeon's success is the firm's driven and often-abrasive chairman, president, and CEO, Garth Drabinsky. It is against the backdrop of the industry's fundamental changes and basic decline that Drabinsky must chart his firm's future actions to insure its continued growth and prosperity.

THE MOTION PICTURE THEATER INDUSTRY

The motion picture theater industry (SIC 783) has undergone a number of radical transformations since its turn of the century beginnings. The first movies were shown in cramped and hastily convened storefront locations called nickelodeons, so named for their 5-cent admission charges. Their numbers grew rapidly because the costs of entering this industry were relatively low and a plentiful supply of films was available in both their legal and pirated versions. By 1907, it was estimated the United States had about 3,000 movie theaters, mainly concentrated in the larger cities. Rural areas were serviced by traveling film shows, which made their presentations in the local town meeting hall.

The typical show lasted only 15 to 20 minutes, augmented by song slides or lectures. As the film medium's novelty declined, audiences began to clamor for more lavish and ambitious productions using recognizable actors and actresses. Feature-length movies replaced one-reel short subjects and comedies in the middle to late 1910s, and the theater industry's greatest building period began. Opulent, specially built structures soon became the focal point of every major city's downtown area.

Often possessing more than 5,000 seats, they came complete with a pit orchestra and vocalists and chorus, baby-sitting facilities, elevators and grand staircases to a heaven-like balcony, numerous doormen, and a watchful and attentive fleet of uniformed ushers.

By the mid-1920s, over 19,000 theaters were in operation, and Hollywood's film producers began what was a continuing attempt to control via acquisitions the first-run exhibitors of their films. The battle was initially waged between Paramount and First National, but soon Loew's (MGM), Fox, and Warner Brothers joined in, with First National being the major loser. By 1935 the twin realities of the Great Depression and the advent of sound films caused the number of theaters to plummet to about 15,000. Because of the nation's bleak economic outlook, many theaters had become too run-down or too costly to convert to the greater demands of sound films. Many Americans also substituted radio's free entertainment for their weekly lemminglike trek to the movies. Surviving theaters introduced the double feature to create more value for the entertainment dollar, while obtaining the major source of their profits from candy, soft drinks, and popcorn sales.

During World War II, motion picture attendance and Hollywood's profits reached their all-time highs; about 82 million people a week going to the nation's 20,400 theaters. This pinnacle did not last long, however, as postwar incomes were spent on new cars, television sets, and homes built in the newly emerging suburbs. Motion picture attendance began its precipitous fall in 1947, with attendance reaching its all-time low of 16 million per week in 1971. The number of theaters followed the same downward trend, although a steady increase in the number of drive-in theaters temporarily took up some of the slack. (See Exhibit 1.)

The postwar period also saw the effects of the government's 1948 Consent Decree. By the early 1940s, Hollywood's five major studios had obtained control or interests in 17 percent of the nation's theaters. This amounted to 70 percent of the important large city first-run theaters. Although certain studios were stronger in different parts of the country—Paramount dominated New England and the South, Warner Brothers the mid-Atlantic region, Loew's and RKO the New York-New Jersey area, and 20th Century-Fox the western states—each controlled all stages of the distribution chain from its studios (manufacturing), its film exchanges (wholesaling), and its movie theaters (retailing). Under the Consent Decree the studios could either divest their studios and film exchanges or get rid of their movie theaters. Hollywood chose to sell the cinemas, and thereby opting to control the supply side of the film distribution system.

In an effort to arrest the decline in attendance and to counter the relatively inexpensive and convenient medium of black-and-white television in the 1950s, the film studios retaliated by offering movies that dealt with subject matter considered too dangerous for home viewing, shown in formats and hues beyond television's technical capabilities. Moviegoers heard the word "virgin" uttered for the first time, women "with child" actually looked pregnant, rather than merely full-skirted, and couples were shown in bed together without having to put one foot on the floor. From 1953 to 1968, about 28 percent of Hollywood's films were photographed and projected in a bewildering array of widescreen processes, such as Cinerama, CinemaScope, RegalScope, SuperScope, Technirama, VistaVision, Panavision, Techniscope, and even three-dimensional color.

As movie attendance stabilized in the mid-1980s to a little more than 20 million patrons per week, two new trends have established themselves in the movie theater business. The first has been the creation of multiple screened theater sites, while the second trend has been Hollywood s reacquisition of theaters and theater chains as part of a general consolidation within the industry. Many theater chains have rediscovered the glitz and glamor of old Hollywood by either subdividing and rejuvenating old

EXHIBIT 1

Number of U.S. Movie Theaters
Selected years, 1929–89

YEAR	THEATERS	DRIVE-INS	TOTAL	SCREENS
1923	15.0		15.0	
1926	19.5		19.5	
1929	23.3		23.3	
1935	15.3		15.3	
1942	20.3	0.1	20.4	
1946	18.7	0.3	19.0	
1950	16.9	2.2	19.1	
1955	14.1	4.5	18.7	
1965	9.2	4.2	13.4	
1974	9.6	3.5	13.2	14.4
1980	9.7	3.6	13.3	17.6
1981	11.4	3.3	14.7	18.0
1984	14.6	2.8	17.4	20.2
1985	15.1	2.8	17.9	20.7
1986	16.8	2.8	19.6	22.8
1987*	17.9	2.8	20.7	23.6
1988*	18.1	2.7	20.8	24.3

* Estimated.

Sources: Joel W. Finler, *The Hollywood Story* (New York: Crown, 1988) p. 288; "The Motion Picture Rides into Town, 1903," *Wall Street Journal*, February 7, 1989, p. B1; *1989 U.S. Industrial Outlook* (Washington, D.C.: U.S. Department of Commerce/International Trade Administration, 1989), p 57-1.

theaters or by constructing multiplexes from scratch in suburban malls and shopping districts. The economies of multiplescreened operations are compelling at the local level. Rather than needing a separate manager and projectionist for each theater, a number of variously sized auditoriums can be combined and centrally serviced. Box office operations and concession stands can also be centrally managed and operated. The availability of a number of screens at one location also yields programming flexibility for the theater operator. A "small" film without mass appeal can often turn a profit in a room seating only 300 people, while it would be unprofitable and would be lost in a larger auditorium. Having a number of screens in operation also increases the likelihood the complex will be showing a hit film, and thereby generating traffic for the other films being shown at the site. Having multiple screens also allows the operator to outfit various rooms with different sound systems (the THX System by Lucasfilm versus the standard 4-track optical stereo system) and projection equipment (at least one 70-mm 6-track magnetic sound projector in addition to the usual 35-mm projector). thereby offering the very finest possible viewing.

The second trend toward consolidation is occurring at all levels of the film distribution chain. A number of studios have recently purchased major theater chains after sensing a relaxation of the enforcement of the Consent Decree (in 1984, the Justice Department offered advance support to any studio financing a lawsuit to reenter the movie theater business) plus their promise to limit their ownership to less than 50 percent of any acquired chain. MCA, owner of Universal Studios, has purchased 49.7 percent of Cineplex Odeon, the Cannon Group has purchased the Commonwealth chain, and United Artists Communications purchased the Georgia Theatre Company, the Gulf States and Litchfield chains, and—in 1988 alone—the Blair, Sameric, Commonwealth (from the Cannon Group), and Moss theater chains. Gulf & Western's Paramount Studios purchased Trans-Lux, Mann Theaters, and Festival Enterprises, while Columbia and Tri-Star (owned by Coca-Cola) bought the Walter Reade and Loews chains. On the retailing side, Cineplex Odeon has purchased the Plitt, RKO, Septum, Essaness, and Sterling chains, and Carmike Cinemas has purchased Stewart & Everett, while AMC Entertainment purchased the Budco Theatres. Through these actions and others the top six chains now own nearly 40 percent of America's screens. This is a 67 percent increase in just three years.

Wholesaling operations have been drastically reduced over the years on a scale unnoticeable to the public but very significant to those in the business. When filmgoing was in its heyday, each studio operated as many as 20 or so film exchanges in key cities across the country. Hollywood's studios have since closed many exchanges, until they are now operating only five to eight branch offices each. Paramount recently merged its Charlotte and Jacksonville branches into its Atlanta office, while Chicago now handles the business once serviced by its Detroit, Kansas City, Des Moines, and Minneapolis branches. As observed by Michael Patrick, president of Carmike Cinemas, "as the geographical regions serviced by these offices increase, the ability of smaller exhibitors to negotiate bookings is diluted relative to the buying power of the larger circuits."[1]

COMPETITIVE CONDITIONS

Despite the glamor associated with Hollywood, its start, its televised Academy Award Show, and such megahits as *Who Framed Roger Rabbit?*, *Rain Man* and *Batman*, theater operators are basically in the business of running commercial enterprises dealing with a very perishable commodity. A movie is a merchandisable product made available by Hollywood and various independent producers to commercial storefront theaters at local retail locations. Given the large degree of concentration in the industry, corporate level actions entail the financing of both acquisitions and

1. Michael W. Patrick, "Trends in Exhibition," *The 1987 Encyclopedia of Exhibition*, ed. Wayne R. Green (New York: National Association of Theatre Owners, 1988), p 109.

EXHIBIT 2

FREQUENCY OF ATTENDANCE BY TOTAL PUBLIC, AGES 12 AND OVER

Attendance	1986	1985	1984
Frequently	21.0%	22.0%	23.0
Occasionally	25.0	29.0	28.0
Infrequently	11.0	9.0	8.0
Never	43.0	39.0	39.0
Not reported	0.0	1.0	2.0

Frequently: At least once a month.
Occasionally: Once in two to six months.
Infrequently: Less than once in six months.

Source: *1988 International Motion Picture Almanac* (New York: Quigley, 1988), p. 29A

EXHIBIT 3

U.S. POPULATION BY AGE GROUP FOR 1980
WITH PROJECTIONS FOR 1990 AND 2000

AGE RANGE	YEAR	NUMBER IN MILLIONS	PERCENT OF TOTAL	PERCENT CHANGE
5–17	1980	47.22	20.7%	
	1990	45.14	18.1	–4.4%
	2000	49.76	18.6	10.2
18–24	1980	30.35	13.2	
	1990	25.79	10.3	–15.0
	2000	24.60	9.2	–4.6
25–44	1980	63.48	27.9	
	1990	81.38	32.6	28.2
	2000	80.16	29.9	–1.5
45–64	1980	44.49	19.5	
	1990	46.53	18.6	4.4
	2000	60.88	22.7	31.1
65 and over	1980	25.71	11.3	
	1990	31.70	12.8	23.3
	2000	34.92	13.0	10.2

1980 total: 227,705,000.
1990 total: 249,675,000.
2000 total: 267,955,000.

Source: Adapted from U.S. Department of Commerce, Bureau of the Census, *Statistical Abstract of the United States, 1985* (Washington, D.C.: Government Printing Office, 1985), pp. 26–27.

new construction, while local operations deal with the booking of films that match the moviegoing tastes of the communities being served.

To the degree a movie house merely retails someone else's product, the theater owner's success lies in the quality and not the quantity of products produced by Hollywood. Accordingly the 1987–88 Christmas season did not produce any blockbusters, while 1987's two big hits were *Beverly Hills Cop II* and *Fatal Attraction*, and 1986's hits were *Top Gun, Crocodile Dundee*, and *The Karate Kid, Part II.* Under these conditions of relatively few real moneymakers, the bargaining power shifts to the studios, leaving the exhibitors with more screens than they can fill with highdrawing films. Although the independent producers (the "indies")—such as the DeLaurentis Entertainment Group, New World, Atlantic, Concorde, and Cannon—are producing proportionally more films every year and the majors are producing fewer,

EXHIBIT 4

AVERAGE OPERATING RESULTS FOR SELECTED MOTION PICTURE THEATER CORPORATIONS, BY ASSET SIZE, 7/84–6/85

OPERATING RESULT	SMALLER-SIZED		MIDDLE-SIZED		LARGER-SIZED	
Revenues	224,171	100.0%	4,476,042	100.0%	151,545,455	100.0%
Cost of operations	93,917	41.9	1,780,066	39.8	54,707,909	36.1
Operating income	130,254	58.1	2,695,976	60.2	96,837,546	63.9
Expenses:						
Compensation of officers	6,788	3.0	150,647	3.4	2,121,636	1.4
Repairs	5,497	2.5	74,134	1.7	2,438,504	1.6
Bad debts	170	.1	4,196	.1	82,661	.1
Rent	32,195	14.4	315,841	7.1	11,489,901	7.6
Taxes (excluding federal tax)	12,904	5.8	179,881	4.0	5,689,843	3.8
Interest	8,045	3.6	117,216	2.6	8,031,999	5.3
Depreciation	8,866	4.0	269,122	6.0	8,954,959	5.9
Advertising	16,004	7.1	152,745	3.4	5,689,843	3.8
Pensions and other benefit plans	—	—	42,662	1.0	771,504	.5
Other expenses	70,971	31.7	1,682,992	37.6	55,245,207	36.5
Net profit before taxes	(31,186)	(13.9)	(293,460)	(6.6)	(3,678,421)	(2.4)
Current ratio	1.0		1.3		0.7	
Quick ratio	0.6		1.0		0.5	
Debt ratio	140.6		52.9		74.2	
Asset turnover	3.0		1.3		1.0	

Source: L. Troy, *Almanac of Business and Industrial Financial Ratios* (Englewood Cliffs, N.J.: Prentice-Hall, 1988), p. 332.

their product is more variable in quality and less bankable. Additionally, theaters often pay a premium for the rights to exclusively show first-run movies in a given area or film zone, such as the May 1989 release of *Indiana Jones and the Last Crusade.* This condition hurts the smaller chains especially hard, because they do not have the resources to outbid the giant circuit.

Marketing research conducted by the industry has consistently found young adults are the prime consumers of motion picture theater entertainment. This group is rather concentrated but not organized. A study by the Opinion Research Corporation in July 1986 found those under the age of 40 accounted for 86 percent of all theater admissions. Frequent moviegoers constitute only 21 percent of the eligible filmgoers, but they account for 83 percent of all admissions. A general downward attendance trend has been occurring, as shown in Exhibit 2, where 43 percent of the population never attended a film in 1986. The long-term demographics also appear to be unfavorable, because America's population moving toward those age categories least likely to attend a movie. Those 40 and over make up only 14 percent of a typical theater's admissions, while they account for 44 percent of the nation's population. Those from 12-29 years of age make up 66 percent of admissions, while accounting for only 36 percent of the population.[2] (See Exhibit 3.)

It appears that certain barriers to entry into the motion picture theater industry exist. Economies of scale are present, with the advantage given to operations concentrated in metropolitan areas. where one omnibus newspaper advertisement covers all the chain's theaters. As shown in Exhibit 4, the largest chains in the United States lost the least during the period of July 1984 to June 1985. Based on these results, scale economies appear to exist in the areas of operating costs, executive compensation, advertising, and rental expenses. Those choosing to enter the industry in recent years have done so through the use of massive conglomerate-backed capital. The possibility that an independent can open a profitable movie theater is very remote. "There's no way the small, independent operator can compete against the large screen owners these days," says John Duffy, cofounder of Cinema 'N' Drafthouse International, of Atlanta, Georgia.[3] (See Exhibit 5.) As a way of carving a niche for himself, Duffy's chain charges $2.00 for an "intermediate run" film but serves dinner and drinks during the movie, and thereby garnering more than $5.00 in food revenue, compared to a theater's average $1.25 per admission.

Despite attempts by various theater owners to make the theater-going experience unique, customers tend to go to the most convenient theater that is showing the film they want to see at the time best for them. Accordingly, a particular theater chain enjoys proprietary product differentiation to the degree it occupies the best locations in any particular market area. Additionally, the cost of building new facilities in the most desirable areas has increased dramatically. Harold L. Vogel, of Merrill Lynch, Pierce, Fenner and Smith, has observed the average construction cost comes to over $1 million per screen in such areas as New York or Los Angeles.[4]

---

2. Presented in *1988 International Motion Picture Almanac* (New York: Quigley, 1988), pp 29A–30A.
3. Quoted by Peter Waldman, "Silver Screens Lose Some of Their Luster," *Wall Street Journal*, February 9, 1989, p. B1.

Just as the motion picture was a substitute for vaudeville shows and minstrels at the turn of the century, radio and now television have been the major somewhat interchangeable substitutes for mass entertainment in America. Most recently cable television, pay-per-view TV, and videocassettes have eaten into the precious leisure

EXHIBIT 5

NORTH AMERICA'S LARGEST THEATER CIRCUITS

CIRCUIT	HEADQUARTERS	SCREENS
United Artists Communications	Denver, Col.	2,677
Cineplex Odeon	Toronto, Canada	1,825
American Multi-Cinema	Kansas City, Mo.	1,531
General Cinema	Chestnut Hill, Mass.	1,359
Carmike Cinemas	Columbus, Ga.	742

Source: 10-Ks and various stockholders' reports for 1988.

time dollar. It has been estimated that 49.2 million homes now subscribe to cable television, 19.0 million homes have pay-per-view capability, and 56.0 million homes have a VCR, with 20.0 percent of those homes having more than one unit. The greatest damage to theater attendance has been accomplished by videocassettes, which deliver over 5,000 titles to viewers, at a relatively low cost, in the comfort of their own living rooms. As Sumner Redstone, owner of the very profitable National Amusements theater chain, says, "Anyone who doesn't believe videocassettes are devastating competition to theaters is a fool."[5]

Although the motion picture medium has been characterized as one that provides visual mass entertainment, those going to movies must ultimately choose between alternative forms of recreation. In that regard skiing, boating, baseball and football games, books, newspapers, and even silent contemplation vie for the consumer's precious time. Exhibit 6 shows the movie theater industry has declined in its ability to capture both America's total recreation dollars or its thirst for passive spectator entertainment. During the period from 1984 to 1987, the greatest increases in consumer recreation expenditures were for bicycles, sports equipment, boats, pleasure aircraft, and television and radio equipment and their repair.

Different marketing strategies are being employed in an attempt to remain viable in this very competitive industry. Some chains, such as Cinemark Theaters and Carmike Cinemas, specialize in $1 or low-price second-run multiplexed theaters in smaller towns and selected markets. In a sense they are applying Wal-Mart's original market strategy of dominating smaller, less-competitive rural towns. Others, such as General Cinema, United Artists Communications, and AMC Entertainment, favor multiplexed first-run theaters in major markets. Within this group, AMC Entertainment has been a pioneer as a multiscreen operator. It opened its first twin theater in

4. Harold L. Vogel, "Theatrical Exhibition: Consolidation Continues," *The 1987 Encyclopedia of Exhibirion*, ed. Wayne R. Green, p. 62.
5. Quoted by Stratford P. Sherman, "Movie Theaters Head Back to the Future," *Fortune*, January 20, 1986. p. 91

EXHIBIT 6

MOTION PICTURE EXHIBITOR'S SHARE OF ENTERTAINMENT EXPENDITURES

Receipts as a Percent of Total for Selected years, 1929–1989

YEAR	CONSUMER EXPENDITURES	RECREATION EXPENDITURES	SPECTATOR EXPENDITURES
1929	0.94%	16.6%	78.9%
1937	1.01	20.0	82.6
1943	1.29	25.7	87.6
1951	0.64	11.3	76.3
1959	0.31	5.6	61.0
1965	0.21	3.5	51.2
1971	0.18	2.7	47.7
1977	0.56	5.8	34.8
1983	0.16	2.4	41.9
1986	0.14	1.9	37.3
1987	0.14	1.8	36.9
1988	0.13	1.8	36.5
1989*	0.13	1.7	36.1

* Estimated by the casewriter

Sources: Finler, *The Hollywood Story*, p. 288; U.S. Bureau of Economic Analysis, *Survey of Current Business*, July issues; and U.S. Bureau of the Census, *Statistical Abstract of the United States: 1989* (109th ed.); Washington, D.C.: U.S. Government Printing Office, 1988).

1963 and its first quadplex in 1969. As of mid-1988, AMC was operating 269 complexes with 1,531 screens, with most of its expansion in the Sunbelt. General Cinema has been diversifying out of the movie theater business through its nearly 60.0 percent interest in the Neiman Marcus Group (Neiman Marcus, Contempo Casuals, and Bergdorf Goodman) and 18.4 percent interest in Cadbury Schweppes. Most recently, General Cinema sold off its soft-drink bottling business to PepsiCo for $1.5 billion to obtain cash for investments in additional nontheater operations.

A great amount of building has occurred in the theater industry in the past few years. Since 1981 the number of screens has increased about 35 percent, but the population proportion attending movies has actually fallen. Additionally, the relatively inexpensive days of "twinning" or quadplexing existing theaters appears to be over, and the construction of totally new multiplexes is much more expensive. Exhibit 7 shows that operating profit margins peaked in 1983 at 11.7 percent, and they have

EXHIBIT 7

PER SCREEN ADMISSIONS, CAPITAL EXPENDITURES, AND OPERATING MARGINS

Selected Years, 1979–1987

ITEM	1979	1981	1983	1985	1987
Tickets sold (000,000)	1,121	1,067	1,197	1,056	1,086
Average admission per screen	65,575	58,422	63,387	49,936	47,797
Capital expenditures (000,000)	$19.0	$57.4	$77.6	$164.0	$515.7
Profit margin	9.3%	9.1%	11.7%	11.6%	8.8%

Source: Peter Waldman, "Silver Screens Lose Some of Their Luster," p. B1.

fallen dramatically since then as the industry has taken on large amounts of debt to finance the construction of more and more screens, now generating 24.6 percent fewer admissions per screen. Many operations are losing money, although certain economies of scale exist and laborsaving devices have allowed industry employment to fall slightly, while the number of screens has increased substantially. The Plitt theaters were money losers before being acquired by Cineplex Odeon, and AMC Entertainment lost $6.0 million in 1987 and $13.8 million in 1988 on theater operations. Carmike was barely profitable in 1986, and General Cinema's earnings from its theater operations have fallen for the past three years, although the operation's assets and sales have been increasing. Generally speaking, about half the nation's motion picture theaters and chains have been unprofitable in the 1980s, while numerous chains have engaged in the illegal practice of "splitting," wherein theater owners in certain markets decide which one will negotiate or bid for which films offered by the various distributors available to them.

THE CINEPLEX ODEON CORPORATION

Today's exhibition giant began in 1978 with an 18-screen complex in the underground garage of a Toronto shopping center. Garth Drabinsky, a successful entertainment lawyer and real estate investor (see Exhibit 8), joined with the Canadian theater veteran Nathan Aaron (Nat) Taylor in this enterprise. After three years and dozens of new theaters, Cineplex entered the U.S. theater market by opening a 14-screen multiplex in the very competitive and highly visible Los Angeles Beverly Center. Despite the chain's growth, it was only marginally profitable. When the fledgling chain went public on the Toronto Stock Exchange in 1982, it lost $12.0 million on sales of $14.4 million.

Cineplex nearly went bankrupt but not through poor management by Drabinsky or Taylor. Canada's two major theater circuits, Famous Players (Paramount Studios) and the independent Odeon chain, had pressured Hollywood's major distributors into keeping their first-run films from Cineplex. But in 1983, Drabinsky, who as a lawyer had written a standard reference on Canadian motion picture law, convinced Canada's version of the U.S. Justice Department's antitrust division that Famous Players and Odeon were operating in restraint of trade. Armed with data gathered by Drabinsky, the Combines Investigative Branch forced the distributors to sign a consent decree,

EXHIBIT 8

CINEPLEX ODEON THEATER ACQUISITIONS

Odeon
Plitt Theatres
RKO Century Warner Theaters
Walter Reade Organization
Circle Theatres
Septum
Essaness
Sterling Recreation Organization
Maybox Movie Centre, Ltd.

thus opening all films to competitive bidding. Ironically, without the protection provided by its collusive actions, the 297 screen Odeon circuit soon began to lose money, whereupon Cineplex purchased its former adversary for $22 million. The company subsequently changed its name to Cineplex Odeon.

In its development as an exhibition giant, the chain has always been able to attract a number of smart, deep-pocketed backers. Early investors were the since-departed Odyssey Partners, and, with a 30.2 percent stake, the Montreal-based Claridge Investments & Company, which is the main holding company of Montreal financier Charles Bronfman. The next major investor was the entertainment conglomerate MCA Incorporated, of Universal City, California. MCA purchased 49.7 percent of Cineplex's stock (but is limited to a 33.0 percent voting stake because of Canadian foreign-ownership rules) in January 1986 for $106.7 million. This capital infusion gave Cineplex the funds to further pursue its aggressive expansion plans. As Drabinsky said at the time, "There's only so much you can do within the Canadian marketplace. It was only a question of when, not where, we were going to expand."[6] In short order the company became a major U.S. exhibitor by acquiring six additional chains. Some rival and fearful exhibitors, because of Drabinsky's quest for growth via the acquisition route, have been tempted to call him Darth Grabinsky. (See Exhibit 8.)

Despite these rumblings, Cineplex Odeon has reshaped the moviegoing experience for numerous North Americans. Many previous theater owners had either let their urban theaters fall into decay and disrepair or they had sliced their larger theaters into unattractive and sterile multiplexes. Others had built new but spartan and utilitarian facilities in suburban malls and shopping centers. When building their own theaters from either the ground up or when refurbishing an acquired theater, Cineplex pays great attention to making the patron's visit to the theater a pleasurable one. When the Olympia I and II Cinemas in New York City were acquired, a typical major renovation was undertaken. Originally built in 1913, the theater seated 1,320 and was billed as having "the world's largest screen." New owners subsequently remodeled it in 1939 in an art deco style, and in 1980 it was renovated as a triplex, with a fourth screen added in 1981. As part of Cineplex's renovation, the four smaller auditoriums were collapsed into two larger 850-seat state-of-the-art wide-screened theaters featuring Dolby stereo sound systems and 70-mm projection equipment. Its art deco design was augmented by postmodern features, such as marble floors, pastel colors. and neon accents.

Whether through new construction or the renovation of acquired theaters, many Cineplex cinemas feature entranceways made of terrazzo tile, marble, or glass. The newly built Cinema Egyptien in Montreal has three auditoriums and a total seating capacity of 900. It is replete with mirrored ceilings and hand-painted murals rendered in the traditional Egyptian colors of Nile green, turquoise, gold, lapis lazuli blue, and amber red. Historically accurate murals measuring 300 feet in length depict the daily life and typical activities of the ancient Egyptians. Toronto's Canada Square office complex features a spacious, circular art deco lobby, with a polished granite floor and recessed lighting highlighted by a thin band of neon encircling the high domed ceiling.

6. Quoted by David Aston in "A New Hollywood Legend Called—Garth Drabinsky? *Business Week*, September 23, 1985, p. 61.

On the lobby's left side, moviegoers can snack in a small cafe outfitted with marble tables, bright red chairs, and thick carpeting. In New York City the chain restored the splendor and elegance of Carnegie Hall's Recital Hall as it was originally conceived in 1981. The plaster ceilings and the original seats were completely rebuilt and refinished in the gold and red velvet colors of the great and historic Carnegie Hall.

Just to make the evening complete, and to capture the high profits realized from concession operations, patrons of a Cineplex theater can typically sip *cappuccino* or taste any of the 14 different blends of tea served in Rosenthal china. Those wanting heavier fare can nibble on croissant sandwiches, fudge brownies, carrot cake, or a *latte macchiato,* while freshly popped popcorn is always served with real butter. In-theater boutiques selling movie memorabilia to add to the dollar volume obtained from the moviegoer were created but discontinued, due to unnecessarily high operating costs.

EXHIBIT 9

CINEPLEX ODEON'S REVENUE SOURCES, 1985–88

REVENUE SOURCE	1988	1987	1986	1985
Admissions	51.1%	62.5%	64.5%	68.0%
Concessions	16.5	18.2	20.0	20.0
Distribution and other	22.5	11.1	8.6	7.0
Property sales	9.9	8.0	6.8	5.0

Source: Various 10-Ks and the 1988 annual report.

EXHIBIT 10

SELECTED SUMMARY FINANCIAL DATA
(in millions, except when presented as percents)

	1989[1]	1988	1987	1986	1985	1984
Revenue	$710.0	$695.8	$520.2	$357.0	$124.3	$67.1
Net profit	43.0	40.4	43.6	22.5	9.1	3.5
Net profit %	6.1%	5.8%	6.6%	6.3%	7.3%	5.3%
Long-term debt	$720.0	$600.0	$464.3	$333.5	$40.7	$36.1
Interest	52.6[2]	40.2[2]	33.8	16.4	3.9	2.1
ROE (%)	10.2%	10.3%	11.0%	18.1%	40.7%	30.5%

1. Estimated by Value Line.
2. Estimated by the case writer.

Source: Various 10-Ks and the 1988 annual report.

EXHIBIT 11

1988 PER CAPITA ATTENDANCE RATES

United States	4.4%	Canada	2.8%	West Germany	1.9%
Great Britain	1.4	France	1.9	Italy	1.6

Source: "Movies 'Held Firm' Last Year," *Tulsa Tribune*, February 16, 1989, p. 9C.

EXHIBIT 12

CINEPLEX ODEON CORPORATION
UNAUDITED FIRST QUARTER CONSOLIDATED STATEMENT OF INCOME
(in thousands of U.S. dollars)

	1989	1988
Revenue:		
Admissions	$ 85,819	$ 80,389
Concessions	26,657	24,082
Distribution, post production and other	70,033	28,782
Sale of theater properties	5,731	1,600
Total revenue	188,240	134,853
Expenses:		
Theater operations and other expenses	133,158	97,733
Cost of concessions	5,085	4,466
Cost of theater properties sold	5,837	550
General and admin. expenses	8,035	6,310
Depreciation and amortization	11,207	7,923
Total expenses	163,322	116,982
Income before the undernoted	24,918	17,871
Interest on long-term debt and bank indebtedness	12,257	9,138
Income before taxes	12,661	8,733
Minority interest	978	—
Income taxes	968	727
Net income	$ 10,715	$ 8,006

Source: *First Quarter Report 1989*, pp. 12–13.

This glamor does not come cheaply, because the chain usually charges the highest prices in town. For those in a financial bind, the American Express credit card is now honored at many of the chain's box offices. Cineplex broke New York City's $6 ticket barrier by raising its prices to $7, thus incurring the wrath of Mayor Ed Koch, who marched in picket lines with other angry New Yorkers. Cineplex's action also caused the New York state legislature to pass a measure requiring all exhibitors to print admission prices in their newspaper advertisements. When justifying the increased ticket price, Drabinsky said the alternative was "to continue to expose New Yorkers to filthy, rat-infested environments. We don't intend to do that."[7] Instead of keeping prices low, $30 million was spent refurbishing Cineplex Odeon's 30 Manhattan

7. Quoted by Richard Corliss, "Master of the Movies' Taj Mahals," *Time*, January 25, 1988, pp. 60–61.

theaters to attract better-paying customers. Another unpopular and somewhat incongruous action, given the upscale image engendered by each theater's trappings, is the running of advertisements for Club Med and California raisins before its films. Regardless of the anger and unpopularity created among potential patrons, Cineplex is not interested in catering to the "average" theater patron. Rather than trying to attract the mass market, the theater chain aims its massive and luxurious theaters at the aging Baby Boomers, who are becoming a greater portion of the United States's population.

Over the years, Cineplex Odeon and Garth Drabinsky have received high marks for their creative show business flair. As observed by theater industry analyst Paul Kagan, "Garth Drabinsky is both a showman and a visionary. There were theater magnates before him, but none who radiated his charisma or generated such controversy."[8] These sentiments are reiterated by Roy L. Furman, president of Furman Selz Mager Dietz & Birney, Inc., one of Drabinsky's intermediaries in the Plitt acquisition. "Too many people see the [theater] business as just bricks and mortar. Garth has a real love for the business, a knowledge of what will work and what won't."[9] When a new Cineplex Odeon theater opens, it begins with a splashy by-invitation-only party, usually with a few movie stars on hand. Besides his ability to attract smart investors, Drabinsky believes moviegoers want to be entertained by the theater's ambience as well as by the movie it shows. Accordingly, about $2.8 million (about $450,000 per screen) is spent when building one of the chain's larger theaters, as opposed to the usual $1.8 million for a simple no-frills sixplex. "People don't just like coming to our theaters," says Drabinsky. "They linger afterward. They have another cup of *cappuccino* in the cafe or sit and read the paper. We've created a more complete experience, and it makes them return to that location."[10] He later expanded on this observation, saying, "This company has attempted to change the basic thinking. We've introduced the majesty back to picture-going."[11]

Drabinsky dates his fascination with the silver screen to his childhood bout with polio, which left him bedridden much of the time from the ages of 3 to 12. His illness also imbued him with a strong sense of determination, and this resolution has helped to drive Cineplex Odeon forward. No one speaks for the company except Drabinsky, and he logs half a million miles a year visiting his theaters and otherwise encouraging his employees. The energetic CEO likes to drop by his theaters unannounced to talk with ushers and cashiers, and he telephones or sees 20 to 25 theater managers a week. His standards are meticulously enforced, often in a very personal and confrontational manner. He has been known to exemplify his penchant for detail by stooping in front of one of his ushers to pick up a single piece of spilled popcorn.

The combative nature that helped Drabinsky break the Famous Players and Odeon cartel in the early 1980s still resides in him. When Columbia pictures temporarily pulled its production of *The Last Emperor* out of distribution, he retaliated by

8. Ibid., p. 60.
9. Aston, "A New Hollywood Legend," p. 62.
10. Quoted by Alex Ben Block, "Garth Drabinsky's Pleasure Domes," *Forbes*, June 2, 1986, p. 93.
11. Mary A. Fischer, "They're Putting Glitz Back into Movie Houses," *U.S. News & World Report*, January 25, 1988, p. 58.

EXHIBIT 13

CINEPLEX ODEON CORPORATION
CONSOLIDATED STATEMENT OF INCOME
(in thousands of U.S. dollars)

	1988	1987	1986	1985
Revenue:				
Admissions	$355,645	$322,385	$230,200	$ 84,977
Concessions	114,601	101,568	71,433	24,949
Distribution, post production and other	156,372	61,216	30,846	7,825
Sale of theater properties	69,197	34,984	24,400	6,549
	695,815	520,153	356,989	124,300
Expenses:				
Theatre operations and other expenses	464,324	371,909	258,313	89,467
Cost of concessions	21,537	18,799	13,742	5,980
Cost of theatre properties sold	61,793	21,618	11,690	2,736
General and admin. expenses	26,617	17,965	15,335	5,701
Depreciation and amortization	38,087	23,998	14,266	3,678
	612,358	454,289	313,346	107,562
Income before the undernoted	85,457	65,864	43,643	16,738
Other income	3,599	—	—	(330)
Interest on long-term debt and bank indebtedness	42,932	27,026	16,195	3,961
Income before taxes, equity earnings, preacquisition losses and extraordinary item	44,124	38,838	27,148	13,107
Income taxes	3,728	4,280	6,210	3,032
Income before equity, earnings, preacquisition losses and extraordinary item	40,396	34,558	21,138	8,075
Add back: Preacquisition losses attributable to 50% interest Plitt not owned by the corporation	—	—	1,381	—
Equity in earnings of 50% owned companies	—	—	—	1,021
Income before extraordinary item	40,396	34,558	22,519	10,374
Extraordinary item	—	—	—	9,096
Net income	$40,396	$34,558	$22,519	$10,374

Source: Company annual reports for 1987 and 1988.

canceling 140 play dates of the studio's monumental bomb *Leonard Part 6* starring Bill Cosby. "Some people are burned by his brashness," says Al Waxman (formerly of the television series "Cagney & Lacey"). "There is no self-denial. He stands up and says, 'Here's what I'm doing.' Then he does it."[12] As his long-time mentor Nat Taylor has observed, "He's very forceful, and sometimes he's abrasive. I think he's so far ahead of the others that he loses patience if they can't keep up with him."[13] This nature may have long-term negative consequences for Cineplex, however, because Drabinsky has recently had heated arguments with Sidney Jay Sheinberg. president of MCA Incorporated, the head of the circuit's largest shareholder group.

In addition to its motion picture theater operations, Cineplex Odeon has been engaged in other entertainment-related ventures, such as television and film production, film distribution. and live theater. In the latter area, the company is restoring the Pantages Theatre in downtown Toronto into a legitimate theater for the housing of its $5.5 million production of Andrew Lloyd Webber's *The Phantom of the Opera*, scheduled for Fall 1989.

Cineplex also began a 414-acre motion picture entertainment and studio complex in Orlando, Florida, as a joint venture with MCA Incorporated, but sold its stake to a U.S. unit of the Rank Organization PLC for about $150 million in April 1989 after having invested some $92 million in the project. Various industry observers felt Cineplex withdrew from the potentially profitable venture to help reduce its bank debt, which had grown to $640 million.

The firm also created New Visions Pictures as a joint venture with a unit of Lieberman Enterprises, Inc., in August 1988, to deliver 10 films over a two-year period. Cineplex Odeon also owns Toronto International Studios, Canada's largest film center. This operation licenses its facilities to moviemakers and others for film and television production. In a related motion picture production move, Cineplex acquired The Film House Group, Inc., in 1986, to process 16-mm and 35-mm release prints for Cineplex Odeon Films, another one of the company's divisions, and for other distributors. After doubling and upgrading its capacity in 1987, it sold 49 percent of its interest to the Rank Organization in December 1988 for $73.5 million. Rank exercised its one-year option to buy the remaining portion of The Film House Group shortly thereafter.

The company has also engaged in various motion picture distribution deals and television productions, none of which have been commercially successful. Cineplex has distributed such films as Prince's *Sign o' the Times*, Paul Newman's *The Glass Menagerie*, *The Changeling* with George C. Scott, and *Madame Sousatzka* starring Shirley Maclaine, while its television unit contracted 41 new episodes of the revived "Alfred Hitchcock Presents" series for the 1988–89 television season. The series, however, was canceled. For future release, Cineplex is financing five low-budget ($4 million to $5 million each) films joint-ventured with Robert Redford's Wildwood Enterprises through Northfork Productions, Inc. The five movies will be distributed through Cineplex Odeon Films.

12. Block, "Garth Drabinsky's Pleasure Domes," p. 92.
13. Aston, "A New Hollywood Legend," p. 63.

EXHIBIT 14

CINEPLEX ODEON CORPORATION
UNAUDITED FIRST QUARTER CONSOLIDATED BALANCE SHEET
(in thousands of U.S. dollars)

	1989	1988
Assets:		
Current assets:		
Accounts receivable	$ 229,961	$ 151,510
Advances to distributors and producers	18,334	26,224
Distribution costs	9,695	10,720
Inventories	7,781	7,450
Prepaid expenses and deposits	6,756	5,505
Properties held for disposition	23,833	25,557
Total current assets	296,360	226,966
Property equipment, and leaseholds	844,107	824,836
Other assets:		
Long-term investments and receivables	35,169	130,303
Goodwill (less amortization of $3,545; 1988–$2,758)	53,589	53,966
Deferred charges (less amortization of $8,456; 1988–$7,724)	30,222	27,100
Total Other assets	118,980	211,369
Total assets	$1,259,447	$1,263,171
Liabilities and Shareholders' Equiity		
Current Liabilities:		
Bank indebtedness	$37,185	$21,715
Accounts payable and accruals	98,876	107,532
Deferred income	38,167	21,967
Income taxes payable	3,726	5,651
Current portion of long-term debt and other obligations	12,174	10,764
Total current liabilities	190,128	167,629
Long-term debt	625,640	663,844
Capitalized lease obligations	14,213	14,849
Deferred income taxes	10,920	10,436
Pension obligations	6,847	6,326
Stockholders' equity:		
Capital stock	284,533	283,739
Translation adjustment	12,473	13,348
Retained earnings	88,571	77,856
Total equity	385,577	374,943
Total liabilities and shareholders' equity	$1,259,447	$1,263,171

Source: *First Quarter Report 1989*, pp. 14–15.

Garth Drabinsky's financial dealings and his ability to attract capital to his firm have always been very important to its success. Serious questions have been raised, however, about the propriety of some of Cineplex's financial reporting methods. Charles Paul, a vice-president of MCA Incorporated and a Cineplex board member, says various members are very concerned about the company's financial reporting practices and procedures. In a highly critical report distributed by Kellogg Associates, a Los Angeles accounting and consulting firm, a number of questionable practices were noted. Most frequently cited is Cineplex's treatment of the gains and losses associated with asset sales, with the overall effect being an overstatement of operating revenues. As an example, Cineplex treated its gain of $40.4 million from the sale of The Film House Group as revenue, rather than as extraordinary income. The report also criticized (1) Cineplex's $18.7 write-off on the value of its film library, thereby "postponing" losses on the sale of U.S. theaters, and (2) its inclusion of the proceeds from the sale of theaters as nonoperating income in its cash flow statement but calling it operating revenue in its profit-and-loss statement. In 1988 alone, Cineplex reported a profit of $49.3 million from the sale of certain theater properties.

Also of concern is the role that asset sales play in the company's revenue and cash flow picture. Jeffrey Logsdon, a Crowell Weedon analyst, believes Cineplex has been selling its assets just to keep operating, citing as evidence the sale of both The Film House Group and its 50 percent stake in MCA's Universal Studios tour project to the Rank Organization. Exhibit 9 demonstrates how Cineplex's revenue sources have changed since 1985, with box office receipts constantly falling and property sales constantly rising. Over the period shown, the sale of theater assets has increased 98 percent as a source of corporate revenues. Additionally, the return on those sales, based on selling price over acquisition costs, has fallen every year from a high of 139.1 percent in 1985 to a low of 13.3 percent in 1988.

There is also a question about whether Cineplex can continue its current growth rate via acquisitions and debt financing. The cost of acquisitive growth may become more expensive, because many of the bargains have already been obtained by Cineplex or other chains. The early purchase of the Plitt Theater chain in November 1985 cost about $125,000 per screen, although the bargain price for Plitt may have been a one-time opportunity, because it had just lost $5 million on revenues of $111 million during the nine months ending June 30, 1985. To get into the New York City RKO Century Warner Theaters chain in 1986, Cineplex had to pay $1.9 million per screen, while it paid almost $3.0 million a screen in 1987 for the New York City-based Walter Reade Organization. Overall, Cineplex Odeon paid about $276,000 each for the screens it acquired in 1986, and some are questioning the prices being paid for old screens, as well as the wisdom of expanding operations in what many see is a declining and saturated industry. A past rule of thumb has been that a screen should cost 11 times its cash flow; but some experts feel a more reasonable rule should be 6 to 7 times its cash flow, given the glut of screens on the market. The changing effects of Cineplex's acquisition and debt structure since 1984 have been summarized in Exhibit 10.

Given the nature of the North American market and Cineplex Odeon's penchant for growth, it is currently implementing a planned expansion into Europe. Cineplex is scheduled to build 100 screens in 20 movie houses throughout the United Kingdom by 1990, and it has further plans in Europe and Israel for the early 1990s. Exhibit 11

EXHIBIT 15

CINEPLEX ODEON CORPORATION
CONSOLIDATED BALANCE SHEET
(in thousands of U.S. dollars)

	1988	1987	1986
Assets:			
Current assets:			
Accounts receivable	$151,510	$42,342	$20,130
Advances to distributors and producers	26,224	10,704	4,671
Distribution costs	10,720	10,593	4,318
Inventories	7,450	8,562	6,978
Prepaid expenses and deposits	5,505	4,683	4,027
Properties held for disposition	25,557	22,704	16,620
Total current assets	226,966	99,588	56,744
Property, equipment, and leaseholds	824,836	711,523	513,411
Other assets:			
Long-term investment and receivables	130,303	49,954	14,292
Goodwill (less amortization of $2,758; 1987—$1,878)	53,966	52,596	40,838
Deferred charges (less amortization of $7,724; 1987—$1,771)	27,100	12,015	6,591
	211,369	114,565	61,721
Total assets	$1,263,171	$925,676	$631,876
Liabilities and Shareholders' Equity:			
Current liabilities:			
Bank indebtedness	$21,715	$20,672	$30
Accounts payable and accruals	107,532	74,929	47,752
Deferred inncome	21,967	755	—
Income taxes payable	5,651	4,607	1,926
Current portion of long-term debt and other obligations	10,764	5,965	6,337
Total current liabilities	167,629	106,173	55,945
Long-term debt	663,844	449,707	317,550
Capitalized lease obligations	14,849	14,565	15,928
Deferred income taxes	10,436	13,318	11,142
Pension obligations	6,326	4,026	3,668
Minority interest	25,144	—	—
Stockholders' equity			
Capital stock	283,739	289,181	212,121
Translation adjustment	13,348	1,915	(3,591)
Retained earnings	77,856	46,791	19,113
	374,943	337,887	227,643
Total liabilities and shareholders' equity	$1,263,171	$925,676	$631,876

Source: Company annual reports for 1987 and 1988.

lists the comparative per capita motion picture attendance rates found in various European countries. Other exhibitors are also interested in bringing multiscreened theaters to Europe. In addition to Cineplex's plans, Warner Brothers, American Multi-Cinema, Odeon, and National Amusements have announced their intentions of opening a total of more than 450 screens in the United Kingdom, with further theaters scheduled for later dates.

While few deny the attractiveness of the theaters owned and operated by Cineplex, the firm may have overextended itself both financially and operationally. (See Exhibits 12, 13, 14, and 15). Is Cineplex Odeon on the crest of a new wave of creative growth in North America and Europe, or does it stand at the edge of an abyss? Is consolidation or a thorough review of past actions in order? What next moves should Garth Drabinsky and Cineplex make to continue the firm's phenomenal success story?[14]

SELECTED REFERENCES

Finler, Joel W. *The Hollywood Story.* New York: Crown, 1988.

Gertner, Richard, ed. *1988 International Motion Picture Almanac.* New York: Quigley, 1988.

Green, Wayne R., ed. *Encyclopedia of Exhibition.* New York: National Association of Theatre Owners, 1988.

Hall, Ben M. *The Best Remaining Seats: The Story of the Golden Age of the Movie Palace.* New York: Bramhall House, 1961.

Harrigan, Kathryn Rudie. *Managing Mature Businesses.* Lexington, Mass.: Lexington, 1988

———. "Strategies for Declining Industries." *Journal of Business Strategy.* Vol. 1 no. 2 (Fall 1980), pp. 20-34.

Musun, Chris. *The Marketing of Motion Pictures.* Los Angeles: Chris Musun Company, 1969.

1988 International Motion Picture Almanac. New York: Quigley, 1988.

1989 U.S. Industrial Outlook. Washington, D.C.: U.S. Department of Commerce/International Trade Administration, 1989.

Tromberg, Sheldon. *Making Money, Making Movies.* New York: New Viewpoints/Vision Books, 1980.

Troy, L. *Almanac of Business and Industrial Financial Ratios.* Englewood Cliffs, N.J.: Prentice-Hall, 1988

U.S. Bureau of the Census. *Statistical Abstract of the United States: 1989.* 109th ed. Washington, D.C.: U.S. Government Printing Office, 1988.

14. This case was published in John Pearce II and Richard B. Robinson, Jr. (eds), *Cases in Strategic Management*, second edition (Homewood, Illinois: Irwin, 1991). Reprinted with permission.

U.S. Department of Commerce. Bureau of the Census. *Statistical Abstract of the United States: 1985*. Washington, D.C.: Government Printing Office, 1985.

Waldman, Peter. "Silver Screens Lose Some of Their Luster." *Wall Street Journal*. February 9, 1989, p. B1.

QUESTIONS

1. Based on the data in this case, did the number of movie theaters per thousand people in the U.S. decline during 1950-88? If so, why?
2. According to Exhibit 6, the share of movie theater receipts in consumer expenditures has fallen dramatically in the past 50 years. Is this because movie attendance is an inferior good? If not, what have been the primary reasons?
3. Exhibit 6 also indicates that the share of movie theater receipts in spectator expenditures has fallen less than the share of movie theater receipts in all recreation expenditures. How do you explain this?
4. During 1984-85, would you expect that there would have been entry or exit in the movie theater industry? Why?
5. If many movie theaters were losing money in the 1980s, why did other theater chains buy them? Also, why were more screens built?
6. Would you expect economies of scale in advertising by movie theaters? Why or why not? Is there any evidence on this score in this case? If so, what is this evidence?
7. What proportion of Cineplex Odeon's income seems to have come from the gross profit from concessions?
8. How can microeconomic concepts and techniques be useful to Cineplex Odeon in forging its strategies for the future?

Part 5 Markets for Inputs

Chapter 15 Price and Employment of Inputs

Case Study: Welfare Payments

The welfare system in the United States has been criticized by both conservatives and liberals. To understand some of the effects of welfare payments, it is instructive to examine the indifference map of a (hypothetical) low-income worker, Jean Lapp, shown below.

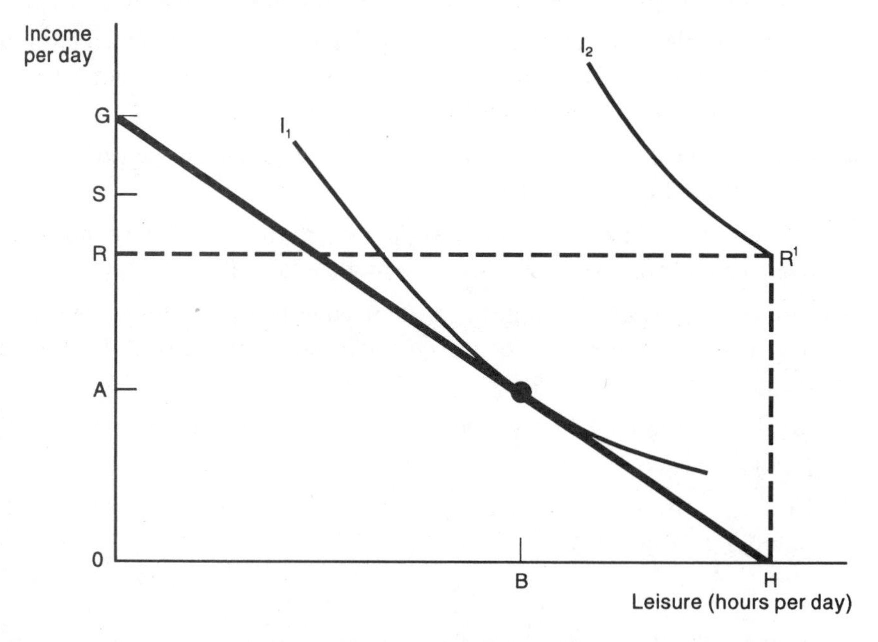

If she holds a job, her aftertax wage rate is –1 times the slope of the line *GH*, where *OH* equals 24 hours of leisure. For simplicity, suppose that, if her earnings are less than *OR*, the welfare system makes up the difference, so she is assured an income per day of *OR*.

a. If she were to work 24 hours a day, how much would she earn?

b. Given that her indifference curves are I_1 and I_2, how many hours per day would she work in the absence of the welfare program, and how much would she earn per day?

c. Given that the welfare program exists, how many hours per day will she work, and how much will she earn per day?

d. Suppose that this welfare system is replaced by a negative income tax which has the following provisions: A minimum income of *OR* is guaranteed; if the worker's income exceeds *OS*, he or she pays taxes; if income is less than *OS*, he or she receives (negative) taxes. Will this have an effect on the amount of work some workers will do? (Their indifference curves need not be the same as those shown above.)

Completion Questions

1. In competitive firm A, the marginal product of the first unit of labor is 4 units of output, the marginal product of a second unit of labor is 3 units of output, and the marginal product of the third unit of labor is 2 units of output. If the price of a unit of output is $20, the firm will demand ______________ units of labor when the price of unit of labor is $75, ______________ units of labor when the price of a unit of labor is $59, and ______________ units of labor when the price of a unit of labor is $45.
2. If the ratio of each input's marginal product to its price is not the same for all inputs, the firm can ________ its costs by using some other input combination.
3. If a firm is maximizing profit under perfect competition, the ratio of the marginal product to the price of each input equals ______________.
4. The value of the marginal product of an input equals the marginal product of the input times ______________.
5. The price of an input in fixed supply is called ______________.
6. A quasi-rent is paid to an input which is temporarily ______________ in ______________.
7. The price elasticity of demand for a nondurable input is generally greater in the ______________ run than in the ______________ run.
8. If an input's supply curve is backward-bending, beyond some point increases in its price bring forth ______________ amounts supplied.
9. Firms frequently use education as an important ______________ of a person's productivity.
10. Once a worker is hired, a ____________ problem must be considered because the worker may pursue his or her interests, not those of the employer.

True or False

____ 1. If an innovation occurs which increases the marginal productivity of labor, and if at the same time the supply curve of labor shifts to the right, the price of labor must fall.

____ 2. A backward-bending supply curve for labor implies that labor is an inferior good.

____ 3. So long as the supply of land in a country is inelastic with regard to its price, economic efficiency does not require payments for land.

____ 4. If more than one input is variable in quantity, the value-of-marginal-product curve will be the demand curve for an input.

____ 5 Changes in the amounts used of other inputs will generally change the value-of-marginal-product curve for an input.

____ 6. Under perfect competition, the supply of an input to an individual firm is perfectly inelastic.

____ 7. Euler's theorem states that, if there are constant returns to scale, the total physical output of a firm will be identically equal to the sum of the amount of each input used multiplied by the input's marginal product.

____ 8. Euler's theorem holds even if there are increasing returns to scale.

____ 9. The more easily other inputs can be substituted for a certain input, the less price elastic is the demand for this input.

____ 10. In the case of a backward-bending supply curve for labor, the income effect more than offsets the substitution effect.

____ 11. An intermediate good is a good that is produced neither by agriculture nor by manufacturing.

____ 12. The price elasticity of demand for an input depends on the price elasticity of the good the input produces.

____ 13. The efficiency wage often is less than the perfectly competitive wage.

____ 14. There frequently is asymmetric information in the labor market.

Multiple Choice

1. The production function of firm Z, a perfectly competitive firm, is as follows:

Number of workers employed per day	*Output per day*
1	0
2	5
3	20
4	35
5	38

If the value of the marginal product of a worker is $150 when between 2 and 3 workers are employed per day, the price of a unit of output must be
a. $1,000.
b. $100.
c. $20.
d. $10.
e. $5.

2. The traditional factors of production are
a. land and labor.
b. land and capital.
c. land, labor, and capital.
d. air, land, and capital.
e. none of the above.

3. (*Advanced*) If one minimizes cost subject to the constraint that a certain output is produced, the Lagrangian multiplier turns out to equal
a. marginal revenue.
b. average cost.

c. marginal cost.
d. all of the above.
e. none of the above.

4. A principal-agent problem often exists because the worker's goals may differ from the firm's goals. To induce workers to work hard, firms often use
a. bonus payments.
b. profit sharing.
c. fixed wages.
d. *a* and *b* only.
e. all of the above.

Review Questions

1. What factors determine the distribution of income? What role do input prices play?

2. Show that the cost-minimizing perfectly competitive firm will set

$$\frac{P_X}{MP_X} = \frac{P_Y}{MP_Y} = \ldots = \frac{P_Z}{MP_Z} = MC$$

where P_X, P_Y, and P_Z are prices of inputs, MP_X, MP_Y, and MP_Z are marginal products of inputs, and MC is marginal cost.

3. Show that a profit-maximizing perfectly competitive firm will set an input's price equal to the value of its marginal product.

4. Suppose that labor is the only variable input, and that the marginal product of labor is as follows:

Amount of labor	*Marginal product of labor*
1	9
2	8
3	7
4	6
5	5
6	4

(Figures concerning the marginal product pertain to the interval between the indicated amount of labor and one unit less than this amount.) If the price of the product is $2 per unit and the firm is perfectly competitive, plot the firm's demand curve for labor in the graph on the following page. (Assume that labor can only be hired in integer amounts.)

5. Describe and discuss the factors that influence the price elasticity of demand for an input.

6. Describe the conditions that can result in a backward-bending supply curve for an input.

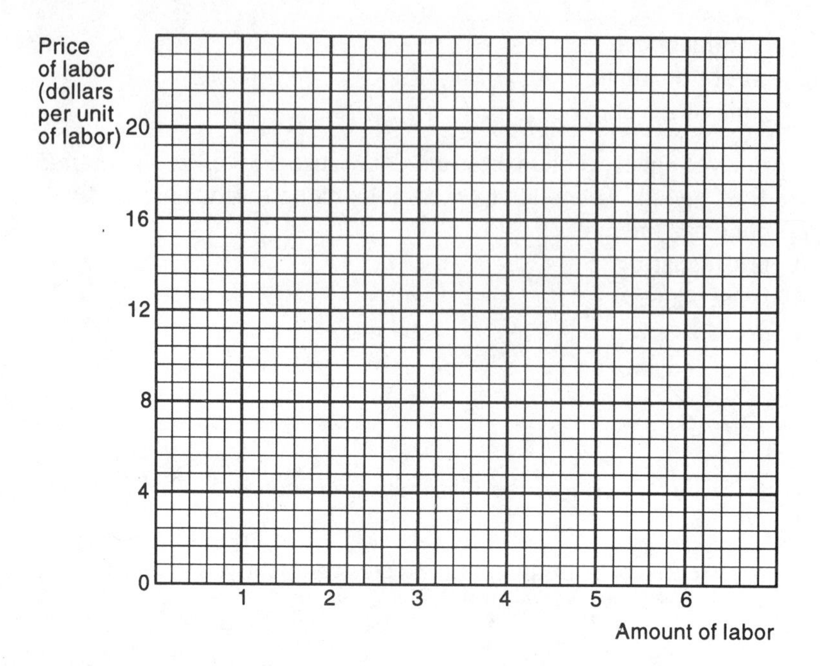

7. Why is a price of OP_0 not an equilibrium price? If OP_0 is the price, what forces will be set in motion to change it? Similarly, why is OP_1 not an equilibrium price? What is the equilibrium price for this input? (*DD′* and *SS′* are the demand and supply curves of the input.)

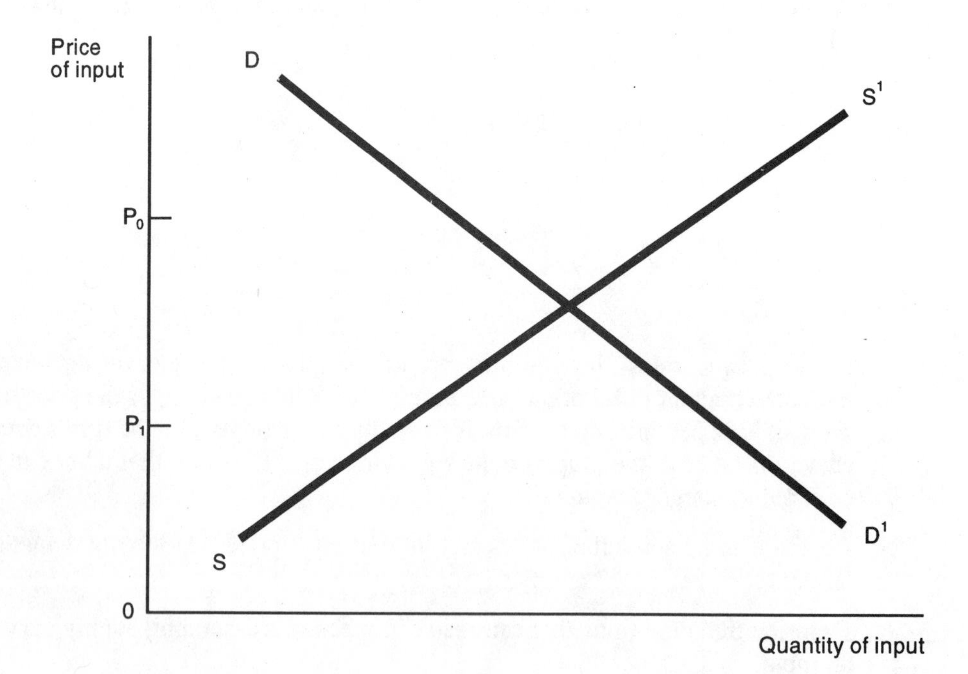

8. Show that a reduction of a payment to an input will not influence the availability and use of the input if the payment is a rent. Why is this important?
9. Describe what is meant by a quasi-rent.

Problems

1. The production function for firm A is

$$Q = 10L - L^2$$

where Q is the number of units of output produced per hour and L is the number of units of labor used per hour. The price of each unit of output is \$1, and the hourly wage rate per unit of labor is \$8.

a. To maximize profit, how many units of labor should firm A hire per hour?

b. If labor is the only input, what will be firm A's hourly profit?

c. Would it make sense for firm A to hire 6 units of labor per hour? Why or why not?

2. The Krakow Manufacturing Company is a member of a perfectly competitive industry. The relationship between various amounts of labor input and the firm's output is shown below.

Product price (dollars)	*Units of labor*	*Units of output*	*Marginal product of labor*	*Value of Marginal product (dollars)*
20	0	0		
			_____	_____
20	1	2½		
			_____	_____
20	2	5		
			_____	_____
20	3	7		
			_____	_____
20	4	8		

a. Fill in the blanks.

b. If you are told that the Krakow Manufacturing Company is hiring 3 units of labor, you can establish a range for the value of the wage rate prevailing in the labor market (assuming that the firm maximizes profit). What is this range? Specifically, what is the minimum value that the wage (for a unit of labor) may be? What is the maximum value? Why?

3. Henry George, an influential nineteenth-century economist, argued that land rents should be taxed away by the government. In his view, owners of land were receiving substantial land rents simply because their land happened to be well situated, not because they were doing anything productive. Since this land rent was unearned income and since the supply of land would not be influenced by such a tax, George felt that it was justifiable to tax away such land rent. Indeed, he argued that a tax of this sort should be the only tax imposed by the government. What weaknesses can you detect in his views?

4. Since various studies indicated a relationship between cigarette smoking and lung cancer, there has been considerable discussion of ways to reduce cigarette smoking in the United States. Suppose that it was decided to impose a tax of $1 on each ounce of tobacco grown by farmers. In your answers to the following questions, assume that all inputs are supplied by competitive input markets, that land has alternative uses, and that the land which can be used for tobacco production is not fixed.
   a. Will the tax have any effect on the equilibrium price of tobacco land? Explain.
   b. Will the tax have any effect on the equilibrium price of machinery used to raise tobacco? Explain.
   c. Is the return to tobacco land an economic rent from the point of view of the tobacco industry? Explain.
   d. Suppose that the tax was imposed only on one small tobacco farmer, rather than on all farmers. What would be its effect on the equilibrium price of tobacco land and the equilibrium price of tobacco machinery?
5. A textile mill uses labor and capital. The price of a unit of labor is $5 and the price of a unit of capital is $6. The marginal product of a unit of labor is the same as the marginal product of a unit of capital. Is this firm (which is a perfect competitor) maximizing its profit?
6. A perfectly competitive firm can hire labor at $30 per day. The firm's production function is as follows:

Number of days of labor	*Number of units of output*
0	0
1	8
2	15
3	21
4	26
5	30

   If each unit of output sells for $5, how many days of labor should the firm hire?
7. Suppose that you own a car wash and that its production function is
$$Q = -0.8 + 4.5L - 0.3L^2$$
   where Q is the number of cars washed per hour and L is the number of persons employed. Suppose that you receive $5 for each car washed and that the wage rate for each person you employ is $4.50. How many people should you employ to maximize profit? What will your hourly profit amount to? (Assume that labor is the only input.)
8. Suppose once again that you own the car wash described in the previous problem. How many people will you employ if the wage rate is $1.50? How many if it is $7.50? Based on these results, draw three points on your firm's demand curve for labor in the grid on the next page.

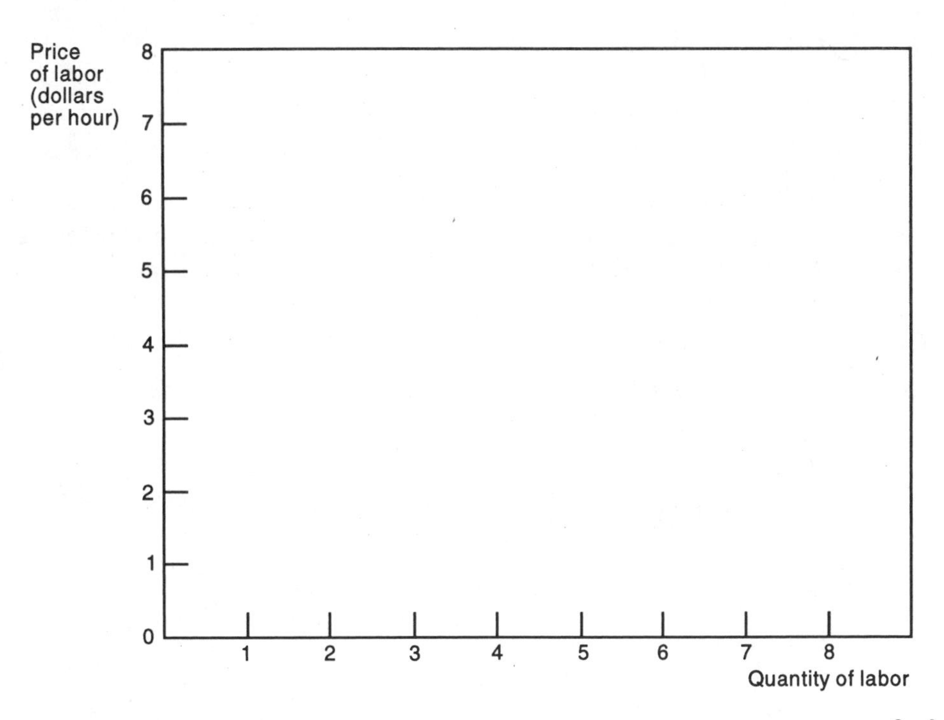

9. (*Advanced*) Suppose that a textile plant's production function in 1994 is $Q = L^{.8}K^{.2}$. If it takes the product price as given and the input prices as given, show that total wages paid by the firm will equal 80 percent of its revenues.
10. Economists used microeconomic analysis to estimate the costs to the Department of Defense of an all-volunteer army. Can you guess how they went about doing this?
11. According to Elliot Berg, "the quantity of wage labor offered by the individual African tends to be inversely related to changes in village income and changes in wage rates." What does this imply about the shape of the supply curve for labor? How do you explain this shape?

Key Concepts for Review

Land
Labor
Capital
Rent
Wages
Profits
Firm's demand curve for an input
Value of marginal product
Market demand curve for an input
Bonus payments
Derived demand
Price elasticity of demand for an input
Market supply curve for an input
Intermediate goods
Backward-bending supply curve
Quasi-rents
Signaling
Principal-agent problem
Efficiency wage
Shirking

ANSWERS

Case Study: Welfare Payments

a. *OG*.
b. She would work *BH* hours per day, and she would earn *OA* per day.
c. Under the welfare program, she can attain point *R′* without working. Since this point lies on her highest attainable indifference curve, she will choose it; that is, she will not work, but receive the assured daily income of OR.
d. Yes.

Completion Questions

1. 1; 2; 2
2. reduce
3. the reciprocal of the price of the product
4. the price of the product
5. rent
6. fixed; supply
7. long; short
8. smaller
9. signal
10. principal-agent

True or False

1. False	2. False	3. False	4. False	5. True	6. False
7. True	8. False	9. False	10. True	11. False	12. True
13. False	14. True				

Multiple Choice

1. *d* 2. *c* 3. *c* 4. *d*

Review Questions

1. A person's income depends largely on the quantity of inputs he or she owns and the prices of these inputs. Input prices play a very important role.
2. It follows from the fact that the marginal rate of technical substitution must equal the input price ratio that

$$\frac{P_x}{MP_x} = \frac{P_y}{MP_y} = \dots = \frac{P_z}{MP_z}$$

Also, the extra cost of producing an extra unit of product, if the extra production comes about by increasing use of any input, must equal the ratio of the price of the input to its marginal product. Consequently, each of the ratios shown above equals marginal cost.

3. This follows from the previous question. Since MC = Price under perfect competition, it follows that

$$\frac{P_X}{MP_X} = \frac{P_Y}{MP_Y} = \ldots = \frac{P_Z}{MP_Z} = \text{Price}$$

Thus, MP_X times Price = P_X; MP_Y times Price = P_Y; and so forth.

4. The demand curve for labor is:

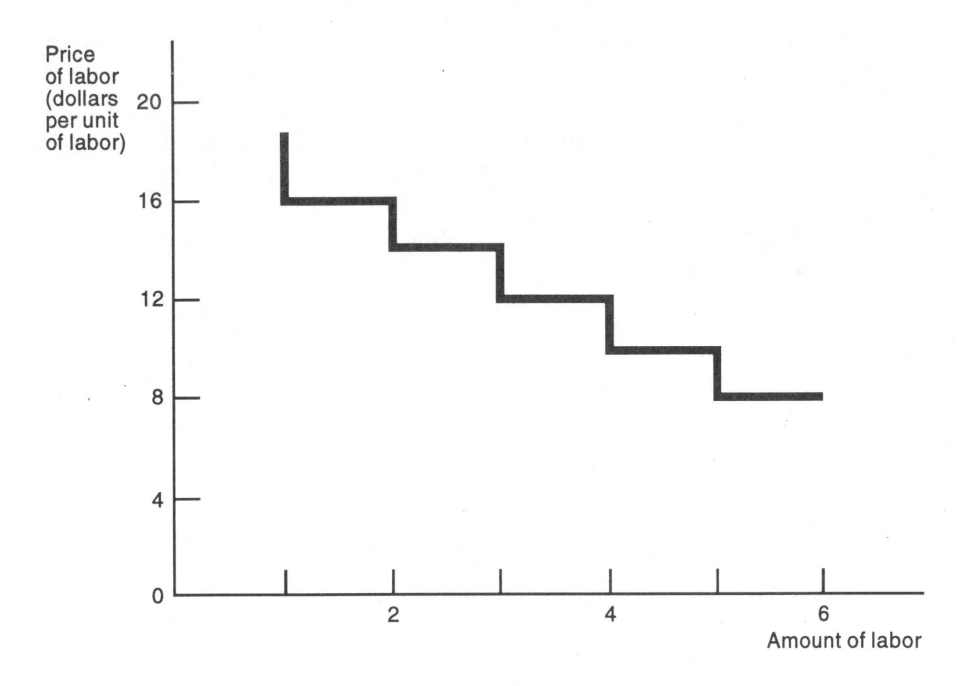

5. First, the more easily other inputs can be substituted for a certain input, the more price-elastic is the demand for this input.

   Second, the larger the price elasticity of demand for the product that the input helps to produce, the larger the price elasticity of demand for the input.

   Third, the greater the price elasticity of supply of the other inputs, the greater is the price elasticity of demand for this input.

   Fourth, the price elasticity of demand for a nondurable input is likely to be greater in the long run than in the short run.

6. A backwàrd-bending supply curve can occur when the person supplying the input can use it himself. As the price of the input increases, the person supplying the input becomes richer. If the income effect more than offsets the substitution effect, the result will be that the person supplies less of the input.

7. OP_0 is not an equilibrium price because at this price the quantity supplied exceeds the quantity demanded. If the input is a good (like coal), inventories of the input would build up and sellers would begin to reduce the price.

   OP_1 is not an equilibrium price because at this price the quantity demanded exceeds the quantity supplied. Since this is the case, inventories would fall and shortages might develop, with the result that the price would be bid up. The equilibrium price is at the intersection of DD' and SS'.

8. This is because a rent is, by definition, a payment to an input that is in fixed supply. It is important because, if the government imposes a tax on rents, there will be no effect on the supply of resources to the economy.
9. A quasi-rent is a payment to any input that is in temporarily fixed supply.

Problems

1. *a.* The value of the marginal product is $1(10 - 2L)$. Thus,

$$1(10 - 2L) = 8$$
$$L = 1$$

   Firm A should hire one unit of labor per hour.

   b. \$9 – \$8 = \$1.

   c. No.

2. *a.* 2½ \$50
   2½ 50
   2 40
   1 20

   b. \$20 to \$40.

3. Critics of George's views pointed out that land of many kinds can be improved, with the result that the supply is not completely price inelastic. The supply of such land really is not fixed, and it may be influenced by the tax. (The return from such land, although called "land rent," is not really rent, as we have defined it.) Moreover, they argued that if land rents are "unearned" so are many other types of income. Further, they claimed that it is unrealistic to expect such a tax to raise the needed revenue.

4. *a.* Because the tax is likely to reduce the output of tobacco, it may shift the demand curve for tobacco land to the left, thus reducing its equilibrium price.

   b. Because the tax is likely to reduce the output of tobacco, it may shift the demand curve for such machinery to the left, thus reducing its equilibrium price.

   c. No, because the supply curve for such land is not vertical.

   d. This farmer would be driven out of business, and there would be little or no effect on the equilibrium price of tobacco land and the equilibrium price of such machinery.

5. No. It should set the marginal product of labor divided by \$5 equal to the marginal product of capital divided by \$6.

6. The value of the marginal product is shown at the top of the next page.

Number of days of labor	*Output*	*Marginal product*	*Value of Marginal product (dollars)*
0	0		
		8	40
1	8		
		7	35
2	15		
		6	30
3	21		
		5	25
4	26		
		4	20
5	30		

Thus, if the daily wage of labor is \$30, the firm should hire 2 or 3 days of labor.

7. Your hourly profit equals $5Q - 4.5L$. Substituting for Q, your profit equals

$$5(-0.8 + 4.5L - 0.3L^2) - 4.5L = -4 + 18L - 1.5L^2$$

If you plot profit against L, you will find that it is a maximum at $L = 6$. Thus, you should employ 6 persons.

\$50.

8. 7 persons.

5 persons.

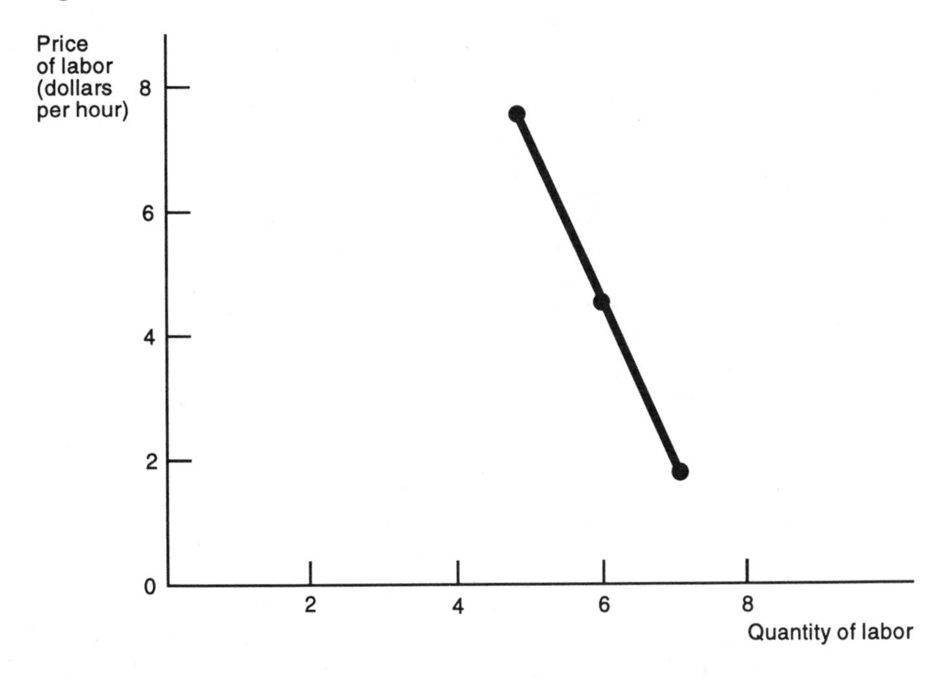

9. The wage, P_L, will equal the price of the product, P, times the marginal product of labor, which equals

$$\frac{\partial Q}{\partial L} = 0.8L^{-.2}K^{.2} = \frac{0.8Q}{L}$$

Thus $P_L = (0.8Q/L)P$; this means that

$$\frac{P_L L}{PQ} = 0.8$$

Since $P_L L$ equals the total wages paid by the firm and PQ equals its revenues, this completes the proof.

10. On the basis of cross-section data, Stuart Altman and Alan Fechter estimated the supply curve for manpower to the defense establishment. Then on the basis of independent estimates of the necessary size of the armed forces, they estimated the wage required to obtain this much labor. The total cost to the Department of Defense was this wage times the number of people in the armed forces.

11. It is backward-bending. See answer to Review Question 6.

Chapter 16 Price and Employment of Inputs under Imperfect Competition

Case Study: A Statistical Study of the Salaries of Baseball Players

Economist Gerald Scully[1] estimated the contribution of baseball players of varying quality to a team's gross revenues. For hitters, his results were as follows:

Lifetime batting average of player	*Gross*	*Net*	*Salary*
255	$121,200	–$39,100	$9,700
283	135,000	–25,300	20,000
338	256,600	128,300	29,100
427	405,800	290,500	42,200

The gross marginal revenue product is an estimate of the player's gross contribution to the team's revenues. The net marginal revenue product deducts the costs of player development and training, as well as other costs, from the gross marginal revenue product. The average 1969 salary of players with each batting average is shown in the last column. These data pertain to the period when the reserve clause was in effect.

a. According to these results, were all baseball players exploited?

b. When the courts allowed players to become free agents, the salaries of star players skyrocketed. Why?

Completion Questions

1. To a monopsonist in the labor market, the cost of hiring an additional laborer (increases, decreases) ________________ as it hires more and more labor. The cost of hiring an additional laborer (exceeds, equals, is less than) ___________ the wage that must be paid this worker. If the monopsonist can hire 10 workers at a daily wage of $40 and 11 workers at a daily wage of $45, the daily cost of hiring the eleventh worker is _________________, as compared with the daily wage of the eleventh worker, which is ______________.

2. The condition for profit maximization under imperfect competition is that the firm should set the price of the input equal to _____________.

3. The marginal revenue product of an input differs from the value of the marginal product in the following way: _________________.

4. If the marginal revenue is equal to price, the marginal-revenue-product schedule becomes precisely the same as the _________________.

5. If all firms are monopolists in their product markets, the market demand curve for an input would simply be the _________________ of the demand curves of the individual firms.

1. G. Scully, "Pay and Performance in Major League Baseball" *American Economic Review.*

6. Monopsony is a case where there exists ______________.
7. The classic case of monopsony is ______________.
8. The marginal expenditure curve for an input lies ______________ its supply curve, if its supply curve slopes upward to the right.
9. If a monopsonist maximizes profit, it sets the marginal expenditure on an input equal to ______________.
10. The monopsonist employs ____________ of the input than if the input market were perfectly competitive.
11. The monopsonist pays a ______________ price for the input than if the input market were perfectly competitive.

True or False

____ 1. If a union shifts the supply curve of labor to the left, and if the price elasticity of demand for labor is infinite, the price of labor will rise.

____ 2. If a union wants to maximize the total payments to labor, it will operate in the inelastic portion of the demand curve for labor.

____ 3. Under perfect competition, the marginal expenditure for an input equals its price.

____ 4. A monopsonist will hire inputs so that the ratio of the marginal product to the marginal expenditure for each input is the same for all inputs.

____ 5. If a labor union tries to maximize its wage bill, it will try to operate at the point where its marginal revenue is zero.

____ 6. The marginal-revenue-product curve for an input often has a positive slope.

____ 7. The marginal revenue product of an input is equal to marginal revenue divided by marginal product.

____ 8. Oligopsony occurs when there are few buyers.

Multiple Choice

1. It makes perfectly good sense for a labor union to want to
   a. maximize the wage rate.
   b. maximize its marginal revenue.
   c. keep its members fully employed
   d. all of the above.
   e. none of the above.

2. If the price of labor is $1 a unit, its marginal product is 2 tons of output, and the firm's marginal revenue is $5 per ton, the firm (which is not a monopsonist) is
   a. minimizing cost.
   b. maximizing profit.
   c. not maximizing profit.
   d. at a corner solution.
   e. a monopolist.

3. If there is only one variable input, the firm's demand curve for an input is obtained by multiplying
   a. marginal cost and marginal revenue curves.
   b. marginal product and marginal revenue curves.
   c. average cost and marginal revenue curves.
   d. average cost and average revenue curves.
   e. none of the above.

4. Under imperfect competition in the product market, the equilibrium price of an input is given by
   a. the intersection of the product demand and supply curves.
   b. the intersection of the input demand and supply curves.
   c. the intersection of the product demand and input supply curves.
   d. all of the above.
   e. none of the above.

5. Monopsony often stems from the fact that
   a. an input is not mobile.
   b. an input is extremely mobile.
   c. an input is an intermediate good.
   d. an input has a wide variety of uses.
   e. none of the above.

6. The marginal expenditure for an input will always be
   a. higher than the input's price.
   b. higher than the input's price if its supply curve is upward-sloping.
   c. higher than the input's price if its supply curve is horizontal.
   d. all of the above.
   e. none of the above.

Review Questions

1. A craft union is formed which forces employers to hire only union members. The union membership is restricted by high initiation fees and a variety of other devices. What are the effects on the demand curve, supply curve, and price of labor?

2. "A union will almost always insist on maintaining the current wage even at the cost of severe contraction in employment, whereas it would not insist on increasing the money wage if the consequences for employment were anything like as severe." If true, how can this fact be incorporated into our models of union behavior? What is the implication of this statement for classical theories of unemployment, which assumed that prices and wages are flexible?

3. A union sometimes presents demands that it does not expect to be acted on immediately. "It is the normal opening move in the chess game." Give some examples of demands that were first met by outraged opposition by management, but which ultimately came to be accepted.

4. Describe how a profit-maximizing firm under imperfect competition will deter-

mine how much of each input to employ. Prove that this does indeed result in maximum profit.

5. Prove that the marginal revenue product is the product of the marginal product and marginal revenue.
6. Discuss what is meant by monopsony. Give some cases in the real world where monopsony seems to be present.
7. If you were a laborer, would you prefer the labor market to be perfectly competitive or monopsonistic?
8. If you wanted your area to become a low-wage area, would you prefer the labor market to be perfectly competitive or monopsonistic? Why?
9. Describe the conditions determining how much of each input a profit-maximizing monopsonist will employ.
10. How can unions influence the wage rate paid to their members? Describe at least three ways that they can influence the wage.
11. Discuss the nature of union objectives. Give a few plausible kinds of union motivation and describe their implications for the wage rate and employment.

Problems

1. The Pennsylvania Widget Company uses capital and labor to produce widgets. The price of a unit of capital is $10 per unit and the price of labor is $5 per unit. The marginal product of capital is 20 units and the marginal product of labor is 10 units. Is this firm maximizing profit if it is
   a. a perfectly competitive firm in perfectly competitive input markets?
   b. a monopolist that buys inputs in perfectly competitive input markets?
   c. a monopsonist? Explain your answer in each case.
2. Firm M, a monopsonist, confronts the supply curve for labor shown below.
   a. Fill in the blanks.
   b. Firm M's demand curve for labor is a horizontal line at a wage rate of $8 per hour. What quantity of labor will firm M demand?

Wage rate per hour (dollars)	*Number of units of labor supplied per hour*	*Cost of adding an additional unit of labor per hour (dollars)*
3	1	
		______
4	2	
		______
5	3	
		______
6	4	
		______
7	5	
		______
8	6	

c. What will be the equilibrium wage rate?

3. *a.* Suppose that the market supply schedule for input *Y* is as follows:

Quantity of Y	*Price of Y (dollars)*
10	1
11	2
12	3

Plot the marginal expenditure curve for *Y* in the relevant range in the graph below.

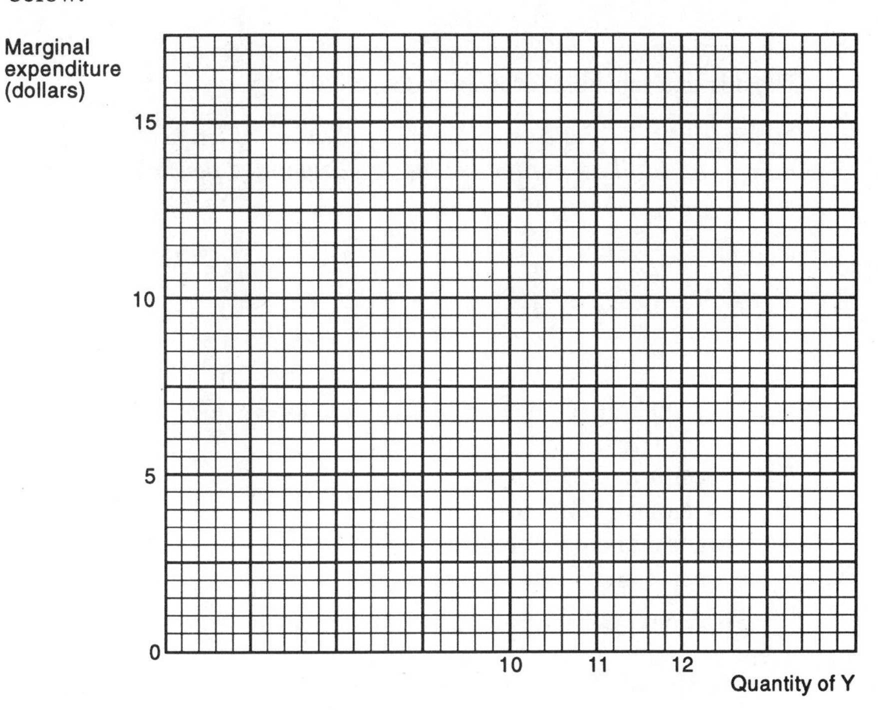

b. If the market supply curve for an input is a horizontal line, will the marginal expenditure curve for this input differ from the market supply curve? Explain.

4. Suppose that labor is the only input used by the Bijou Manufacturing Company, a hypothetical producer of tea kettles. This firm's production function is

Quantity of labor (workers)	*Number of tea kettles produced per hour*
0	0
1	5
2	10
3	15
4	19
5	22

The demand schedule for its product is:

Price (dollars per tea kettle)	*Number of tea kettles demanded per hour*
1	22
2	19
3	15
4	10
5	5
6	0

What is the marginal revenue product of labor when
a. between 0 and 1 laborers are hired?
b. between 1 and 2 laborers are hired?
c. between 2 and 3 laborers are hired?
d. between 3 and 4 laborers are hired?
e. between 4 and 5 laborers are hired?

5. In the previous problem, how many laborers will the Bijou Manufacturing Company hire if
a. the wage of a laborer is \$30 per hour?
b. the wage of a laborer is \$25 per hour?
c. the wage of a laborer is \$15 per hour?
d. laborers can be obtained free of charge?

6. *a.* Suppose that the demand curve for a firm's product is as follows:

Output	*Price of good (dollars)*
23	\$5.00
32	4.00
40	3.50
47	3.00
53	2.00

Also, suppose that the marginal product and total product of labor at this firm is

Amount of labor	*Marginal product of labor*	*Total output*
2	10	23
3	9	32
4	8	40
5	7	47
6	6	53

(Note that the figures regarding marginal product pertain to the interval between the indicated amount of labor and one unit less than the indicated amount of labor.) Given these data, how much labor would the firm employ if labor costs $12 a unit?

b. How much labor will the firm employ if labor costs $13 a unit?

c. How much labor will the firm employ if labor costs $10 a unit?

d. How much labor will the firm employ if labor costs $1 a unit?

e. Assuming labor is the only variable input, plot the firm's demand curve for labor in the graph below.

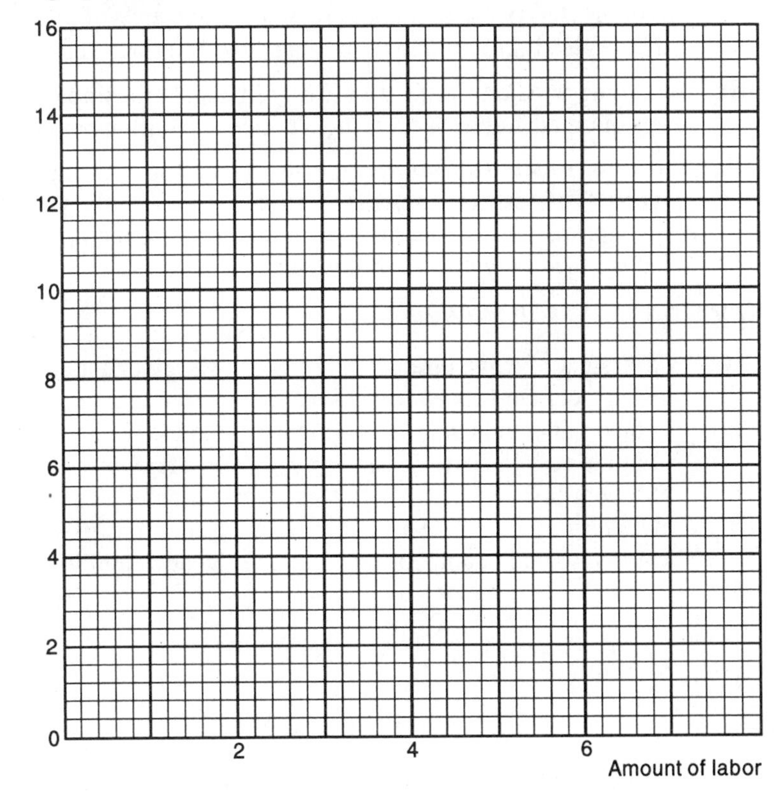

7. According to Ruth Greenslade, the ratio of average hourly earnings in the bituminous coal industry to that in general manufacturing rose from 1:1 at the

turn of the century to about 2:1 at the peak of post-World-War-I union strength in the industry in 1922. Then as union strength declined in the industry, the ratio declined to 1.07 in 1933. But as union strength increased subsequently, the ratio increased too, and reached 1.7:1 in 1957. What do these findings seem to indicate about the economic effects of unionism?

8. When an economist studied the behavior of wages in 34 manufacturing industries, he found that wages were directly related to degree of concentration of output, changes in output per hour of labor, and extent of unionism. But the extent of unionism seemed least important and was not statistically significant. Comment on these findings.

9. The market for a particular input is a monopsony. For this input,

$$MRP = 8 - 2Q_D$$
$$P = 2Q_S$$
$$ME = 4Q_S$$

where MRP is the input's marginal revenue product, P is its price, ME is marginal expenditure for this input, Q_D is the quantity demanded of this input, and Q_S is the quantity supplied of this input. (MRP, P, and ME are expressed in dollars per unit of input; Q_D and Q_S are expressed in thousands of units per day.) How large is the deadweight loss due to monopsony? Interpret your result.

10. According to Allan Cartter and F. Ray Marshall, the impact of unionism on wage levels of organized workers is most noticeable during periods of recession. Why?

11. According to Simon Rottenberg, in his discussion of the labor market for baseball players: "The reserve rule, which binds a player to the team that contracts him, gives a prima facie appearance of monopsony to the market." [2] Do you agree? What are the likely effects of this rule?

Key Concepts for Review

Marginal revenue product
Firm's demand curve for an input
Market demand curve for an input
Market supply curve for an input
Monopsony
Oligopsony
Monopsonistic competition
Marginal expenditure curves
Labor unions
Collective bargaining
Bilateral monopoly
Union objectives
Economic effect of unions
Deadweight loss due to monopsony

2. Simon Rottenberg, "The Baseball Players' Labor Market," *Journal of Political Economy*.

ANSWERS

Case Study: A Statistical Study of the Salaries of Baseball Players

a. No. The players with low batting averages seemed to be paid more than the net contribution they made to their team's profit.

b. According to Scully's estimates, the star hitters contributed much more to their team's profit than they were being paid when the reserve clause was in effect. This was also true of the star pitchers.

Completion Questions

1. increases; exceeds; $95; $45
2. the input's marginal revenue product
3. $MRP = MR \cdot MP$, whereas $VMP = P \cdot MP$
4. value-of-marginal-product schedule
5. horizontal summation
6. a single buyer
7. a company town where a single firm is the sole buyer of labor.
8. above
9. the input's marginal revenue product
10. less
11. lower

True or False

1. False	2. False	3. True	4. True
5. True	6. False	7. False	8. True

Multiple Choice

1. *c*	2. *c*	3. *b*	4. *b*	5. *a*	6. *b*

Review Questions

1. The supply curve for labor is shifted to the left, with the result that the price of labor will increase. There may be no effect on the demand curve for labor, unless employers are forced to hire more labor at a given wage than they otherwise would have.

2. If true, this proposition implies that the current wage is a floor that almost never will be penetrated in the course of collective bargaining. It implies that an assumption of downward wage flexibility is unrealistic. However, during the 1980s, many unions did grant significant concessions.

3. The Auto Workers and Steel Workers asked for a "guaranteed" annual wage in the 1940s. Also, pensions and health and welfare funds are other examples.

4. The firm will set the marginal revenue product of each input equal to the input's price. Since the former measures the extra revenue derived from an extra unit of input and the latter measures the extra cost of an extra unit of input, profit cannot be a maximum if the latter is less than the former—or more than the former. It can only be a maximum when they are equal.

5. By definition, the marginal revenue product is $\Delta R \div \Delta I$,where ΔR is the change in total revenue and ΔI is the change in the quantity of the input. Since marginal revenue (MR) equals $\Delta R \div \Delta Q$, where ΔQ is the change in output, it follows that the marginal revenue product equals $MR\Delta Q \div \Delta I$. And since marginal product (MP) equals $\Delta Q \div \Delta I$, it follows that the marginal revenue product equals $MR \cdot MP$.
6. Monopsony is a case where there is a single buyer. Company towns or mill towns are sometimes examples of monopsony with respect to labor.
7. Perfectly competitive, because the wage will be higher.
8. Monopsonistic, because the wage will be lower.
9. A profit-maximizing monopsonist will hire an input up to the point where its marginal product times the firm's marginal revenue equals the marginal expenditure for the input.
10. They can shift the supply curve for labor to the left, try to get the employer to pay a higher wage while allowing some of the supply of labor forthcoming at this higher wage to find no opportunity for work, or try to shift the demand for labor upward and to the right.
11. The union might want to maximize the wage, subject to the constraint that a certain number of its members be employed. Or it might want to maximize the wage bill. It is difficult to summarize union goals adequately and simply.

Problems

1. *a.* and *b.* Under the conditions set forth in parts *a* and *b* of this question, a necessary condition for profit maximization is that the price of each input be proportional to its marginal product. This condition is fulfilled in this case. However, this is not a sufficient condition for profit maximization. Even though this condition is satisfied, the firm may not be maximizing profit.
   c. Unless the marginal product of each input is proportional to the marginal expenditure for this input, the firm is not maximizing profit. If they are proportional, the firm may or may not be maximizing profit. See the answer to parts (*a*) and (*b*).
2. *a.* 5.
   7.
   9.
   11.
   13.
   b. 3 units of labor per hour.
   c. $5 per hour.

3. *a.*

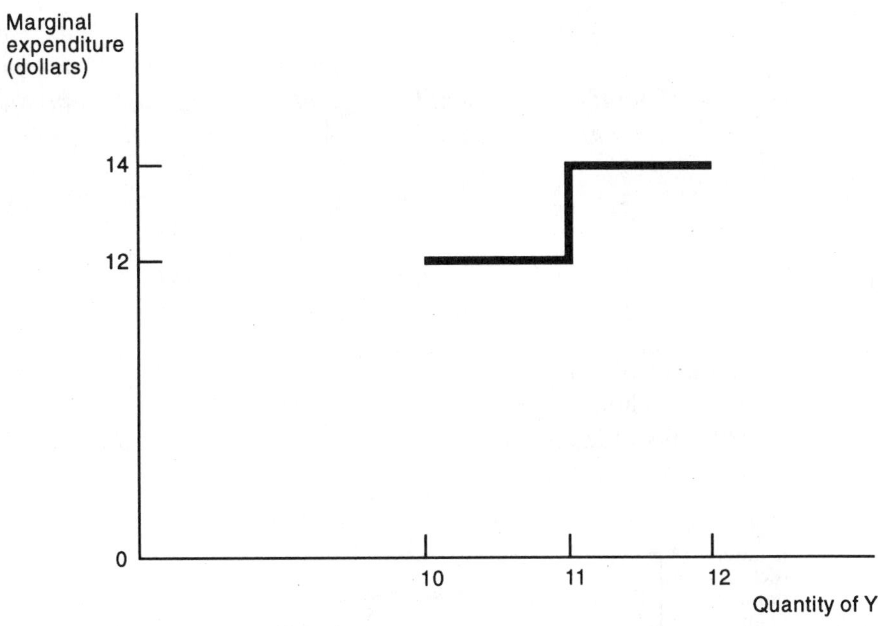

b. No. The additional expenditure for an additional unit of the input equals its price under these circumstances.

4.

Quantity of labor	*Number of tea kettles*	*Price (dollars)*	*Total revenue (dollars)*	*Marginal revenue product (dollars)*
0	0	6	0	
1	5	5	25	25
2	10	4	40	15
3	15	3	45	5
4	19	2	38	–7
5	22	1	22	–16

a $25.
b. $15.
c. $5.
d. –$7.
e. –$16.

5. *a.* None.
b. Zero or 1.
c. 1 or 2.
d. 3.

6. *a.* 3 or 4 units of labor. For reason, see below.

Amount of labor	*Total output*	*Marginal product*	*Price of good (dollars)*	*Total revenue (dollars)*	*Marginal revenue product (dollars)*
2	23	10	5.00	115	n.a.
3	32	9	4.00	128	13
4	40	8	3.50	140	12
5	47	7	3.00	141	1
6	53	6	2.00	106	–35

b. 2 or 3 units of labor.
c. 4 units of labor.
d. 4 or 5 units of labor.
e. The demand curve is as follows:

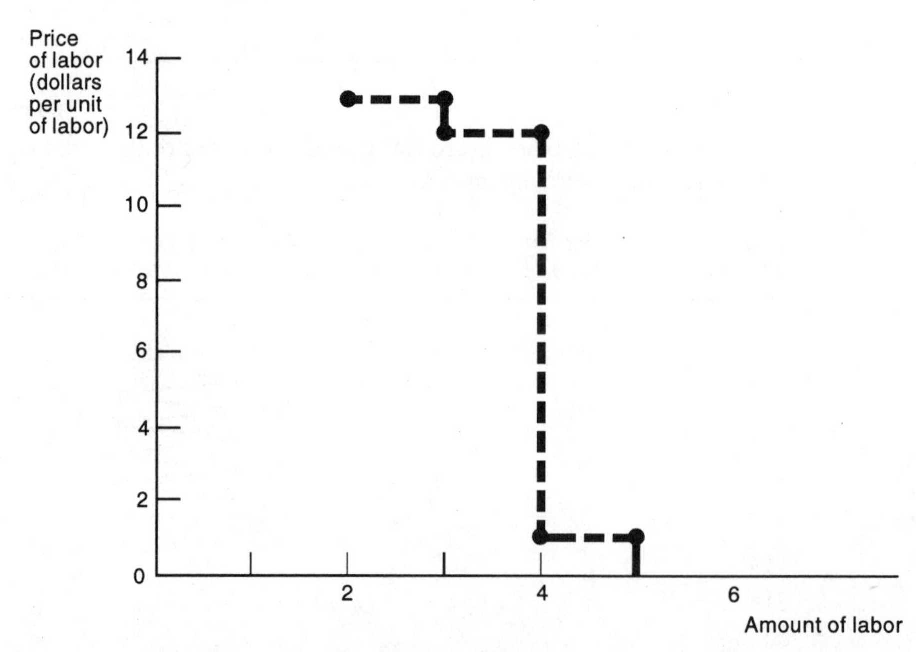

7. They seem to indicate that the strength of the union was directly related to the relative height of wages in the industry.

9. The deadweight loss is the shaded area in the graph below.

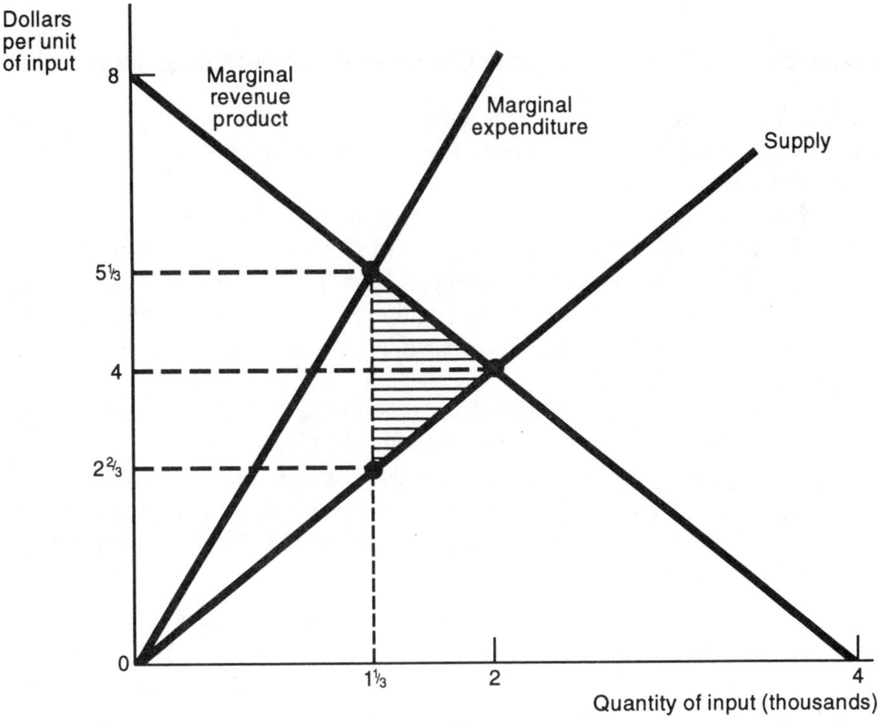

The deadweight loss equals ½(⅔ thousand units times ⁴⁄₃ dollars per unit) + ½(⅔ thousand units times ⁴⁄₃ dollars per unit) = ⁸⁄₉ thousands of dollars per day.

10. Because union contracts tend to introduce rigidity in wage structures.
11. Wages of baseball stars may be lower than in a competitive market.

Chapter 17 Investment Decisions and Risk

Case Study: How One Firm Made Investment Decisions

The following excerpt from congressional testimony by executives of the Armstrong Cork Company describes the procedure it used to decide which investment products to undertake:

Effective evaluation of capital appropriation requests is essential to maintaining a satisfactory return on investment, both short-term and long-term. Such evaluation should be planned to follow the concept and terminology in use for budgetary controls and operating reports: otherwise the evaluation will fail to answer the questions that arise in the minds of top management when considering the advisability of granting an appropriation request.

Faced with these considerations, a system for evaluating capital appropriation requests has been developed by the Armstrong Cork Company which, we believe, gives management the information it needs to make sound decisions. This system is based on the same concept of cost used in reporting budgets and operating results.

The more important aspects of the Armstrong Cork Company system can be summarized as follows:

1. The economics of proposed capital appropriation requests are examined by the controller's office in consultation with the interested staffs like central engineering, research, industrial engineering, economic and marketing research, purchasing, and so forth, before presentation to top management.
2. The basic evaluation is made in return on investment terms.
3. An evaluation of risk is made, based on the length of time required to recover the cash to be expended.
4. Management relies on these valuations in making appropriation decisions.
5. Follow-up reports of the in-process status and actual results of capital appropriations granted are made annually by the controller's office to top management, which, in turn, discusses the results with the accountable persons.

"Return on investment," or as we use the phrase, "return on capital employed" (ROCE), is nothing more than the ratio of net profit after tax to total book assets. In the Armstrong Cork Company, ROCE is accepted as the basis for measuring operating management performance; the success or failure of all our individual and collective efforts to improve operations is reflected in our ROCE results.

Our internal ROCE concept is designed around our operations general managers. These people are responsible for the production and sale of all products within the major market areas assigned to them. They are expected to use the capital they are allocated by the board to obtain the best possible return on it. Furthermore, and this is important, these line people are also responsible for recommending opportunities for the employment of additional capital to improve or expand existing businesses and to enter new businesses. The operations managers, therefore, occupy key positions in our company. The principal function of our staff departments is to help these line people fulfill their responsibilities; that is, to obtain higher returns on more capital, or, in other words, to obtain superior performance as measured by ROCE, coupled with growth in and into businesses having high ROCE potentials.

All matters related to financing are reserved to a staff vice-president. The determination of what shall be financed is reserved, depending on the amounts involved, for the executive committee of our board of directors or the board itself. In practice, the board of directors and its executive committee control "what shall be financed" almost entirely through our system of capital appropriation requests. This control of capital expenditures automatically controls the basic level of working capital. The general level of cash, accounts receivable, and inventories required by operations is determined basically by the types of businesses in which we operate. Since we are primarily a manufacturing company, the types of businesses we operate are established by the plant, property, and equipment we buy. Therefore, the board of directors and its executive committee, in controlling such purchases, indirectly but effectively control the types of businesses we operate and, consequently, the basic levels of cash, accounts receivable, and inventories.

Our procedures for reviewing capital appropriation requests are designed specifically to fit the needs of our organization. They do this by profit and loss statements expressed in terms of ROCE. We place our emphasis on accuracy of perspective rather than on accuracy of detail. Our aim is to discover the best of the many potential expenditures. We don't attempt to provide the figure that makes approval or disapproval of a request automatic. The figures we develop are relative, not absolute. One of the most valuable results of our system will never be measured since it is the elimination of projects of low ROCE potential from consideration before an undue amount of valuable technical and management effort is expended upon them.[1]

a. Does this procedure attempt to accomplish the same objective as the net-present-value rule? Explain.

b. What differences can you detect between this procedure and the net-present-value rule?

Completion Questions

1. The decision maker should carry out any investment project with a ________ net present value.

2. The ____________ is the premium received by the consumer one year hence if he or she lends a dollar for a year.

3. The ______________ of the consumer's intertemporal budget line is equal to $-(1 + r)$, where r is the interest rate.

4. The investment demand curve slopes ____________ to the right.

5. The _____________ is the relationship between the quantity of loanable funds supplied and the interest rate.

---

1. Testimony before the Joint Economic Committee of Congress, 89th Congress, 2nd Session.

6. The __________ of the interest rate is the level where the quantity of loanable funds demanded equals the quantity supplied.

7. If the interest rate is .10, the _______________ of a dollar received four years hence is about 68 cents.

8. If a decision maker's von Neumann-Morgenstern utility function is ________, the decision maker maximizes expected monetary value.

9. The Monongahela Company stores spare parts at two warehouses, one in Scranton and one in Cleveland. The number of defective and acceptable spare parts at each warehouse is given below.

	Number of Spare Parts		
Warehouse	*Defective*	*Acceptable*	*Total*
Scranton	200	800	1,000
Cleveland	50	450	500

If one of the 1,500 spare parts kept by the firm is chosen at random (i.e., if each spare part has a 1/1,500 chance of being chosen), the probability that it will be defective is _____________.

10. In the previous question, the probability that the chosen spare part will be at the Scranton warehouse is _____________.

11. In Question 9, the probability that the chosen spare part will *both* be defective and at the Scranton warehouse is _____________.

12. In calculating the expected value of perfect information, it is important to recognize that the decision maker (does, does not) _____________ know what this information will be.

13.[2] The problem of _______________ occurs if a person's or firm's behavior changes after buying insurance so as to increase the probability of a loss.

True or False

____ 1. If an investment yields $100 per year indefinitely into the future and the interest rate is .10, the present value of this investment is $500.

____ 2. Increases in the interest rate will always result in increases in the amount saved.

____ 3. If you receive X dollars n years from now, the present value of this amount is $X(1 + r)^n$, where r is the interest rate.

____ 4. To take risk into account when calculating the net present value of an investment project, firms often use as a discount rate the interest rate on risk-free government bonds minus 1 or 2 percentage points.

____ 5. No matter whether expected utility is maximized or minimized, you get the same result if the von Neumann-Morgenstern utility function is linear.

2. This question pertains to the chapter appendix.

____ 6. According to the subjective definition of probability, the probability of a particular outcome is the proportion of times that this outcome occurs over the long run if this situation exists over and over again.

____ 7.[3] Sellers of health insurance often offer lower prices to members of group plans than to individuals.

____ 8.[4] To construct James Johnson's von Neumann-Morgenstern utility function, we begin by setting the utility attached to three monetary values arbitrarily.

____ 9. If James Johnson is indifferent between the certainty of a $50,000 gain and a gamble where there is a .4 probability of a $40,000 gain and a .6 probability of a $60,000 gain, the utility he attaches to a $50,000 gain equals $.4U(40) + 0.6U(60)$, where $U(40)$ is the utility he attaches to a $40,000 gain and $U(60)$ is the utility he attaches to a $60,000 gain.

____ 10. If a firm accepts a particular gamble, there is a .3 chance that it will gain $1 million and a .6 chance that it will lose $2 million. Based on this information, one cannot calculate the expected monetary value of this gamble to the firm.

Multiple Choice

1. Firm R buys a painting for $1,000 in 1994. If maintenance of the painting is costless, and if the interest rate is 0.10, how much must it sell the painting for in 1997 if the present value of the investment in the painting is to be nonnegative?
   a. At least $(1.10)^3 \times \$1,000$.
   b. At least (1.30) x $1,000.
   c. At least $1,000.
   d. At least $\$1,000 + (.30)^3 \times \$1,000$.
   e. None of the above.

2. The present value of a dollar received 10 years hence is ______ cents, if the interest rate is .04. (use Table 17.2)
   a. 96.154
   b. 97.007
   c. 92.456
   d. 67.556
   e. none of the above

3. If the present value of a dollar received six years hence is 70.496 cents, the interest rate (based on Table 17.2) must be
   a. 4 percent.
   b. 6 percent.
   c. 8 percent.
   d. 3 percent.
   e. none of the above.

3. This question pertains to the chapter appendix.
4. This question pertains to the chapter appendix.

4. The Monongahela Company will receive $952 in three years. If the interest rate is 8 percent, the present value of this amount is
   a. $812.
   b. $792.
   c. $683.
   d. $701.
   e. $756.

5. If a decision maker is risk neutral, and if the expected value of perfect information is $40,000, the decision maker should be willing to pay which of the following for perfect information?
   a. $35,000.
   b. $45,000.
   c. $55,000.
   d. $65,000
   e. $75,000.

6. If a decision maker is a risk averter for values of monetary gain above $10,000, then he or she
   a. must be a risk averter for values of monetary gain below $10,000.
   b. must be risk neutral for values of monetary gain below $10,000.
   c. must be a risk lover for values of monetary gain below $10,000.
   d. must be either *a* or *b*.
   e. none of the above.

7. John Jerome rolls a true die. If the die comes up a 2, he receives $6,000; if it does not come up a 2, he pays $1,000. The expected monetary value of this gamble to him is
   a. $1,167.
   b. $833.
   c. $167.
   d. –$167.
   e. –$833.

8. John Jerome's von Neumann-Morgenstern utility function is $U = 10 + 3M$, where U is utility and M is monetary gain (in dollars). He will prefer the certainty of a gain of $20 over a gamble where there is a .3 probability of a $9 gain and a .7 probability of
   a. a $31 gain.
   b. a $29 gain.
   c. a $27 gain.
   d. a $25 gain.
   e. none of the above.

9. In the previous question, if John Jerome has a .4 probability of receiving $100 and a .6 probability of losing $200, the expected utility of this gamble is
   a. –70.
   b. –80.
   c. –230.

d. –250.
e. none of the above.

10. Because of adverse selection, the following is likely to occur:
a. an increase in insurance prices.
b. an increase in the proportion of insurance buyers who are high-risk.
c. a reduction in the proportion of insurance buyers who are low-risk.
d. all of the above.
e. none of the above.

11. A firm must decide whether or not to increase its price. If its competitors will match the price increase, it will gain $2 million. If they do not match the increase, it will lose $4 million. The expected monetary value if it increases its price is –$1.6 million. The probability that its competitors will match the price increase is
a. 0.
b. .4.
c. .6.
d. 1.
e. none of the above.

12. In the previous question, the firm will neither gain nor lose if it does not increase its price. Thus the expected monetary value if the firm does not increase price is
a. $800,000
b. –$1,600,000.
c. –$2,400,000.
d. zero.
e. none of the above.

Review Questions

1. Ann Millaway's budget line is as follows:
a. What is the interest rate? Why?

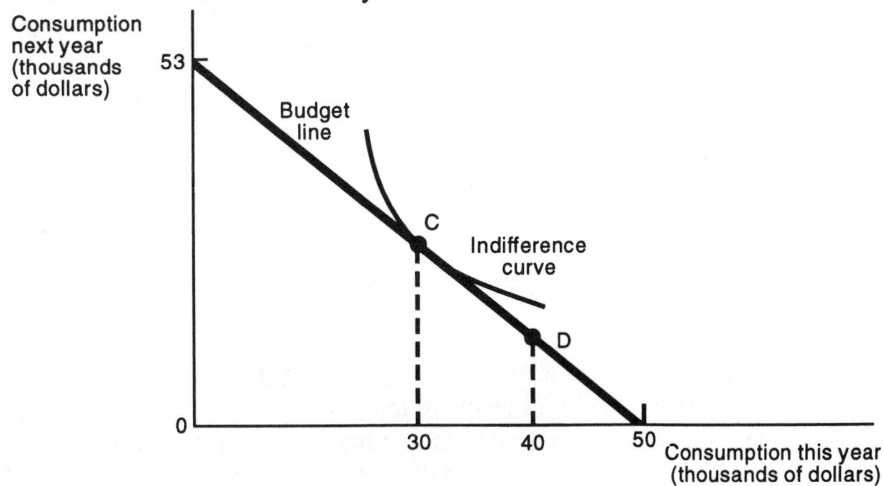

b. Ms. Millaway's endowment position is at point *D*. How much will she save this year? Why?

c. If the interest rate were 10 percent, would her budget line still go through point *D*? Why, or why not?

2. The supply and demand for loanable funds are as shown below. What is the equilibrium level of the interest rate?

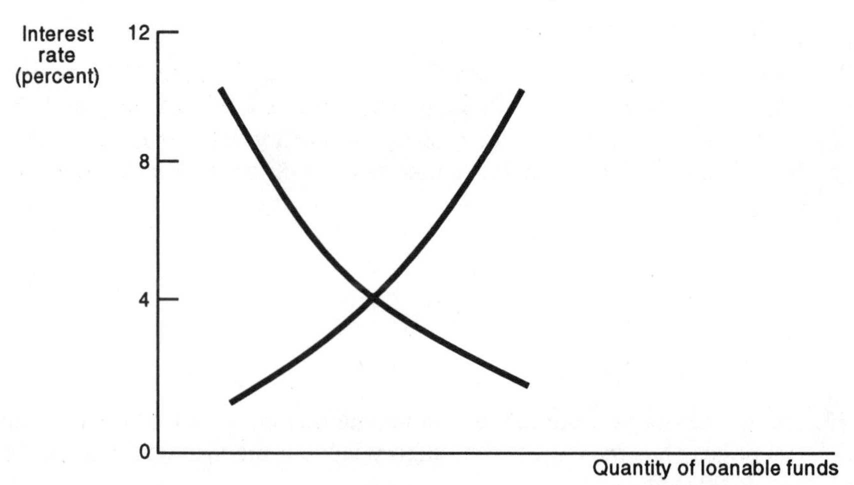

3. The Jackson Company will receive $1,000 two years hence and $2,000 three years hence. If the interest rate is 10 percent, what is the present value of this stream of income?

4. John Jones's utility function is shown below.

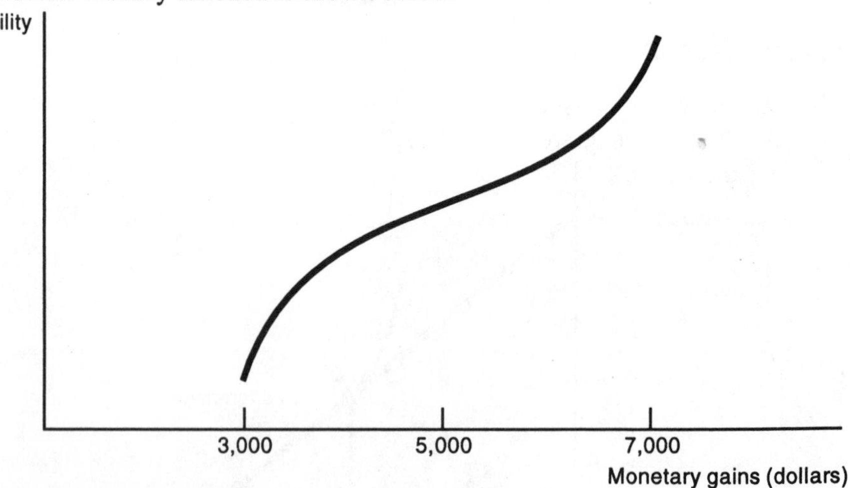

a. Is he a risk averter at all levels of monetary gain?

b. Is he a risk lover at all levels of monetary gain?

c. Does he prefer the certainty of gaining $4,000 over a gamble where there is a .5 probability of gaining $3,000 and a .5 probability of gaining $5,000?

5. If each card in a standard 52-card deck of playing cards has the same probability of being drawn, what is the probability that a single card that is drawn will be
   a. a red 9?
   b. a black ace?
   c. a black card?
   d. neither a red card nor a queen?
   e. a queen, king, or ace?
6. Margaret Murphy is selling her house. She believes that there is a .2 chance that each person who inspects the house will purchase it. What is the probability that no more than two people will have to inspect the house before she finds a buyer? (Assume that the probability that each person who inspects the house would buy it is unaffected by the decisions of persons who inspected it previously.)
7. In the previous problem, what is the probability that more than two people will have to inspect the house before Ms. Murphy finds a buyer?
8. A true die is rolled twice.
   a. What is the probability of getting a total of 6?
   b. Is this a subjective probability?

Problems

1. A government bond pays $10 per year and has no due date; that is, the interest payment of $10 per year goes on forever. If the interest rate is 0.05, how much would you be willing to pay for the bond?
2. A government program involves the training of disadvantaged teenagers. The cost of training each person is $8,000. With this training, he or she makes $500 more each year than he or she otherwise would. If this income differential is an adequate measure of the social benefit, and if the interest rate is 10 percent, is this program worthwhile?
3. John Miller is trying to decide whether to install aluminum siding on his house. As matters stand, his house has wood siding, which must be painted every year at an annual cost of $500. If he installs aluminum siding now, it will cost S2,000, and the need for painting will be eliminated. If he does not install aluminum siding, he will paint the house now, a year from now, 2 years from now, 3 years from now, and 4 years from now. Then he plans to sell the house. He believes that he will be able to get no more for the house if he installs the aluminum siding than if he does not. If Mr. Miller can get 8 percent on alternative investments, should he install the aluminum siding?
4. The Milwaukee Manufacturing Company is trying to determine whether to buy a type A or type B machine tool. The price of a type A machine tool is $80,000, while the price of a type B machine tool is $50,000. Each machine tool will last 6 years, after which its scrap value will be zero. Each machine tool produces the same quantity and quality of output, but the type A machine tool requires 2,000 hours of labor per year, whereas the type B machine tool requires 5,000 hours of labor per year.

a. If the interest rate is 10 percent, how big must the hourly wage of labor be for the present value of the savings in labor costs of the type A machine (relative to the type B machine) to exceed the extra investment in the type A machine (relative to the type B machine)?
b. Based on the above data, can we be sure that the firm should buy either machine tool?

5. If firm Z buys a particular machine in 1994, the effect will be that (1) the firm's cash inflow will be reduced by $10,000 in 1994, and (2) the firm's cash inflow will be increased by $2,000 annually in 1995–2001. If the interest rate is .10, should the firm buy this machine?

6. Firm X must determine whether or not to add a new product line. If the new product line is a success, firm X will increase its profits by $1 million; if it is not a success, its profits will decrease by $0.5 million. Firm X feels that the probability is .6 that the new product line will be a success and .4 that it will not be a success.
   a. If Firm X is risk neutral, should it add the new product line?
   b. How would you determine whether firm X is risk neutral?
   c. What is the expected value of perfect information to Firm X?

7. A newspaper publisher in a small town must decide whether or not to publish a Sunday edition. The publisher thinks that the probability is .7 that a Sunday edition would be a success and that it is .3 that it would be a failure. If it is a success, he will gain $200,000. If it is a failure, he will lose $100,000.
   a. If the publisher is risk neutral, should he publish the Sunday edition?
   b. What is the expected value of perfect information?
   c. How would you go about trying to determine whether the publisher is in fact risk neutral?

8. A Honolulu restaurant owner must decide whether or not to expand his restaurant. She thinks that the probability is .6 that the expansion will prove successful and that it is .4 that it will not be successful. If it is successful she will gain $100,000. If it is not successful, she will lose $80,000. What should her decision be if she is risk neutral?

9. In the previous problem, would the restaurant owner's decision be altered if she felt that:
   a. the probability that the expansion will prove successful is .5, not .6?
   b. the probability that the expansion will prove successful is .7, not .6?
   c. What value of the probability that the expansion will prove successful will make the restaurant owner indifferent between expanding and not expanding the restaurant?

10.[5] Mary Malone says that she is indifferent between the certainty of receiving $10,000 and a gamble where there is a .5 chance of receiving $25,000 and a .5 chance of receiving nothing. In the graph on the next page, plot three points on her utility function.

5. This problem pertains to the chapter appendix.

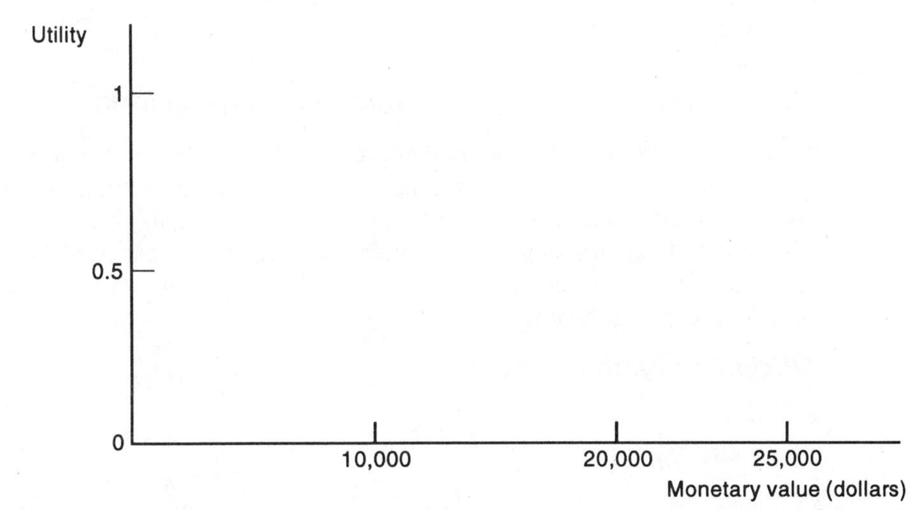

11. The Tremont Corporation is considering the purchase of a firm that produces tools and dies. Tremont's management feels that there is a 50–50 chance, if Tremont buys the firm, that it can make the firm into an effective producer of auto parts. If the firm can be transformed in this way, Tremont believes that it will make $1 million if it buys the firm; if it cannot be transformed in this way, Tremont believes that it will lose $2 million. What is the expected monetary value to Tremont of buying the firm?

12. In fact, the Tremont Corporation decides to purchase the firm described in the previous question. Does this mean that Tremont's management is (a) risk neutral, or (b) risk averting?

13. Banker William Martin says that he is risk neutral. Suppose that we let zero be the utility he attaches to $100,000 and 1 be the utility he attaches to $200,000. If what he says is true, what is the utility he attaches to
    a. $300,000?
    b. $50,000?
    c. –$10,000?

Key Concepts for Review

Rate of interest	Probability
Lending	Expected monetary value
Borrowing	Expected value of perfect information
Endowment position	von Neumann-Morgenstern utility function
Saving	Expected utility
Investment demand curve	Risk averter
Supply curve of loanable funds	Risk lover
Demand curve of loanable funds	Risk neutral
Present value	* Adverse selection
Discount rate	* Moral hazard
Net-present-value rule	

* This concept pertains to the chapter appendix.

ANSWERS

Case Study: How One Firm Made Investment Decisions

a. In crude terms, both seem to attempt to promote the wealth (or something like it) of the firm. The firm's executives are supposed "to use the capital they are allocated by the board to obtain the best possible return on it."

b. This firm used the ratio of net profit (after tax) to investment as a measure of the desirability of an investment project, whereas the net-present-value rule uses the net present value of the project.

Completion Questions

1. positive
2. interest rate
3. slope
4. downward
5. supply curve for loanable funds
6. equilibrium level
7. present value
8. linear
9. 250/1,500 = 1/6
10. 1,000/1,500 = 2/3
11. 200/1,500 = 2/15
12. does not
13. moral hazard

True or False

1. False	2. False	3. False	4. False	5. False	6. False
7. True	8. False	9. True	10. True		

Multiple Choice

1. *a*	2. *d*	3. *b*	4. *e*	5. *a*	6. *e*
7. *c*	8. *e*	9. *c*	10. *d*	11. *b*	12. *d*

Review Questions

1. *a.* 6 percent, because the slope of the budget line equals –1.06 (and it must equal minus 1 times 1 plus the interest rate).
   b. \$10,000, because she receives \$40,000 this year and wants to spend only \$30,000.
   c. Yes, because the budget line always goes through the point representing the endowment position.

2. 4 percent.

3. $$\frac{1{,}000}{(1.10)^2} + \frac{2{,}000}{(1.10)^3} = .82645(1{,}000) + .75131(2{,}000) = \$2{,}329.07$$

4. *a.* No.

b. No.
c. Yes.

5. *a.* 1/26.
   b. 1/26.
   c. 1/2.
   d. 6/13.
   e. 3/13.

6. The probability that the house is bought by the first person who inspects it equals .20. The probability that the first person will not buy but that the second person will buy it equals (.8)(.2) = .16. Thus, the probability that either the first or second person will buy it equals .20 + .16 = .36.

7. 1 − .36 = .64.

8. *a.* 5/36.
   b. No.

Problems

1. $200.

2. If the benefit continued indefinitely, its present value would be $500 ÷ .10, or $5,000. This is an overestimate, since the benefit will not continue indefinitely. Since the cost is $8,000, the investment is not worthwhile.

3. The present value of painting costs is

$$\$500(1 + 1/1.08 + 1/(1.08)^2 + 1/(1.08)^3 + 1/(1.08)^4) =$$
$$\$500(1 + 0.926 + 0.857 + 0.794 + 0.735) = \$500(4.312) = \$2{,}156$$

This must be compared with $2,000, the cost of installing aluminum siding now. Since the latter is less than the former, it is cheaper to install the aluminum siding.

4. *a.* The annual saving in labor costs with the type A machine equals $3{,}000W$, where W is the hourly wage rate. If the annual saving in labor costs occurs 1, 2, . . ., 6 years hence, and if the extra $30,000 (that must be paid for the type A machine) must be paid now, the extra investment in the type A machine is less than the present value of the savings in labor costs if

$$\$30{,}000 < 3{,}000W\left(\frac{1}{1.1} + \frac{1}{1.1^2} + \frac{1}{1.1^3} + \frac{1}{1.1^4} + \frac{1}{1.1^5} + \frac{1}{1.1^6}\right)$$
$$< 3{,}000(.9091 + .8264 + .7513 + .6830 + .6209 + .5645)W$$
$$< 13{,}065.6W$$

or if $W > \$30{,}000/13{,}065.6 = \$2.30$. Thus, if the wage rate exceeds $2.30, this will be true.

   b. No. No data are presented to indicate the profitability of the investment in either machine tool, relative to the profitability of other uses of the funds.

5. The net present value of the investment is

$$-10{,}000 + \frac{2{,}000}{1.10} + \frac{2{,}000}{1.21} + \frac{2{,}000}{1.331} + \frac{2{,}000}{1.464} + \frac{2{,}000}{1.610} + \frac{2{,}000}{1.771} + \frac{2{,}000}{1.948}$$
$$= -10{,}000 + 1{,}818 + 1{,}653 + 1{,}503 + 1{,}366 + 1{,}242 + 1{,}129 + 1{,}027$$

$= -263$ dollars

Thus the firm should not buy the machine.

6. *a.* Yes.
   b. Construct the von Neumann-Morgenstern utility function of the decision maker.
   c. $200,000

7. *a.* He should publish the Sunday edition.
   b. .7($200,000) + .3(0) – $110,000 = $30,000.
   c. Construct his von Neumann-Morgenstern utility function.

8. Since the expected monetary value if she expands is .6 ($100,000) + .4 (–$80,000) = $28,000, she should expand the restaurant.

9. *a.* No.
   b. No.
   c. 4/9.

10. Let zero be the utility of receiving nothing, and 1 be the utility of receiving $25,000. Then, since she is indifferent between the certainty of receiving $10,000 and the gamble described in the problem, the utility of receiving $10,000 must equal

$$.5(1) + .5(0) = .5$$

Thus, three points on her utility function are given below.

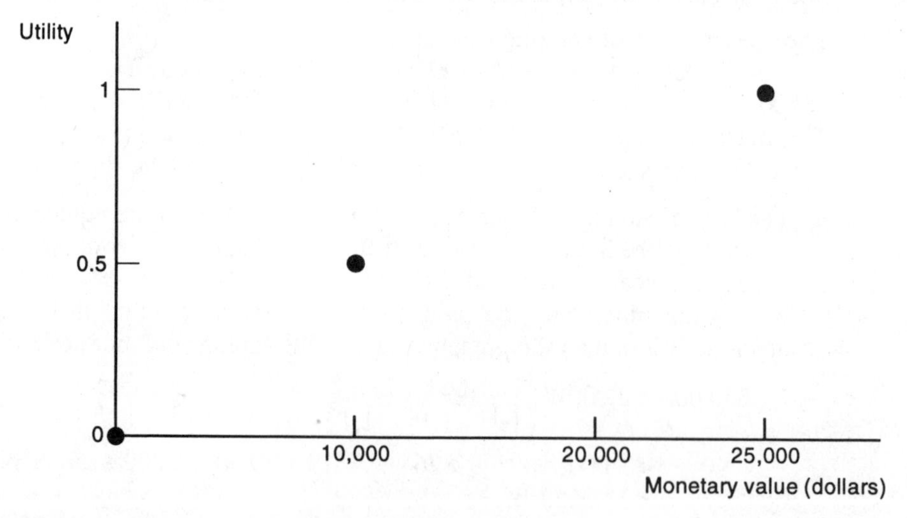

11. .5($1 million) + .5(–$2 million) = –$0.5 million.

12. *a.* No, since it would have chosen not to purchase it if it were risk neutral.
    b. No, since it would have chosen not to purchase it if it were risk averse.

13. *a.* 2.
    b. –0.5.
    c. –1.1.

Case 4 Airbus Industrie: a Wave of the Future

Shaker Zahra
Daniel Hurley, Jr.
John Pearce II

> Like Americans, we Europeans are fiercely proud of our independence. We resent being told what industry we should be in or out of. We do not appreciate being told that we should stick "to building trains and things like that," as was hinted at recently by one of our competitors. Imagine the outcry if a European had remarked that the United States should stick to bottling soft drinks, or producing episodes of "Dallas" or "Dynasty" simply because you are good at "things like that."
>
> — Jean Pierson, CEO & president, Airbus Industrie

HISTORY

In the two decades following World War II, U.S. companies produced most of the free world's commercial aircraft, while British manufacturers made noticeable, but unsustained, inroads into the near U.S. monopoly. During the early 1960s, however, the majority of Europe's aircraft manufacturers began to study projects relating to a high-capacity, short- or medium-range commercial transport. This was potentially a very large market since about 70 percent of the world's air traffic flew routes less than 2,500 nautical miles.

Intent on seizing this opportunity, the aircraft industries of France, West Germany, and Great Britain entered into a preliminary agreement in 1967, defining the design of the airframe and engines of an aircraft to meet the immediate need. One airframe and one engine manufacturer were selected from each country to implement the agreement. The airframe participants were Sub-Aviation (France), Hawker Siddeley Aviation (Great Britain), and Deutsche Airbus (a joint company formed within the German aircraft manufacturing industry). The aircraft engine manufacturers included SNECMA of France, Rolls-Royce of Great Britain, and MTU of Germany. The British government and Rolls-Royce withdrew shortly thereafter to concentrate on participation in Lockheed's L-1011 aircraft. Hawker Siddeley Aviation, however, remained associated with the Airbus consortium.

The initial project, designated the A300 (denoting the objective seating capacity), had the advantage of meeting the needs of a market gap that had been neglected by the U.S. manufacturers. The project also gave the Europeans a decided 10-year technological lead over their major U.S. rivals.

In September 1967, the French government created a new form of business entity, known as a "Groupement d'Intérêt Économique" (GIE). Its purpose was to satisfy the member companies' desires to retain freedom to pursue certain ventures independently while working together on the Airbus project.

Airbus Industrie was formally established under French law on December 18, 1970, using the GIE business arrangement. It reflected the need for European companies to band together, both to design a broad product line of advanced design transport aircraft and to market these aircraft effectively against established competition in a highly competitive market. The members of Airbus Industrie were:

1. Société Nationale Industrielle Aerospatiale SA of France, known as Aerospatiale: 37.9 percent interest.
2. Deutsche Airbus GmbH: 37.9 percent interest.
3. British Aerospace PLC: 20 percent interest.
4. Construcciones Aeronauticas SA (CASA) of Spain: 4.2 percent interest.

EXHIBIT 1

AIRBUS INDUSTRIE MEMBERSHIP

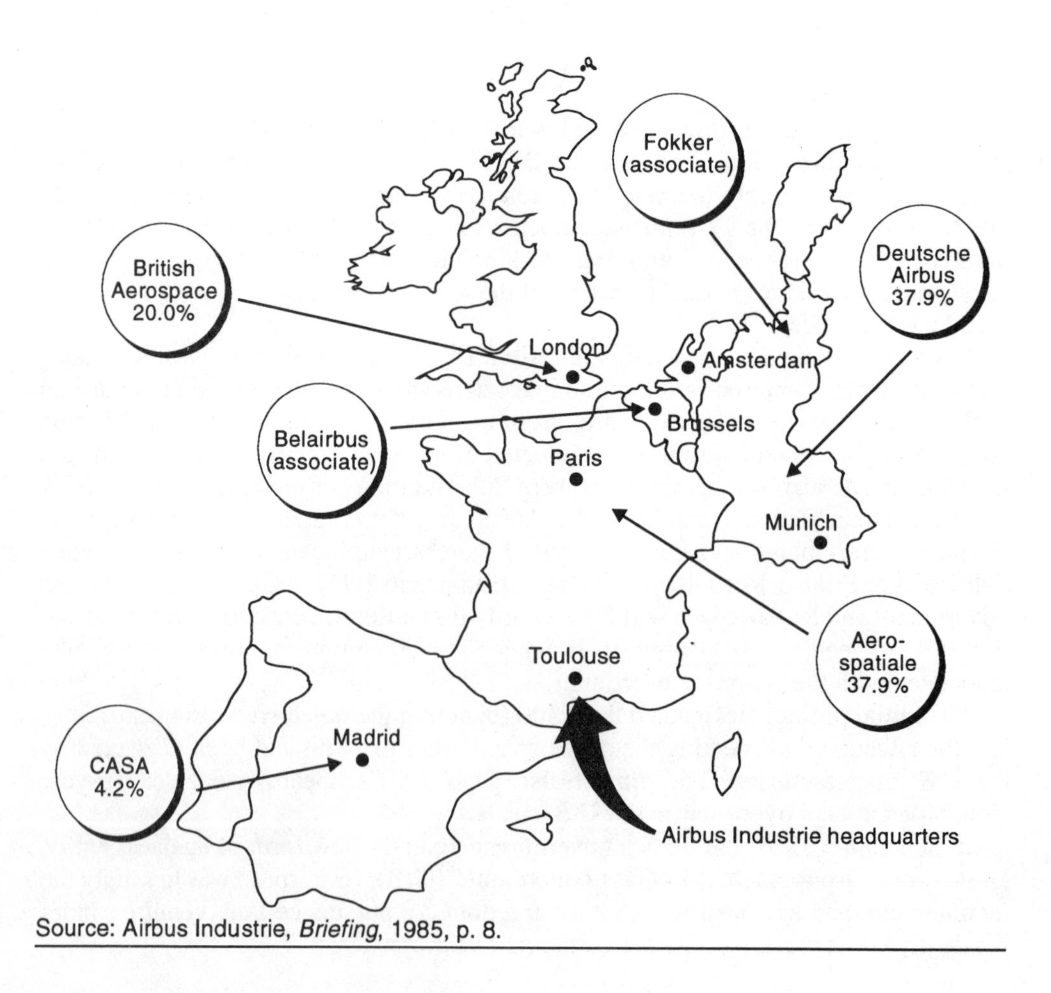

Source: Airbus Industrie, *Briefing*, 1985, p. 8.

EXHIBIT 2
INTERRELATIONSHIPS AMONG THE AIRBUS CONSORTIUM MEMBERS

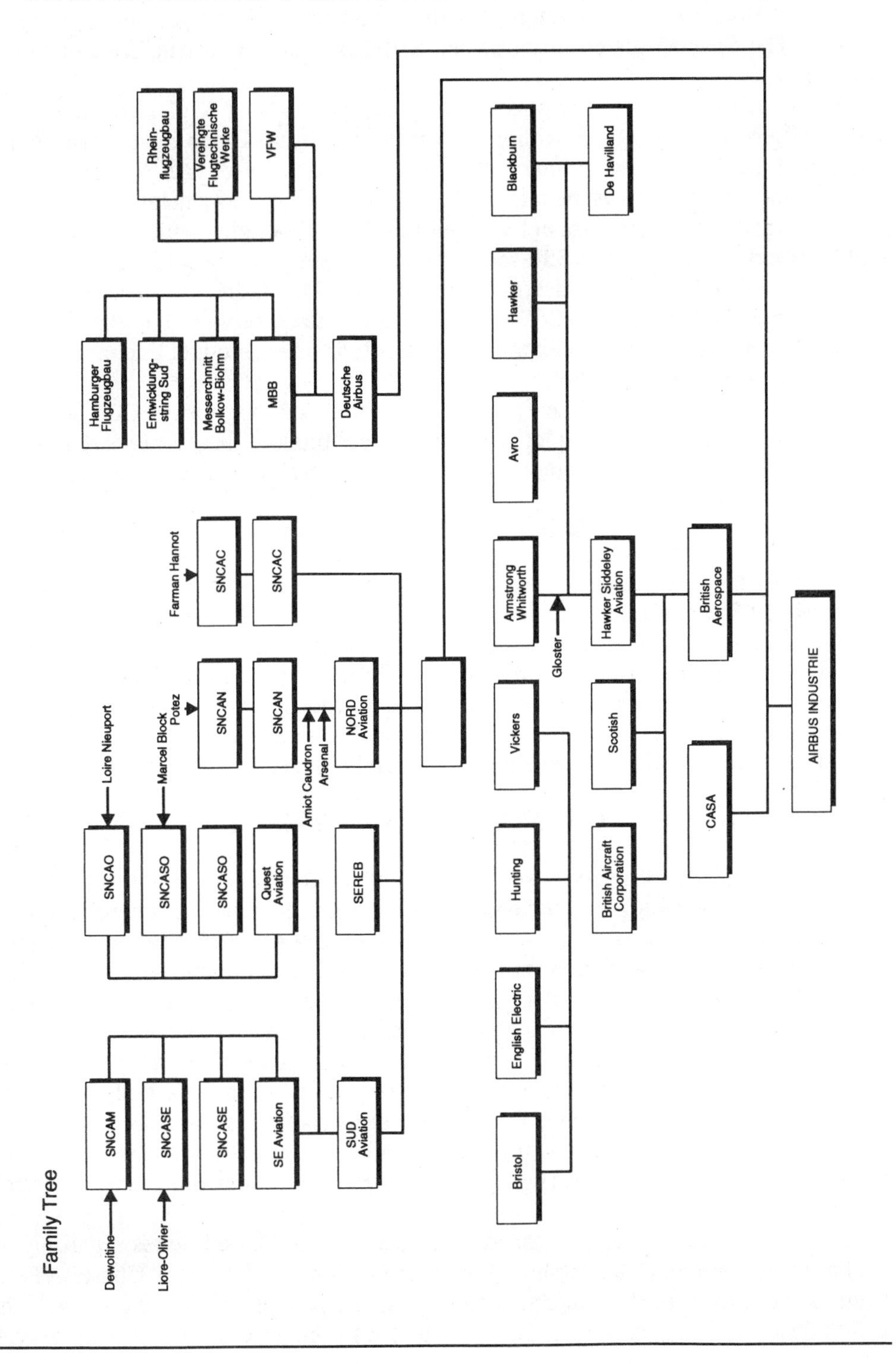

Interestingly, the French government owned over 97 percent of the shares of Aerospatiale. The British government and private investors held equal 48.43 percent portions of British Aerospace interest, with the company's employee holdings at 3.14 percent. The Spanish government owned a 100 percent interest in Construcciones Aeronauticas.

Airbus Industrie also had two associate members: Fokker NV of Holland and The Belairbus group of Belgium, providing for participation by the Dutch and Belgian aircraft industry. Exhibit I depicts primary and associate members of the Airbus Industrie consortium. Exhibit 2 shows the interrelationships among Airbus's membership.

Airbus Industrie's first product was the A300, a twin-engine wide-body commercial transport having a 240- to 350-seat capacity. It went into scheduled service in May 1974. In April 1983, Airbus Industrie's second aircraft, a smaller but more technologically advanced A310 (210- to 265-seat capacity), entered service. The A320 entered scheduled service in 1988 and expanded Airbus Industrie's product line into the 135- to 179-seat, single-aisle market.

With the arrival of the 300 seat capacity, medium- and long-range A330 and A340 models, Airbus Industrie completed its five-model product line. It appeared that by the early 1990s Airbus would compete directly with Boeing's full-range family of aircraft.

MISSION AND GOALS

The long-term objectives of Airbus Industrie centered on achieving a sufficient volume of business to become self-sustaining and profitable. Practically speaking, this translated into a goal of obtaining an overall 30 percent market share worldwide for aircraft with seating capacity of greater than 100 seats by the mid-1990s.

Pursuing these objectives and goals, Airbus Industrie adopted an outlook of designing aircraft to meet the needs of airlines worldwide. As a result, Airbus Industrie placed equal emphasis on revenue generation and reduced operating costs.

CORPORATE STRATEGY

Airbus adopted a strategy based on product differentiation combined with low price. The company sought to differentiate itself from the other commercial aircraft manufacturers by emphasizing technological innovations in airframe design, engine technology, airfoil design, new materials, and advanced digital avionics. In fact, Airbus Industrie developed an impressive list of technological accomplishments. These included twin engine, twin-aisle configuration, advanced rear-loaded airfoil, full flight regime autothrottle, and automatic wind-shear protection. Other innovations were a digital autoflight system and a two-man cockpit on a wide-body, advanced cathode ray tube cockpit displays with an electronic centralized aircraft monitor, and electronic signaling of all secondary control. Plans for future innovations included a second-generation digital autoflight system, fly-through controller, and side-stick controller.

These many achievements intensified concern in the United States regarding its declining technological leadership; For example, former astronaut Charles (Pete) Conrad was quoted in the August 3, 1987, issue of *Time* magazine as saying, "The U.S. industry is falling behind. It disturbs me that the highest technology commercial ship about to fly is the A320—not a U.S.-made plane. In this world, technology sells airliners, and we have to get off our butts fast."

Airbus Industrie has also championed techniques leading to significant reductions in operating costs, using lighter weight, composite materials incorporating the variable camber feature already mentioned and, particularly, in designing aircraft that required only two crew members on the flight deck.

In addition to its strategy of technology-based differentiation, Airbus Industrie simultaneously embarked on a unique pricing system. The company set prices that were based on its costs but were comparable to prices charged by its principal competitor, Boeing. For example, Airbus Industrie sold aircraft for as low as $35 million to $40 million. This price was well below the going market rate of $50 million to $60 million. Consequently, some competitors complained that prices charged by Airbus Industrie did not cover either the development and production costs or any profit. Airbus Industrie explained that its aircraft were priced to show a profit once the particular model's "break-even point" was reached. Airbus Industrie's explanation was challenged because of the practice of some European governments of providing massive subsidies to the consortium.

Unlike its successful strategy of differentiation, Airbus Industrie's pricing created considerable controversy not only within the civil aviation community but also at the international level. These practices prompted the U.S. government to question whether Airbus Industrie's aircraft partners were benefiting unfairly from European government subsidies, in violation of the General Agreement on Tariffs and Trade (GATT).

In response to these allegations, Airbus Industrie introduced new production methods, which resulted in more efficient operations. For example, in contrast to conventional systems employing stationary shipbuilding techniques, Airbus Industrie made use of the excellent production facilities of its members and pioneered a system of "production sharing," involving prefabrication of complete subassemblies in which final assembly represented only a small portion of the total work on the airframe.

Aerospatiale was designed to manufacture the flight deck, the forward fuselage sections, the center fuselage/wing box section below the cabin floor, the lift dampers, and the engine pylons. MBB produced the major fuselage sections, the rudder, and the tailcone. British Aerospace manufactured the main wing box; Belairbus the slats; Fokker the ailerons, wing tips, main landing gear doors, and leg fairings; and MBB the flaps, spoilers, and flap track fairings. MBB also assembled the complete wing. CASA supplied the tailplane and elevators, the nose landing gear doors, and the forward cabin entry doors.

The completed subassemblies were flown by Super Guppy cargo aircraft to Toulouse, France, for final assembly and painting. The completed aircraft was tested at Hamburg, West Germany. There the aircraft was provided with customized internal furnishings. These production innovations saved time, reduced costs, and helped maintain quality. Approximately six weeks elapsed between the arrival of the subassemblies and the departure of the completed aircraft. Nearly 96 percent of the construction work was completed before the subassemblies were flown for final assembly.

Because of the close association with their respective governments and because of the European political tradition of state-supported or socialized industry, the parent governments provided the major portion of the investment capital and developmental financing. The size of this support was estimated at between U.S. $9 billion and U.S. $12 billion for the period from 1970 to 1987. These subsidies occurred in various forms, including direct grants, no-interest loans, deferred-payment loans, and government-guaranteed commercial loans. Most loans were tied to airplane sales, with a loan installment payment being made when each aircraft was sold.

Various European officials defended these subsidies as permissible under the GATT Agreement on Trade in Civil Aircraft. They suggested that these subsidies did not "distort the market," 70 percent of which was controlled by Boeing. Further, they pointed out that subsidies were equivalent to the U.S. government's practice of supporting U.S. manufacturers through military contracts.

Although both Boeing and McDonnell Douglas drafted petitions for relief under the unfair trade provisions of the Trade Act of 1974, neither filed a petition with the U.S. government, because they feared that the sanctions that would result could disrupt the market. For example, more than 30 percent of the construction of a Boeing 767 involved non-U.S. components, and this figure was expected to rise to almost 49 percent in the projected 7J7 program. Similarly, nearly 50 percent of the McDonnell Douglas MD-80 airframe was built outside the United States. Both manufacturers had significant markets in the Western European communities, which were likely to be jeopardized if countersanctions were invoked.

This debate on unfair pricing policies by Airbus was difficult, because the company and its consortium members had not published financial statements that covered company operations. Although pressured to ensure fuller disclosure of company data, Airbus never responded.

MARKETING PRACTICES

In marketing directly to airlines, Airbus Industrie vigorously pursued a "push" approach to aircraft sales. The consortium's member governments played an active role in marketing efforts through the use of export credits, marketing aid, and political influence. As a result, the company established a near monopoly for large aircraft manufacturing in the countries of its major partners.

Airbus often sent mixed market messages, signaling head-to-head competition on one hand, yet stating openly that it sought only 30 percent of the civil transport market segment on the other hand. Exhibit 3 shows changes in the market shares of Boeing, McDonnell Douglas, and Airbus during the 1984–86 period.

ORGANIZATION

Since its inception, the Board of Control, known informally as the Supervisory Board, governed Airbus. Its 17 members were selected to represent equity holdings by member countries. Fokker and Belairbus representatives were permitted to attend meetings of the Supervisory Board as observers only. The board's activities included establishing the general philosophy toward projects, sales strategy, production planning, budget administration, and ensuring that work that could be performed by one of the partners was actually subcontracted to that partner.

The executive director, Administrateur Gerant, reported regularly to the Supervisory Board. His decisions and actions were binding.

Airbus Industrie's general meeting of members established an 81 percent majority vote as a basis for decision making. Because each of the three major Airbus Industrie partners acquired an equity interest of at least 20 percent, the voting requirement ensured that all three must agree on issues decided at the general meeting.

The auditors' annual financial statements were filed with the French tax authorities. The information was also given to Airbus Industrie members, who, in turn, incorporated the financial data on a pro rata basis into their respective individual annual

EXHIBIT 3

MARKET SHARE OF INDUSTRY LEADERS

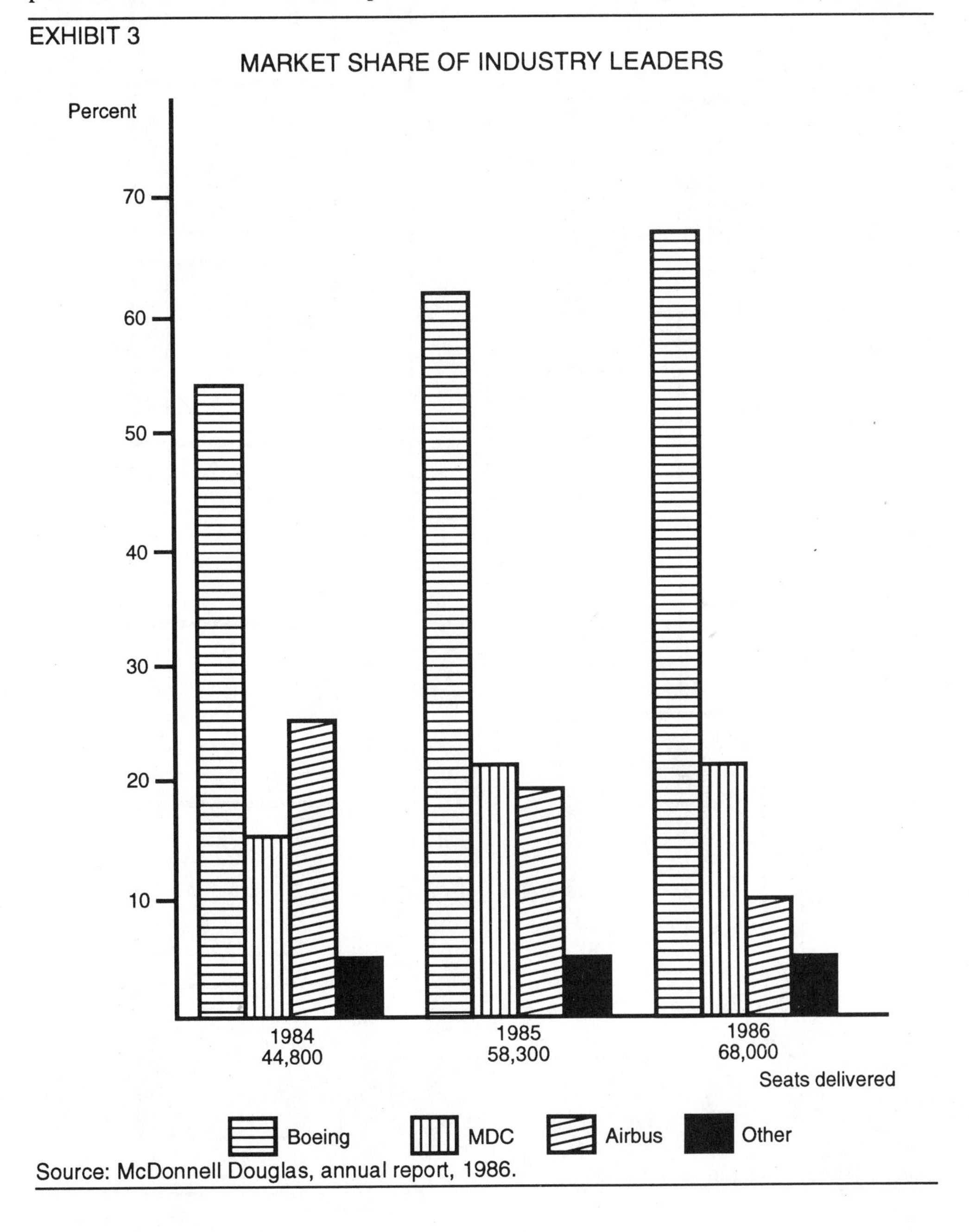

Source: McDonnell Douglas, annual report, 1986.

EXHIBIT 4

AIRBUS ORGANIZATIONAL CHART

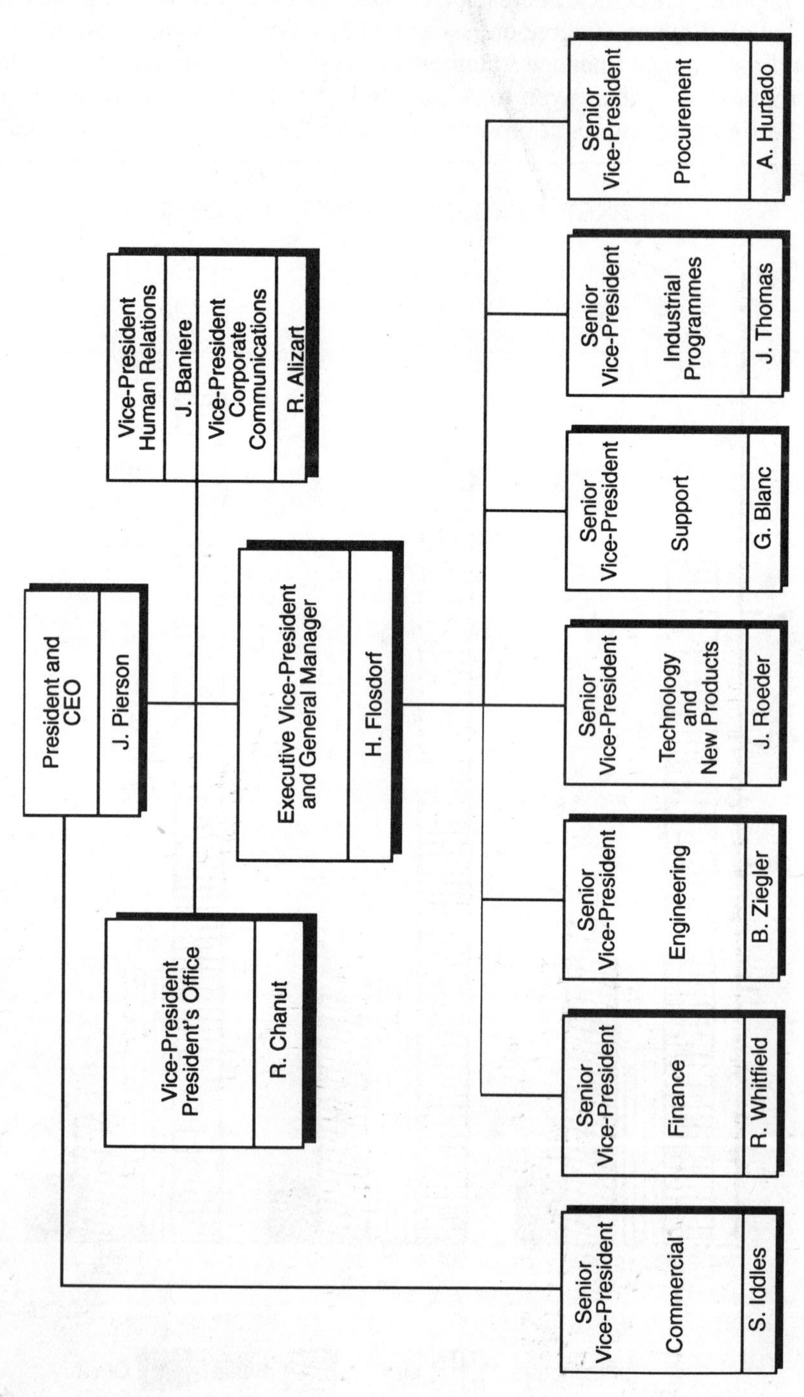

Source: Airbus Industrie, *Briefing*, 1986, p.11

reports, presenting corporate financial performance on a consolidated basis. As a result, financial statements for Airbus were not publicly available. Consequently, competitors and concerned foreign governments did not have access to vital financial data to measure Airbus's performance.

For its day-to-day management, Airbus Industrie employed an organizational structure headed by a chief executive officer and a president, assisted by the senior vice presidents in various functional positions. Exhibit 4 shows Airbus Industrie's 1988 organizational chart.

SUBSIDIARIES

In addition to the partner corporations comprising its ownership, Airbus established three subsidiaries. The first, Aeroformation, was founded in 1972 as a joint venture with Flight Safety International, Inc., of the U.S.A. Aeroformation served as the training center for crews flying Airbus aircraft. There were three training centers, located in Europe at Toulouse, France; in the Far East at Singapore; and in the United States at Dulles International Airport in Reston, Virginia.

The second subsidiary, Airbus Industrie of North America (AINA), was established as a wholly owned U.S. subsidiary incorporated in Delaware. AINA was created to serve as the company's marketing arm in North America, with offices located in New York and Reston, Virginia.

The third subsidiary, Airbus Airspares Center, provided aircraft service and parts delivery. Operating from a hub in Hamburg, West Germany, Airbus Airspares maintained depots in Washington, D.C.; Miami, Florida; and Hong Kong.

THE FUTURE

Airbus Industrie's marketing performance reflected two growing trends: the steady increase in the overall number of Airbus Industrie aircraft delivered or on order, and the relatively high percentage of "export" orders. Orders for Airbus Industrie's 300-seat aircraft models were nearly twice those for the comparable U.S. aircraft model in 1986 and 1987. Moreover, Airbus Industrie outsold the Boeing 767 in the export market by a 5:1 ratio.

TABLE 1

DEMAND FORECAST, 1986–2005

	FORECASTS 1986–2005		
	MDC*	AIRBUS	BOEING*
Average annual world passenger traffic growth (%)	6.4	5.5	5.1
New aircraft required for growth	5,327	3,659	3,913
Aircraft reqquirements	4,437	4,499	3,869
Total aircraft required	9,764	8,158	7,782

* Adjusted to a 1986–2005 basis.

Sources: "1985–1999 Outlook for Commercial Jet Aircraft," Douglas Aircraft Co., September 1985; "Global Market Forecast," Airbus Industrie, March 1986; and "Current Market Outlook—World Travel Demand and Airplane Economic Requirements," Boeing Commercial Airplane Co., 1986.

EXHIBIT 5

AIRBUS SALES FORECAST (MARKET SHARE, 1986–2005)

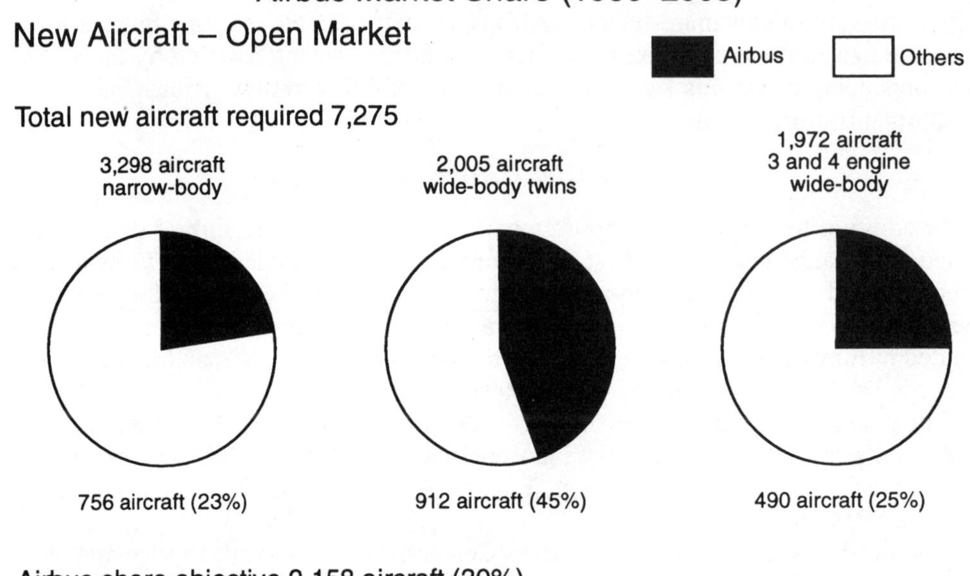

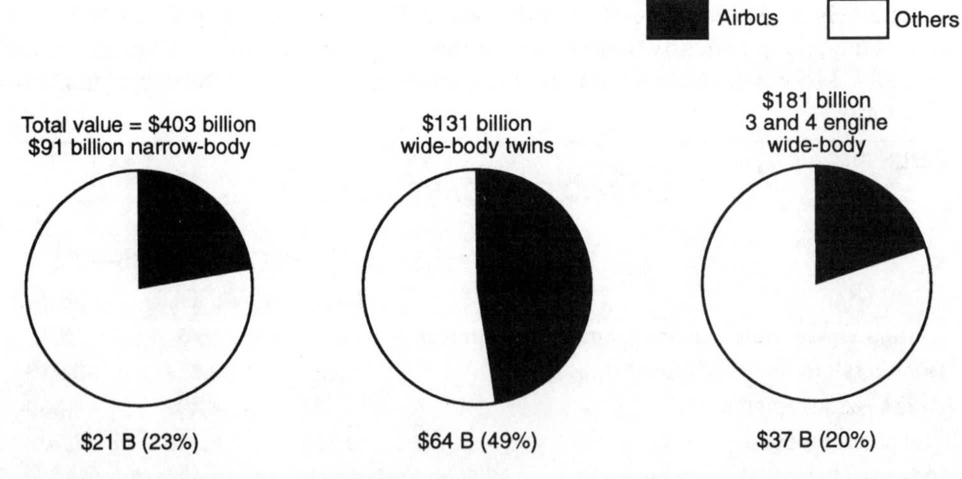

Source: Boeing and McDonnell Douglas, annual reports, 1986.

Airbus Industrie estimated that it should be able to achieve about a 30 percent market share in the markets it entered. Initial indications suggested that Airbus overestimated, by a wide margin, the size of its potential markets for 300- to 200-seat aircraft. The break-even point for the A300 and A310 models was estimated at 850. However, Airbus Industrie had delivered only 350 in almost 14 years of production.

Meanwhile, Boeing and McDonnell Douglas appeared to have accepted Airbus Industrie as a full-fledged member of the commercial aviation market. However, both companies opposed government subsidies to Airbus.

Given the technological leadership of Airbus, the company expected to do well, especially as the world's air carriers selected new and more efficient aircraft to replace their outdated fleets. Airbus Industrie's medium- and long-range aircraft were best suited for this market. In fact, Airbus's strategy revolved around achieving a 30 percent market share of worldwide demand. Table I shows Airbus's forecasts for the period from 1986 to 2005. Exhibit 5 also shows predicted Airbus's sales during the same period.

For this future to become a reality, Airbus Industrie and its partners had to take effective steps to make performance data for Airbus Industrie publicly available. Further, they had to adjust Airbus's pricing structure to more accurately reflect costs.

Airbus Industrie charted a unique strategy in a highly competitive market. Confident of support from its member countries, Airbus appeared poised to penetrate international markets and compete head on with the leading U.S. aircraft manufacturers. Its initial success raised concerns not only among U.S. companies but also among policymakers. As Airbus Industrie pursued its strategy, experts agreed that the company had changed the rules of the competitive game and made competition truly global in scope.[1]

Questions

1. In early 1993, President Bill Clinton made a speech in Seattle, Washington in which he said that unfair European government subsidies were costing jobs and profits in the U.S. aerospace industry. Do you believe that this was true?

2. President Clinton's remarks came after Boeing announced that it would eliminate over 25,000 jobs. Its customers, the world's airlines, had lost a great deal of money, partly because of price wars. In the United States alone, they lost over $9 billion in the previous three years. Thus, many of them had deferred or canceled their orders for aircraft; this prompted Boeing to reduce its production from about 40 airplanes per month in the summer of 1992 to about 21 in mid-1994. To what extent were the job losses at Boeing due to Airbus?

3. Between 1988 and 1993, Boeing had sold 707 aircraft in Europe while Airbus had sold 296 aircraft in the United States. Why did *Business Week* cite this fact as a possible reason why Boeing was less zealous than President Clinton in criticizing Airbus? (*Business Week*, March 15, 1993, p. 30).

4. Boeing's forecasts of aircraft requirements are presented in Table 1. In March 1993, Boeing cut back its forecasts through the year 2010. Estimated demand for

1. This case was published in Pearce and Robinson, op. cit. Reprinted with permission.

all jetliners was reduced by about 5 percent because of ". . . increasing airline productivity, slower traffic growth, and a slowing rate of replacement of older jets by all airlines."[2] Why would increasing airline productivity shift the demand curve for aircraft to the left?

5. Boeing has tried hard to improve efficiency. For example, Ronald Woodard, a division manager, cut inventories by $100 million in 1992. (According to *Business Week*,[3] Boeing traditionally overstocked to prevent shortages.) Woodard claims that "inventory is evil." In what sense is this true?
6. Boeing's announced layoffs caused great concern in the Seattle area where Boeing is the leading employer. In 1993, it employed 98,000 people in the Seattle area; about 1 in 10 jobs in the area were at Boeing.[4] Is Boeing a monopsonist with regard to engineering labor? Why or why not?
7. Suppose that Airbus is the only buyer of a particular item and that the demand, supply, and marginal expenditure curves for this item are such that

$$Q_D = 20 - 1.5P$$

$$P = 2Q_S$$

$$ME = 4Q_S$$

where Q_D is the quantity demanded (in thousands of units per month), Q_S is the quantity supplied (in thousands of units per month), ME is the marginal expenditure (in dollars per unit), and P is the price (in dollars per unit). Calculate the deadweiqht loss from monopsony.

2. *Wall Street Journal*, March 5, 1993, p. A5D
3. *Business Week*, March 1, 1993, p.62
4. *New York Times*, February 13, 1993.

Part 6 Economic Efficiency

Chapter 19 Economic Efficiency, Externalities, and Public Goods

Case Study: The Lackawanna Social Club

The Lackawanna Social Club, which contains 20 resident members, keeps a refrigerator stocked with soda and beer. Each can of soda or beer is obtained from a local distributor for 50 cents per can. Each member has free access to the refrigerator and can consume as many cans of soda or beer as he or she likes. At the end of each month, the total cost of the soda and beer is divided evenly among the members.

a. If one member of the club drinks a can of beer, what is its cost to him or her?
b. What is the cost of this beer to the rest of the members of the club?
c. From the point of view of the club, what is the total social cost of the beer?
d. Is the amount of beer and soda consumed by the club's members likely to depart from the amount that is socially efficient? If so, why?
e. What would be the social advantages and disadvantages of having each member buy and store his or her own soda and beer?
f. What would be the social advantages and disadvantages of having a bartender with a key to the refrigerator who will open it during certain hours and sell soda and beer at 50 cents per can?
g. What would be the social advantages and disadvantages of having the refrigerator open and asking each member to sign on a sheet (and later pay) for each can of soda or beer he or she consumes?
h. If the Lackawanna Social Club contained 400 resident members, rather than 20, do you think that the relative advantages and disadvantages of these and other ways of handling this problem would be unaffected? Explain.
i. Do you think that the sort of problem taken up in this case study arises frequently within households? Explain.

Completion Questions

1. If resources are allocated efficiently, the amount of money a consumer will give up to obtain an extra unit of water must be (different, the same) ____________ for all consumers.

2. Partial equilibrium analysis is adequate in cases where the effect of a change in market conditions in one market has _______________ repercussion on prices in other markets.

3. The contract curve is an efficient set of points in the sense that, if the consumers are off the contract curve, it is always preferable for them to _____________.

4. There is _____________ on which we can validly measure pleasure or pain so that interpersonal comparisons can validly be made.

5. Whether or not one income distribution is better than another is a___________.

6. The marginal conditions for efficient resource allocation imply that consumers are on their ____________.
7. The marginal conditions for efficient resource allocation imply that producers are on their ____________.
8. If there were only one consumer, the marginal conditions for efficient resource allocation would imply that his or her indifference curve be tangent to the ____________.
9. Under perfect competition, the marginal rate of substitution between two goods is the same for any pair of consumers because they pay ____________ for the goods.
10. Under perfect competition, the marginal rate of technical substitution between two inputs is the same for any pair of producers because they pay __________ for the inputs.
11. Public goods generally (are, are not) ____________ sold in the private marketplace.
12. A "free rider" is ____________.
13. The demand curve for a ____________ good will not be revealed.

True or False

____ 1. If commodities are allocated efficiently, the marginal rate of substitution between a pair of commodities must be the same for all consumers that consume both.

____ 2. If commodities are allocated efficiently, the marginal rate of substitution between any pair of commodities must be the same for all pairs of commodities.

____ 3. According to the Pareto criterion, a change that harms no one and improves the lot of at least one person is an improvement.

____ 4. According to the Pareto criterion, a dollar taken from a rich man and given to a poor man is an improvement.

____ 5. Economic efficiency does not require the fulfillment of all of the marginal conditions for efficient resource allocation.

____ 6. Perfect competition results in an efficient allocation of resources. Public goods will be allocated the right amount of resources without government intervention or other nonmarket mechanisms.

____ 7. Tickets to a concert by the Cleveland Symphony Orchestra are an example of a public good.

____ 8. Fifty independent fishing companies attempt to catch fish in the same lake. One of these companies buys up all of the others. The number of fishermen sent out each day by the one company will be the same as the number that was sent out by the fifty companies.

____ 9. Public goods will be provided in the right amount by the market mechanism.

___ 10. Goods that are rival are called public goods.

___ 11. An uncrowded bridge is an example of a nonrival good.

___ 12. National defense is not a public good.

___ 13. For a public good, the market demand curve is the horizontal summation of the individual demand curves.

___ 14. In large part, environmental pollution is due to external economies.

___ 15. The price paid by polluters for water and air is often less than the true social costs.

___ 16. Pollution is tied inextricably to national output.

___ 17. The only way to reduce pollution is to reduce population.

Multiple Choice

1. No market can adjust to a change in conditions without
   a. a change in other markets.
   b. other markets remaining unchanged.
   c. government aid.
   d. violating the assumption of general equilibrium analysis.
   e. none of the above.

2. Compared with a certain point on the contract curve, a point off the contract curve
   a. may be better.
   b. is certainly better.
   c. is never better.
   d. is always equally good.
   e. none of the above.

3. If a point is off the contract curve, we can find
   a. a point on the contract curve that is better.
   b. no point on the contract curve that is better.
   c. a point on the contract curve that is better, if and only if the contract curve is a straight line.
   d. a point on the contract curve that is better, if and only if the contract curve goes through the origin.
   e. none of the above.

4. A road is built which is maintenance-free and which is so wide (relative to the available traffic) that no person or car interferes with the movement of another person or car. The cost of the road is entirely a fixed cost. If marginal cost pricing is adopted
   a. the receipts from the road will be no less than the road's total profit.
   b. the receipts from the road will not depend on the number of autos using the road.
   c. the receipts from the road will be no greater than the road's fixed cost.
   d. all of the above.
   e. none of the above.

5. The marginal conditions for efficient resource allocation
   a. depend on interpersonal comparisons of utility.
   b. help to determine the optimal distribution of income.
   c. are silent concerning the optimal distribution of income.
   d. all of the above.
   e. none of the above.
6. Marginal cost pricing is automatically the rule under
   a. monopoly.
   b. oligopoly.
   c. monopolistic competition.
   d. perfect competition.
   e. none of the above.
7. If a perfectly competitive industry is monopolized, the result is that
   a. the marginal conditions for efficient resource allocation are no longer fulfilled.
   b. the marginal conditions for efficient resource allocation are still fulfilled.
   c. more of the marginal conditions for efficient resource allocation are fulfilled than was formerly the case.
   d. the marginal conditions for efficient resource allocation are no longer applicable.
   e. none of the above.
8. Which of the following is not a public good?
   a. A smog-free environment.
   b. National defense.
   c. Blood donated to the Red Cross.
   d. The Apollo space program.
   e. All of the above.
9. Technological change
   a. has made people more interdependent.
   b. has brought about more and stronger external diseconomies.
   c. has resulted in some harmful ecological changes.
   d. is a powerful positive force in the fight against pollution.
   e. all of the above.
10. If *SS* (on the next page) shows the marginal social cost of a product and *DD* is the product's demand curve, the efficient output of the product is
    a. less than 500.
    b. 500.
    c. 600.
    d. more than 600.
    e. none of the above.
11. In the previous problem, the market mechanism, left to its own devices, would result in an output of
    a. less than 500.

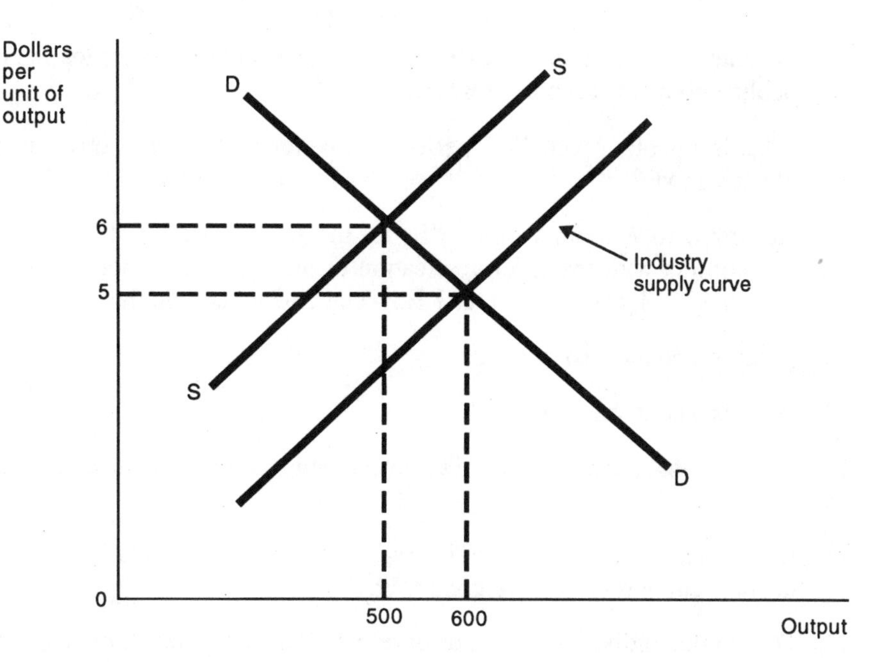

b. 500.
c. 600.
d. more than 600.
e. none of the above.

12. Continuing the previous problem, at the output that would result from the market mechanism, the marginal social cost of the product would be ____________ its marginal social value.
a. greater than
b. equal to
c. less than
d. either less than or equal to
e. none of the above

Review Questions

1. Can we be sure that the three necessary conditions for welfare maximization are satisfied under perfect competition? Are they satisfied under monopoly?

2. Discuss the advantages of marginal cost pricing.

3. According to Thomas Marschak, marginal cost pricing has resulted in savings in the French nationalized electricity industry. For example, he states that there seems to have been a leveling of consumption between the daytime and nighttime periods. Why do you think this occurred?

4. Discuss the meaning and importance of external economies and diseconomies, both of consumption and production. How do they alter the efficiency of perfect competition?

5. Indicate how the theory of external economies and diseconomies sheds light on public policy toward basic research.

6. What is a public good? Will perfect competition result in an efficient allocation of such goods?

7. According to William Vickrey, "By far the most important of the considerations that conflict with the strict application of marginal cost pricing is the need for revenues." Why is this the case? How can one get around this difficulty?

8. What is a nonexclusive good?

9. What is a nonrival good?

10. Why is it that people often fail to reveal their true preferences concerning public goods?

11. Do you sum individual demand curves vertically or horizontally to get the market demand curve for a *private* good? Why?

12. Do you sum individual demand curves vertically or horizontally to get the market demand curve for a *public* good? Why?

13. What are the basic causes of environmental pollution in our society?

14. What is an effluent fee? How can it be used to reduce pollution to more satisfactory levels? Have effluent fees been used elsewhere?

Problems

1. According to some observers, population growth is a tremendous problem and Zero Population Growth is a desirable objective. In their view, population growth is a problem because the private cost of children to parents is less than the social cost imposed on society. To deal with this problem, it has sometimes been suggested that we take one of the following actions. (1) Limit each woman to two children. (2) Issue two tickets to each woman and allow her to sell one or more of them to other women. Each ticket entitles its owner to bear one child. (3) Hold an auction where such tickets are sold by the government to the highest bidders.
   a. Using the Edgeworth box diagram, show, in the graph on the next page, that the first proposal is likely to lead to a less satisfactory outcome than the second proposal.
   b. Do you agree that the private cost of children to parents is less than the social cost imposed on society? Explain.
   c. What factors help to determine whether the third proposal is preferable to the second proposal?
   d. Many religious and other groups disagree with the proponents of Zero Population Growth. Do microeconomic principles prove that Zero Population Growth is the proper objective of our present-day society? Explain.

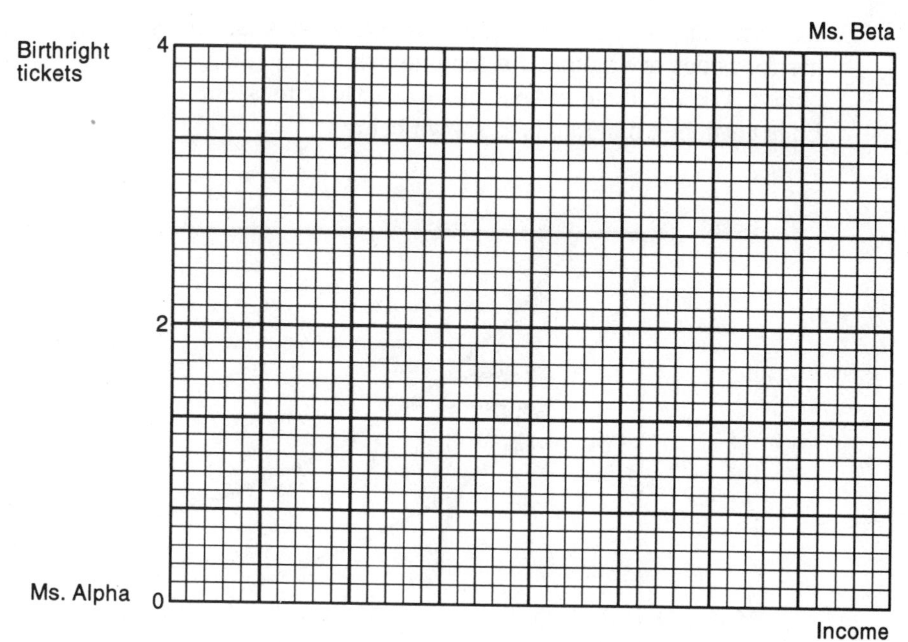

2. John takes a date, Joan, to a Chinese restaurant, and they order a portion of lemon chicken and a portion of sweet and sour pork. When the food arrives, they divide each portion between them.

   a. Indicate below how the Edgeworth box diagram might be used to analyze the way in which they should divide the food.

   b. Is it reasonable to assume that John's satisfaction depends only on the amount of food he consumes and not on the amount Joan consumes too?

3. Suppose that you have six bottles of beer (and no potato chips) and your roommate has four bags of potato chips (and no beer). It is late at night, and the stores are closed. You decide to swap some of your beer for some of her potato chips. The Edgeworth box diagram is as shown below:

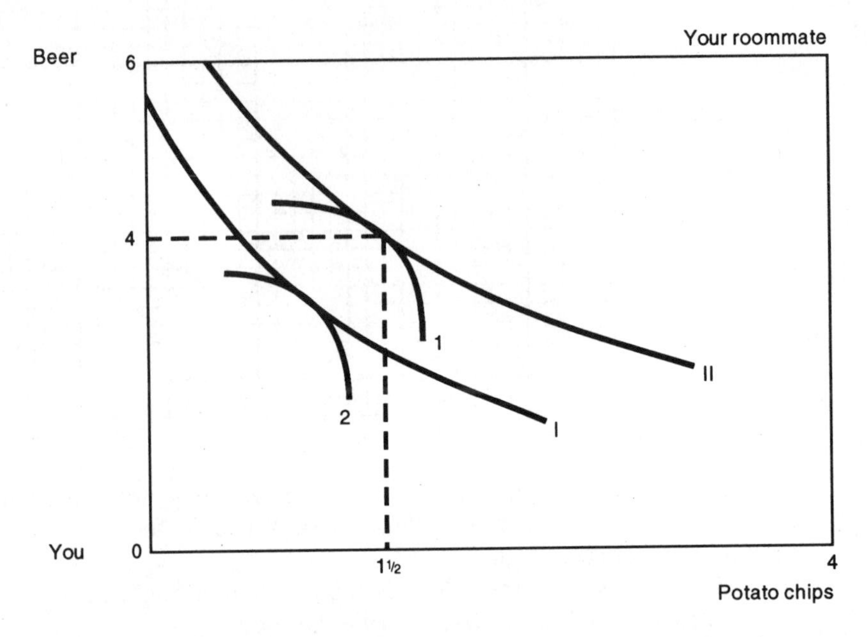

Label the point on the diagram which represents your pretrade situation.

4. In Question 3, suppose that you offer to swap 2 bottles of your beer for 1½ bags of your roommate's potato chips.
   a. Will your roommate agree to this?
   b. Is this a rational offer for you to make; that is, will it make you better off?
5. In the cotton textile industry, there are 100 firms, 30 in the Northeast, 30 in the Southeast, and 40 in the West. The marginal cost of the northeastern firms is 5 percent higher than that of the southeastern firms. Is the cotton textile industry producing efficiently? Why or why not?
6. A perfectly competitive industry produces cotton. The demand and supply curves for cotton are as shown at the top of the next page. If 40 million pounds (rather than 50 million pounds) of cotton are produced per month, how much is the loss to society?
7. The diagram at the bottom of the next page shows the average cost curve, *AA'* of a regulated monopoly. It produces *OQ* units of output and sets a price of *OP*.
   a. Is the firm engaged in marginal cost pricing? Why or why not?
   b. Does existing regulatory practice in the United States generally lead to marginal cost pricing?
8. A poor man receives $10 per day in cash income. In addition, a kind neighbor provides him with three free meals a day, the cost to the neighbor of these three

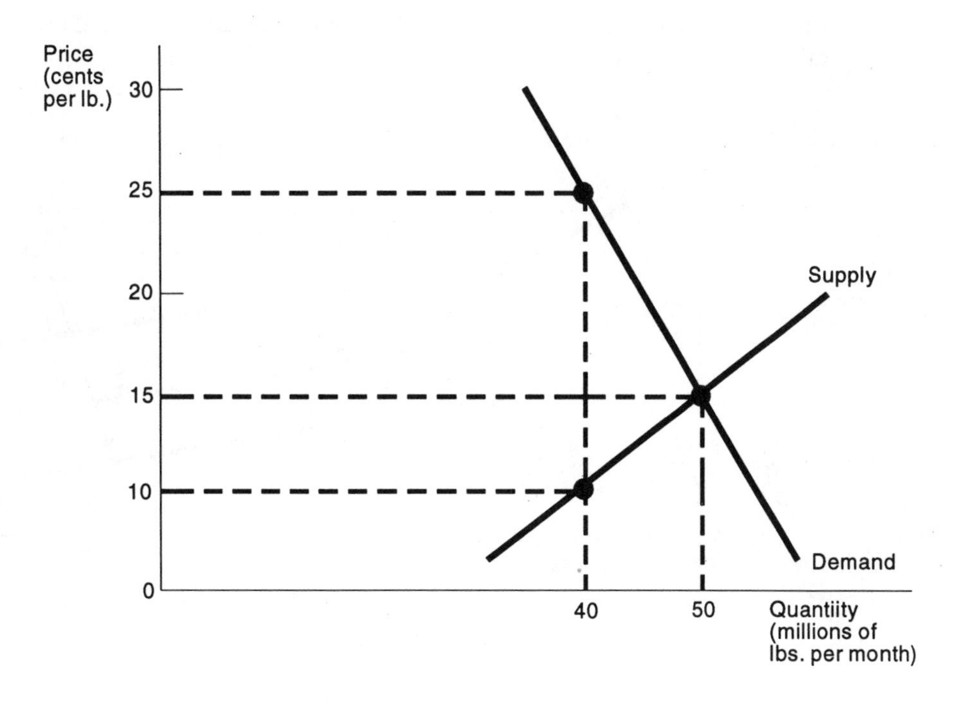
Price
(cents
per lb.)
30
25
20
15
10
0
Supply
Demand
40
50
Quantiity
(millions of
lbs. per month)

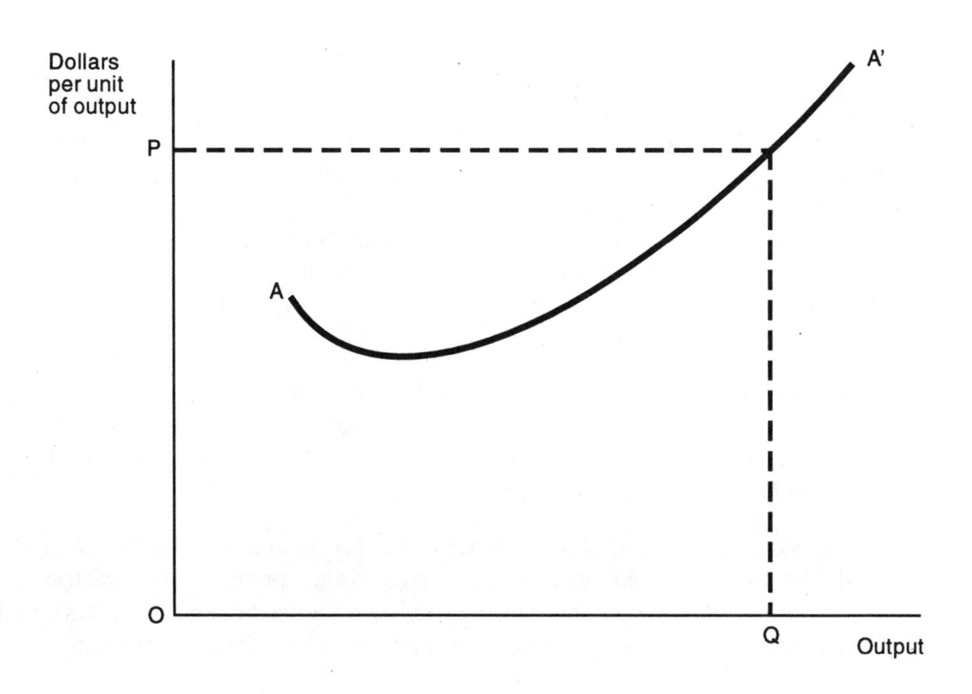
Dollars
per unit
of output
P
A
A'
O
Q
Output

meals being $6. The man's preferences between meals and cash are shown in the indifference map below:

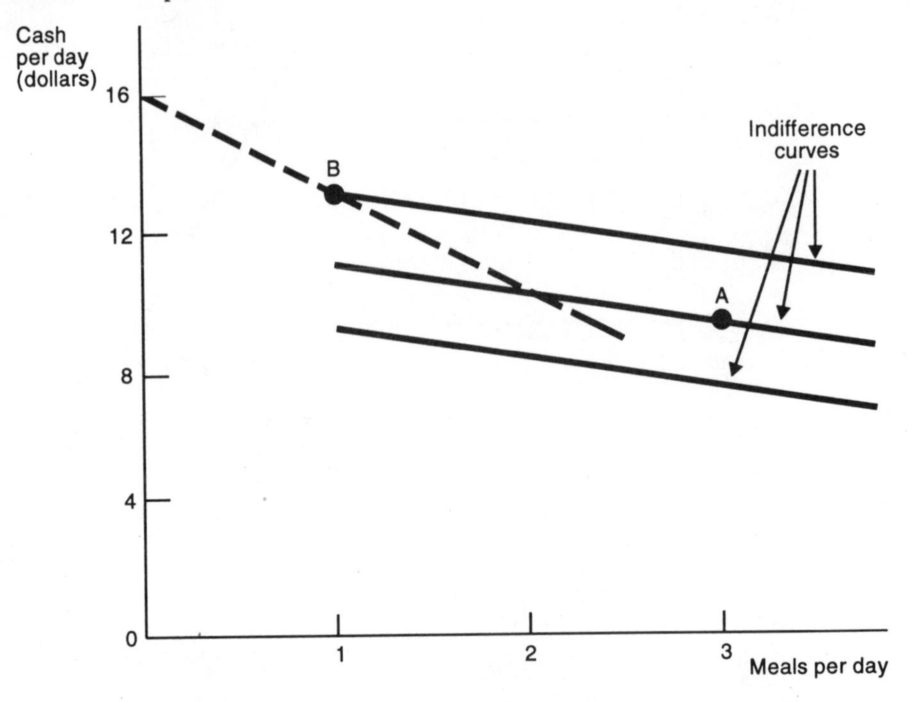

A social worker suggests to the neighbor that she give the man the $6 per day and let him spend it as he pleases. If he can buy a meal for $2.50, would such a change be an improvement, according to the Pareto criterion?

9. For many years, the United States used a draft to obtain military personnel. Did this procedure violate any of the rules for efficient resource allocation? Explain.

10. Suppose that two consumers, after swapping goods back and forth, have arrived at a point on the contract curve. In other words, neither can be made better off without making the other worse off.
    a. Does this mean that neither of them can find a point *off* the contract curve which he or she prefers to the point at which they have arrived?
    b. If it does not mean this, why do economists claim that points on the contract curve are to be preferred?

11. A small private jet lands at Kennedy Airport in New York at the busiest time of day. It pays a nominal landing fee. What divergences may exist between the private and social costs of this plane's landing there at that time? What policies might help to eliminate such divergences?

12. Suppose that the market for videocassette recorders is in disequilibrium; that is, the actual price does not equal the equilibrium price. If all industries in the economy are perfectly competitive (including videocassette recorders), will the necessary conditions for efficient resource allocation be met? Explain.

13. The town of Erewhon is faced with a serious smog problem. The smog can be dispelled if an air treatment plant is installed at an annual cost of $1 million. There is no way to clean up the air for some but not all of the town's population. Each of the town's families acts independently, and no single family can afford to carry out the project by itself. Why doesn't a private firm build the air treatment plant and sell its services to the town's families (acting individually)?

14. About 95 percent of all families in the United States have television sets.
    a. For these families, is a television program a nonrival good?
    b. Is it possible for firms to exclude some of these families from viewing certain programs?
    c. Who pays for ordinary TV programs?
    d. Must there be any commercials on "pay TV"?

15. Discuss the advantages and disadvantages of the following policies for dealing with air pollution due to gas emission from cars:
    a. Ban the internal combustion engine after the year 2000.
    b. Impose a tax on each car which varies with the amount of noxious gas it emits.
    c. Subsidize research to develop better antipollution devices that can be used by cars.

16. "Public goods have marginal costs of zero. In other words, they can be provided as cheaply for one as for all. Consequently, it is foolish to attempt to compare the benefits of public goods with their costs. Any public good is worth producing because it benefits so many people." Comment and evaluate.

17. Suppose that there are only three citizens of a (very small) nation and that the amount of national defense each would demand (at various prices) is as follows:

Price of a unit of national defense (dollars)	*Citizen A*	*Citizen B*	*Citizen C*
1	10	8	12
2	9	7	9
3	8	6	7
4	7	5	5

If the marginal cost of a unit of national defense is $9, what is the efficient amount of national defense for this nation?

Key Concepts for Review

Partial equilibrium analysis
General equilibrium analysis
Edgeworth box diagram
Exchange
Contract curve
Production possibilities curve
Interpersonal comparison of utility
Marginal conditions for resource allocation
Marginal cost pricing
External economies
External diseconomies
Pareto criterion
Public goods
Coase's theorem
Nonexclusive
Nonrival

ANSWERS

Case Study: The Lackawanna Social Club

a. 50 cents ÷ 20, or 2.5 cents.
b. 47.5 cents.
c. 50 cents.
d. It is likely to exceed the efficient amount because the private cost is much less than the social cost.
e. It would result in the private and social costs' being brought into equality, but it would result in more inconvenience (to buy and store and refrigerate the soda and beer) to the members.
f. It would push the private cost toward the social cost, but it would increase cost because a bartender would have to be hired. The price would have to be increased to cover the cost of the bartender.
g. If strictly observed, this procedure would eliminate the gap between private and social costs, and it would eliminate the cost involved in hiring a bartender. But cheating could occur.
h. The bigger the club, the smaller the cost (per can of beer or soda) of having a bartender, and the more difficult it would be to police schemes such as the one in part (*g*).
i. Yes. For example, a family frequently has a refrigerator stocked with food or some other group of goods that the family members can draw on. The cost to a particular family member of consuming some of these items is frequently very small because other members of the family pay for the food or the group of goods. Thus a teenager may consume much more soda or ice cream than if he or she had to pay the full cost of these items.

Completion Questions

1. the same
2. little
3. move to a point on the contract curve
4. no scale
5. value judgment
6. contract curve
7. contract curve
8. production possibilities curve
9. the same prices
10. the same prices
11. are not
12. someone who makes no contribution to a public good, but that uses whatever amount of it is provided
13. public

True or False

1. True	2. False	3. True	4. False	5. False	6. False
7. False	8. False	9. False	10. False	11. True	12. False
13. False	14. False	15. True	16. False	17. False	

Multiple Choice

1. *a*	2. *a*	3. *a*	4. *d*	5. *c*	6. *d*
7. *a*	8. *c*	9. *e*	10. *b*	11. *c*	12. *a*

Review Questions

1. Yes.
   No.
2. See the section of the text on marginal cost pricing.
3. Because of price changes.
4. An external economy occurs when an action taken by an economic unit results in uncompensated benefits to others. An external diseconomy occurs when an action taken by an economic unit results in uncompensated costs to others. Too little is produced of goods generating external economies; too much is produced of goods generating external diseconomies.
5. Basic research results in important external economies. Consequently, there seems to be a good case for the government (or some other agency not motivated by profit) to support basic research. Too little would be supported by a perfectly competitive economy.
6. A public good is a good that one person can enjoy without reducing the enjoyment it gives others. Also, the consumer cannot be made to pay for a public good, because he or she cannot be barred from using it (whether or not he or she pays) or because it is obligatory for everyone to use it.
   No.
7. If average costs decline with output, marginal cost is below average cost, the result being that marginal-cost pricing will result in a deficit. Public subsidy is one possible answer under these circumstances.
8. A good is nonexclusive if people cannot be excluded from consuming the good whether they pay for it or not.
9. A nonrival good can be enjoyed by an extra person without reducing the enjoyment it gives others.
10. Because a person is likely to feel that the total output of the good will not be affected significantly by his or her action. He or she will be likely to make no contribution to supporting the good, although he or she will use whatever output of the good is forthcoming.
11. Horizontally, because each unit of output can be consumed by only one person.
12. Vertically, because each unit of output can simultaneously be consumed by all consumers in the market.
13. External diseconomies from waste disposal.
14. An effluent fee is a fee that a polluter must pay to the government for discharging waste. It brings the private cost of waste disposal closer to the social cost.
    Yes, for example, in the Ruhr Valley in Germany.

Problems

1. *a.* Suppose that two women, Ms. Alpha and Ms. Beta, have indifference maps shown below, where income is plotted along the horizontal axis and birthright tickets are plotted along the vertical axis.

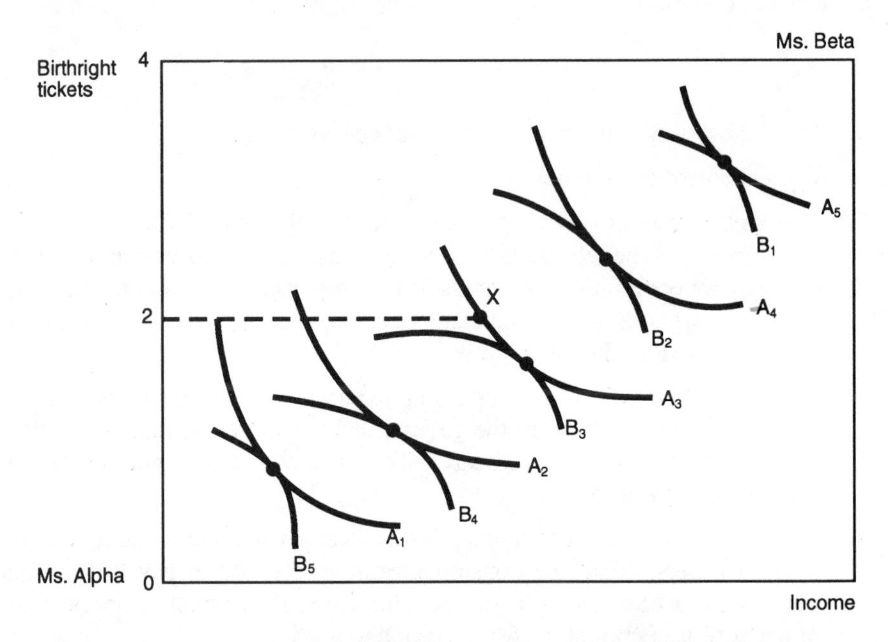

Indifference curves $A_1, \ldots, A_5$ are Ms. Alpha's; indifference curves $B_1, \ldots, B_5$ are Ms. Beta's. The first proposal says that the allocation must be that represented by point X, but in general one or both women can increase utility by moving to the contract curve. The second proposal would enable them to move to the contract curve.

b. This may be true because of external diseconomies; however, there also may be external economies.

c. One important consideration here is what the government does with the proceeds of the auction.

d. No. As stressed repeatedly, economics is no substitute for religion, ethics, or politics. It does not tell us what our objectives or values should be.

2. *a.* The quantity of lemon chicken might be measured along the horizontal axis; the quantity of sweet and sour pork might be measured along the vertical axis. The indifference curves of each person (A_1, A_2, A_3 for Joan; B_1, B_2, B_3 for John) might be inserted, as shown at the top of the next page.
The efficient point would be on the contract curve.

b. Probably not, since he probably gets satisfaction from her well-being (and hopefully she feels the same way).

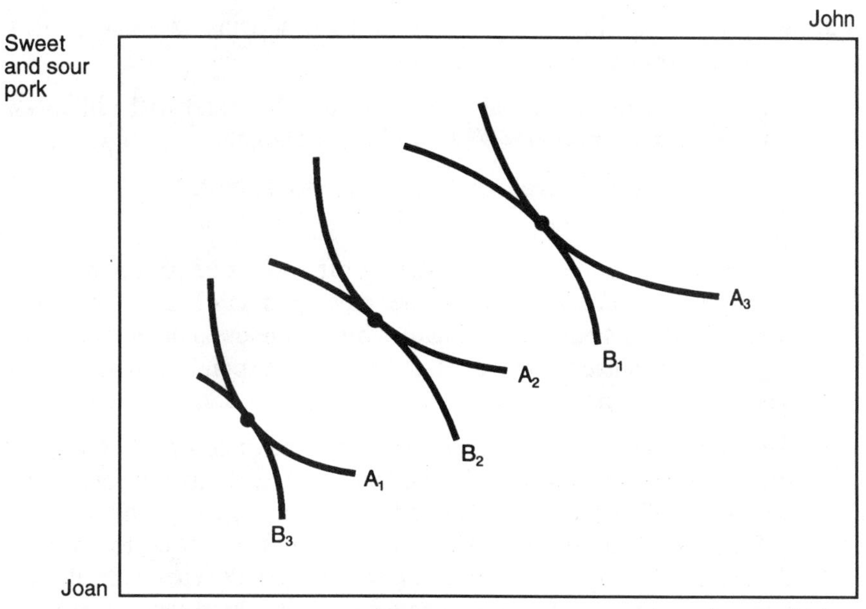

3. Point *Y* represents your pretrade position.

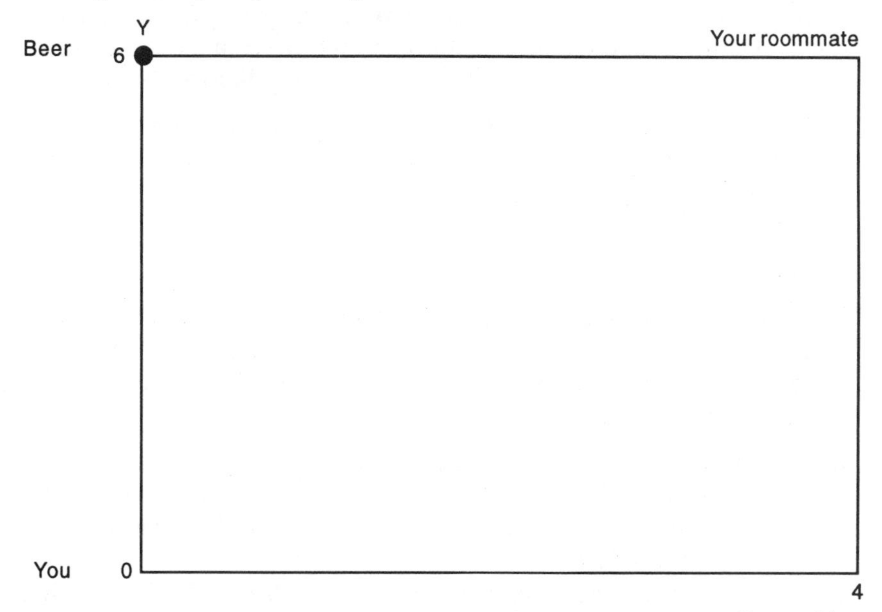

4. *a.* This trade would put her on her indifference curve 1, which is a higher indifference curve than that going through point *Y*. (To identify point *Y*, see the answer to Problem 3) Thus, she should agree to this.
   b. This trade would put you on your indifference curve II, which is a higher indifference curve than that going through point *Y*. Thus, the trade will make you better off.

5. No. Firms with high marginal costs are producing too much; those with low marginal costs are producing too little.

6. The reduction in total surplus is ½(25 cents – 15 cents) x 10 million + ½(15 cents – 10 cents) x 10 million = $0.75 million per month.

7. *a.* No. Price equals average cost, not marginal cost.
   b. No.

8. Yes, because the man would be better off and his neighbor would be no worse off. If the neighbor gives him $6 per day, the man's budget line is shown by the dashed line. This budget line enables him to choose a point like *B*, which is on a higher indifference curve than *A*, the point corresponding to his present situation. The neighbor is no worse off since the cost is the same to her.

9. The draft involved price control, since the maximum price that could be set by the military services was established by law. The number of people of many types who were willing to serve at these prices was less than the number demanded by the defense establishment, and a compulsory draft was used to obtain the number demanded. So long as military manpower of various types were underpriced, there was an incentive to use these manpower in other than an economically efficient way.

10. *a.* No.
    b. Because, if the two consumers arrive at any point off the contract curve, they can find a superior point on the contract curve, in the sense that one of them can be made better off without making the other worse off.

11. The plane's landing at that time may delay large commercial jets and impose substantial costs (in terms of delay and inconvenience) on the passengers carried by these commercial aircraft. Increases in the landing fees paid by small private aircraft could help to eliminate such divergences.

12. No.

13. If any family buys smog-free air, it automatically buys it for others too, whether the latter pay for it or not. And since no family can afford the cost, so long as families act independently, it will be unprofitable for a private firm to carry out this project.

14. *a.* Yes, because once the program is put on the air, an extra family can watch it without depriving others of the opportunity to do so. In other words, once the program is broadcast to some viewers, others can watch it at no (or trivial) extra cost.
    b. Yes. A means has been devised to "scramble" the signal so that, unless a viewer's TV set is equipped to unscramble it, the viewer cannot get the program. Viewers rent this equipment or operate it by means of coins. This is sometimes called "pay TV."
    c. Advertisers pay for the broadcasts in order to air their commercials. In addition, listeners pay something, if they don't like commercials!
    d. No, because the viewers pay directly for the broadcasts.

15. *a.* This is unlikely to be the optimal way to handle the problem, since it would

be very difficult to develop and adopt effective substitutes for the internal combustion engine by 2000.

b. An effluent fee of this sort might be useful but it might be difficult to measure the amount of pollution caused by a particular car and to make sure that each car pays the tax. Also, it might be reasonable to set different levels of the tax, depending on where the car is driven.

c. This step would help to improve matters, but it contains no incentive for people to switch to less polluting transportation techniques in the short run.

16. It is not true that any public good is worth producing. The efficient output of a public good can be determined by summing the individual demand curves vertically and by finding the intersection with the supply curve. In many instances, the efficient output may be zero.

17. 7 units of national defense.

Case 5 California Water Pricing[1]

Dorothy Robyn

Like energy in the East, water has proved to be a limiting natural resource in the western states. The appropriate allocation of this scarce commodity is perhaps nowhere more important than in California, a semiarid state with a gigantic agricultural sector and a burgeoning population. And yet, the state's historical approach to water has established a pattern of consumption far removed from what many people consider ideal.

The evolution of California water law is a story in several chapters. Early settlers adopted the approach prevalent in the eastern states which gave water rights to the owner of land adjacent to a river or stream. That system of "riparian" rights (from the Latin word for bank) sufficed while settlement was sparse, but proved unsuitable as demand grew for nonriparian land. California gold miners were the first to systematically appropriate water for use at off-stream sites. Over time, miners fashioned a sophisticated system of water rights based on the principle of "first in time, first in right," with codes specifying the maximum size of claims, limits on the number of claims per miner, and conditions for forfeiture. Reflecting the special demands of an arid climate, the doctrine of "prior appropriation" served to establish ownership rights that were generally well-defined, enforced, and transferable.

Despite its sophistication, the prior appropriation system began to unravel in the late 1800s. Faced with frontier disputes, eastern-trained judges often enforced riparian precedent, and the resulting mix of riparian and prior appropriation doctrine led to a confusion that stifled the establishment of private property in water. The miners' system was gradually replaced by a system of state control of water, which restricted water sale and transfer in significant ways.

The decline of private water rights also removed a key incentive for conservation. In an effort to reduce waste, a 1928 California constitutional amendment limited water rights to an amount "reasonably required for beneficial use." Conservation was not treated as a beneficial use, however, and so the amendment perversely reinforced farmers' "use it or lose it" mentality toward water. For more than 50 years, the "beneficial use" doctrine—combined with state control of water allocation—punished conservation and discouraged the voluntary transfer of existing allotments to parties who could make better use of the water.

The water system that developed within this legal framework is physically and institutionally complex. California has 1,251 major dams and the two biggest irrigation projects on earth (the gargantuan Central Valley Project, operated by the federal government since 1902, diverts one-third of the Colorado River's flow, primarily to southern California farms and ranches; the State Water Project, begun in 1959, serves

1. This case was written for use at the John F. Kennedy School of Government, Harvard University. It draws heavily on Philip A. Guentert, "Agricultural Water Pricing in California" (Policy Analysis Exercise, Kennedy School of Government, 1983), and on an earlier case by Professor Helen Ladd. Certain institutional and administrative facts have been altered. (0987)

chiefly Los Angeles and Kern County). These engineering feats satisfy only 60 percent of California's enormous thirst for water. The rest comes from under the ground.

Most of the state's water, 85 percent of which goes for irrigation, is delivered by some 2,500 local water agencies. These delivery agencies, or retailers, are supplied by wholesalers who purchase water from federal, state, and local sources at markedly different rates. Federal water goes for the heavily subsidized rate of roughly $10 per acre-foot (volume of water covering an acre to a depth of one foot). Water from state reservoirs costs wholesalers from $30–$60 per acre-foot. Underground supplies cost approximately $100 per acre-foot. Retailers, in turn, are charged a "rolled-in" (i.e., weighted) average of wholesale acquisition costs.

At the retail level, there is no single price for water. Although retailers' revenues cover their costs (water districts are barred from making a profit), in many areas, especially those dominated by farmers, agricultural users pay less than residential users. A recent study done for a major Los Angeles petrochemical firm put the average water price for agriculture at $10 per acre-foot, compared to an average residential water price of approximately $100 per acre-foot. This pricing structure results in such anomalies as the current, and profitable, practice of irrigating hay in Death Valley. According to a 1978 study by the Rand Corporation, the gross inefficiencies created by charging farmers less than the true cost of water result in a $5 billion loss to society.

Recent years have seen a growing interest in alternatives to the current system. A 1982 statewide referendum, defeated with heavy opposition from farm interests (see Exhibit 1), would have encouraged marginal cost pricing by local water districts. That same year, the California Water Code was amended to allow for the temporary sale of "conserved" water without loss of water rights (an earlier bill had defined conservation as a beneficial use). More recently, debate has focused on additional means to expand the ability of those holding water rights—individuals as well as water districts—to buy, sell, and transport surplus water with no danger of losing future rights (see Exhibit 2). While the eventual movement to some kind of market-oriented system of water allocation seems likely, strong political and institutional obstacles remain.

MARGINAL COST PRICING

The nature and degree of potential opposition begin to emerge from some simple tables showing the differential effects a system of marginal cost pricing for water might have: There are 13 major water basins in California (see Exhibit 3), and most

EXHIBIT 1

"NO" VOTE ON WATER REFERENDUM

Region	%	Region	%
Westside San Joaquin	80%	Central Coast	70%
San Joaquin Basin	80%	Delta	69%
Mountain Valley	79%	North Coast	67%
North San Joaquin	78%	North Bay	67%
Sacramento Valley	77%	South Coast	63%
High Desert	75%	South Bay	53%
Imperial Valley	74%		

Source: Compiled from county tabulations in the *Los Angeles Times*, November 4, 1982, p. 19, col. 1.

EXHIBIT 2

PUSH TO LIFT CURBS ON WATER SALES

By Ann Cony
Bee Staff Writer

Legislative attempts to create an open market for the buying and selling of state water have brought together some unlikely bedfellows.

The irony is not lost on Ron Khachigian, senior vice president of Blackwell Land Co. Inc., which operates one of the largest corporate farms in California.

"I'm sitting at a witness table with the Environmental Defense Fund, and here I am a real conservative farmer, and we're both espousing the marketing of water," Khachigian said.

For different reasons, Blackwell and the Environmental Defense Fund are supporting a bill that would require state water agencies to make their canals and aqueducts available to third parties that have concluded water sales.

Sponsored by Assemblyman Richard Katz, D–Sepulveda, the bill (AB 2746) was approved by the Assembly in May. It's one of two water-marketing bills that will be awaiting Senate action when lawmakers return from a recess on Monday.

Those bills and a third one signed recently by the governor are designed to remove institutional, logistical and financial roadblocks to water trading. None is monumental in and of itself.

However, "they're all adding to a policy and a body of law that allows [water marketing] to happen," said Clyde MacDonald of the Assembly Office of Research. "They're like bricks in a building. They're all important."

The Legislature has been working on fundamental changes in water policy for the last seven years or so, pushing water marketing as one means of using a scarce resource more efficiently, MacDonald said.

Four years ago a law was passed to allow farmers and other to sell surplus state water without abandoning their rights to use their water allocations in the future.

But there still are potential barriers to water trading, proponents say.

Katz has said that farmers who support water trading fear that big water districts will block sales by denying access to their canals. His bill would guarantee water traders access to transportation systems as long as space is available and they're willing to pay a fair fee.

The measure has been opposed by some water bureaucracies but supported by a broad coalition that includes environmentalists, the California Farm Bureau and financially strapped irrigation districts where farmers are having trouble paying their water bills.

For farmers, water trading offers flexibility and a new weapon in their survival arsenal. In a free water market, a farmer can sell some or all of his water rights, for any length of time, without selling the land to which the water rights were originally attached.

Environmental groups support water marketing as one way of encouraging water conservation, and hope it will lessen the need for building expensive or environmentally disruptive additions to the state's water delivery system.

Katz' bill will be before the Senate Appropriations Committe on Monday, along with a water marketing bill by Assembly Democrats Jim Costa of Fresno and Phil Isenberg of Sacramento.

The Costa-Isenberg bill (AB 3722) is designed to help overcome bureaucratic inertia. It directs the state Department of Water Resources to "facilitate" the voluntary transfer or exchange of water and serve as an information clearinghouse, maintaining lists of would-be buyers and sellers.

A third bill, sponsored by Assemblyman David Kelley, R–Hemet, was approved by the Legislature last month and has been signed into law by the governor. It removes a seven-year limit on the term of contracts for the transfer of water. Removal of the time limit eliminates an economic roadblock to water transfers that would require construction of canals, pipelines, pumps or other expensive structures. The Farm Bureau and other proponents say the seven year limit on contracts would have made some of those deals impossible to finance.

Observers say that the Katz and Costa-Isenberg bills have a good chance of joining the Kelley bill in becoming law but that none of the measures will trigger immediate wheeling and dealing in water. Numerous farms are reportedly anxious to sell water, but "until we have a situation where there's not enough water, we're not going to have very many buyers," said MacDonald. That is expected to change, however, with the next drought.

Eventually, "I think we will use water marketing," MacDonald said. "I think there's no question about it."

Source: Reprinted with permission from *The Sacramento Bee*, August 10, 1986, pp. D1–2

of the variation in average water prices occurs between basins (see Exhibit 4). Price increases from a marginal cost scheme would differ substantially by region, both because agricultural water prices are closer to the true marginal cost in some regions and because the true marginal cost is lower in some regions. (Costs within basins are assumed to be uniform.) Since certain regions tend to produce certain crops, the shift to a marginal cost pricing scheme would affect some crops more than others (see Exhibit 5).

Water is also much more important for some California crops than others. The "factor share" of water—the percentage of total costs accounted for by water—ranges from 1.0 percent for celery to 35.8 percent for grain hay (see Exhibit 6). Moreover, the additional value created by marginal applications of water differs markedly across crops. For example, the marginal application of an acre-foot of water produces $62.00 worth of lemons on average across the state, but only $8.83 worth of lima beans (see Exhibit 7).

In addition, farmers in different regions use different amounts of water to produce a given crop. The coefficient of variation—which represents the degree of interregion variation in the ratio of water to nonwater inputs—is much greater for onions and pears, for example, than for rice and lima beans (see Exhibit 8).

EXHIBIT 3

CALIFORNIA'S WATER BASINS

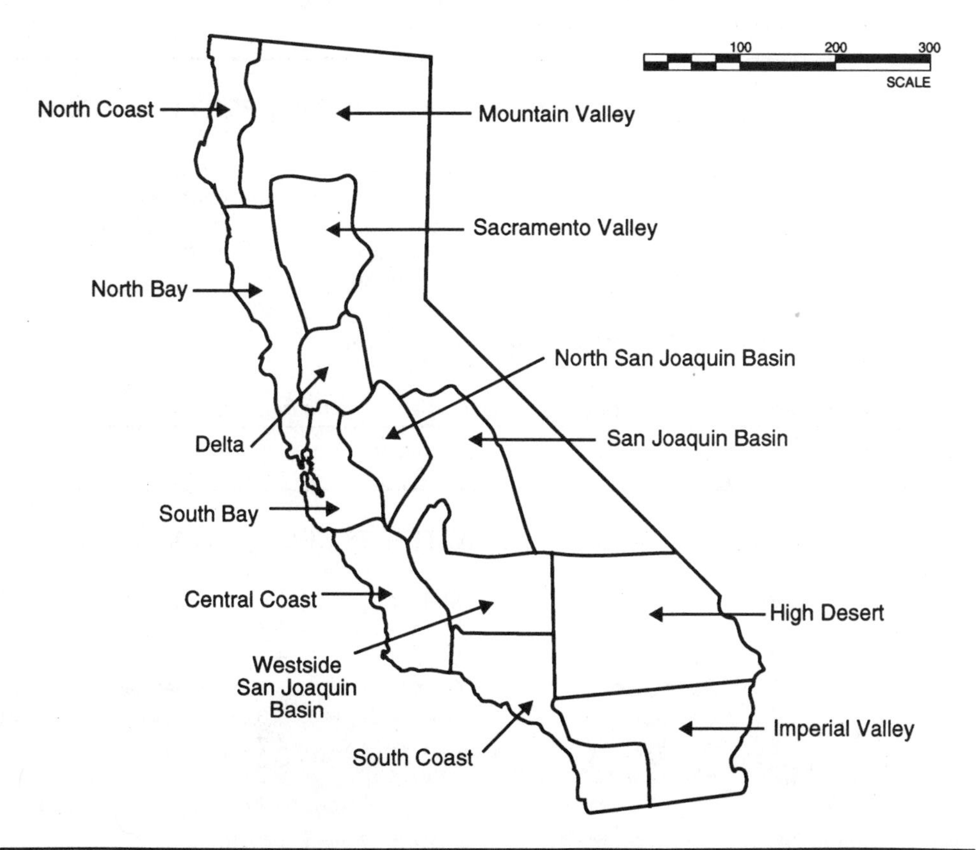

EXHIBIT 4

WATER PRICE INCREASES FROM MARGINAL COST PRICING BY WATER BASIN ($ PER ACRE-FOOT)

BASIN	CURRENT AVERAGE PRICE	MARGINAL COST	CHANGE
North Coast	20	50	30
Sacramento Valley	30	40	10
North Bay	20	80	60
Delta	20	50	30
South Bay	20	20	0
North San Joaquin	10	70	60
Mountain Valley	20	50	30
San Joaquin	20	90	70
Central Coast	10	100	90
Westside San Joaquin	20	100	80
South Coast	30	90	60
High Desert	40	80	40
Imperial Valley	10	90	80

Source: Derived from Philip Guentert, *Agricultural Water Pricing in California*, Policy Analysis Exercise, Kennedy School of Government, Harvard University, 1983.

EXHIBIT 5

AVERAGE PRICE INCREASE FOR WATER BY CROP ($ PER ACRE-FOOT)

Lettuce	80	Melons	60	Almonds	40	Tomatoes	40
Carrots	70	Sorghum	60	Apricots	40	Irrigated pasture	30
Celery	70	Peaches	50	Field corn	40	Rice	30
Grapes	70	Pears	50	Grain hay	40	Walnuts	30
Oranges	70	Onions	50	Lima beans	40		
Barley	60	Wheat	50	Prunes	40		
Cotton	60	Alfalfa hay	40	Sugar beets	40		

Source: Derived from Philip Guentert, *Agricultural Water Pricing in California.*

EXHIBIT 6

AVERAGE FACTOR SHARE OF WATER BY CROP (% COSTS)

Grain hay	35.8	Wheat	17.8	Melons	6.7	Lima beans	4.0
Irrigated pasture	34.7	Sorghum	14.2	Carrots	6.5	Apricots	3.5
Alfalfa hay	25.9	Oranges	14.2	Walnuts	6.3	Pears	3.0
Field corn	21.7	Sugar beets	13.6	Grapes	5.9	Peaches	2.3
Barley	20.2	Onions	12.5	Tomatoes	5.3	Celery	1.0
Rice	19.7	Cotton	11.3	Lettuce	4.7		
		Almonds	6.8	Prunes	4.3		

Source: Derived from Philip Guentert, *Agricultural Water Pricing in California.*

Farmers of a given crop also incur different costs of production, largely because some land is more fertile than others. No consistent data are available with which to estimate the precise size of the cost variations, but it is clear that some farmers operate on marginal land while others farm land that returns large profits to owners.

Finally, crops differ with respect to their place in the domestic and international market, which has implications for a marginal cost pricing scheme. For example, California farmers produce 99.9 percent of the U.S. almond crop, which represents 20.7 percent of the total world production. About 60.7 percent of California's almonds are exported outside the state. In contrast, California only produces 2.9 percent of the nation's wheat, but 78.1 percent of it is exported (see Exhibit 9).

URBAN WATER ACQUISITION STRATEGIES

The various metropolitan water districts of California (MWDs), water wholesalers that represent primarily urban and often rapidly growing regions of the state, are eager to secure rights to additional supplies of water. At present, these districts negotiate with water authorities in agricultural areas for medium and long-term supply contracts.

EXHIBIT 7

MEAN MARGINAL VALUE ($ PER ACRE-FOOT)

Crop	Value	Crop	Value	Crop	Value	Crop	Value
Lemons	62.00	Wheat	37.74	Sorghum	32.78	Lettuce	16.84
Cotton	55.98	Barley	37.17	Prunes/plums	28.49	Peaches	10.59
Melons	53.53	Alfalfa hay	36.37	Irrigated pasture	20.93	Artichokes	10.00
Carrots	53.27	Sugar beets	32.78	Celery	20.77	Broccoli	10.00
Onions	51.50	Tomatoes	32.75	Asparagus	17.40	Apricots	9.41
Oranges	46.86	Walnuts	32.45	Field corn	17.27	Lima beans	8.83
Sweet corn	40.00	Almonds	32.26	Rice	16.89		
Grapes	39.61	Grain hay	31.37				

Source: Derived from Philip Guentert, *Agricultural Water Pricing in California.*

EXHIBIT 8

COEFFICIENT OF VARIATION OF INPUT RATIOS[1]

Crop	Value	Crop	Value	Crop	Value	Crop	Value
Onions	.829	Almonds	.393	Apricots	.235	Sugar beets	.138
Pears	.532	Peaches	.302	Walnuts	.230	Melons	.123
Grain hay	.475	Tomatoes	.266	Cotton	.219	Rice	.119
Sorghum	.445	Field corn	.265	Celery	.195	Lima beans	.005
Prunes	.419	Wheat	.250	Oranges	.187		
Lettuce	.403	Grapes	.244	Alfalfa hay	.181		
Carrots	.395	Irrigated pasture	.242	Barley	.163		

1. Defined as the standard deviation of the set of ratios divided by their mean.
Source: Derived from Philip Guentert, *Agricultural Water Pricing in California.*

EXHIBIT 9

CALIFORNIA CROPS IIN THE WORLD MARKETPLACE

	CALIF. PROD. (TONS)	CALIF. % OF U.S. PROD.	CALIF. % OF WORLD PROD.	U.S. IMPORTS (TONS)	U.S. EXPORTS (TONS)	% CALIF. PROD. EXPORTED
Almonds	159,000	99.9	20.7	688		60.7
Walnuts	195,000	99.4	27.1	145	49,887	30.8
Artichokes	174,000	98.0	19.5		9,795	18.7
Broccoli	180,000	94.9				10.8
Grapes	3,924,000	90.4	N.B.[1]	27,240[2]	192,130	18.1[3]
Tomatoes	7,270,600	85.5	14.0	368,720[4]	167,840	2.1[5]
Plums[6]	273,000	79.2	10.2	3,229	206,500	25
Lemons	844,000	75.5	27.4	12	198,913	31.4
Lettuce	1,912,500	73.2	23.6			5.1
Celery	530,600	66.2				7.2
Peaches	914,500	64.7	19.1		99,312	10
Melons	306,500	63.8	5.7			
Carrots	520,800	54.0	7.6			5.2
Asparagus	53,500	50.3				16.3
Lima beans	46,300	48.2				
Pears	300,400	39.3	5.4	8,159	35,667	6.9
Onions	545,100	34.7	2.8	34,800	87,900	29.1
Sugar beets	8,476,000	29.0	N.B.[7]	3,867,000	203,000	
Rice	1,504,400	23.6	0.4	1,102	2,287,949	79.0
Cotton[8]	465,700	23.3	1.8	29,970	827,750	87.5
Oranges	2,062,500	20.2	5.8	30,980	529,743	25.8
Barley	1,450,100	15.8	0.9	337,998	545,472	
Apples	230,000	6.1	0.8	42,953	120,978	
Hay[9]	7,642,000	5.8				4.1[10]
Wheat	1,866,800	2.9	0.5	43,130	25,179,022	78.1
Sorghum[11]	417,300	2.0			6,663,916	
Corn	775,200	0.5	0.2	49,117	27,397,696[12]	

1. 24,943,687 acres worldwide; 524,420 acres in California.
2. Raisins and fresh equivalent.
3. Fresh only (no wine).
4. All products.
5. Processing tomatoes only (no fresh).
6. Including prunes.
7. 11,506,910 tons sugar, cane and beet. Imports in tons cane and beet sugar. Exports in tons centrifugal sugar only.
8. Lint only (not seed).
9. Of total California production, 6,608,000 is alfalfa.
10. Alfalfa only.
11. Grain only.
12. Corn and cornmeal.

Source: Derived from Guentert, *Agricultural Water Pricing in California*.

When the MWDs can induce farm regions to forego the use of water, it becomes available for purchase under the provisions of the California Water Code, as amended in 1982:

> §1011. Reduction of water use due to conservation efforts.
>
> (a) When any person entitled to the use of water under an appropriative right fails to use all or any part of the water because of water conservation efforts, any cessation or reduction in the use of such appropriated water shall be deemed equivalent to a reasonable beneficial use of water to the extent of such cessation or reduction in use. No forfeiture of the appropriative right to the water conserved shall occur upon the lapse of the forfeiture period applicable to water appropriated pursuant to the Water Commission Act or this code or the forfeiture period applicable to water appropriated prior to December 19, 1914. . . .
>
> For purposes of this section, the term "water conservation" shall mean the use of less water to accomplish the same purpose or purposes of use allowed under the existing appropriate right. Where water appropriated for irrigation purposes is not used by reason of land fallowing or crop rotation, the reduced usage shall be deemed water conservation for purposes of this section.
>
> (b) Water, or the right to the use of water, the use of which has ceased or been reduced as the result of water conservation efforts as described in subdivision (a), may be sold, leased, exchanged, or otherwise transferred pursuant to any provision of law relating to the transfer of water or water rights, including, but not limited to, provisions of law governing any change in point of diversion, place of use, and purpose of use due to the transfer.

Needless to say, when negotiating with agricultural water districts, the MWDs find farmers reluctant to part with water acquired at prices so favorable to agriculture. Accordingly, the metropolitan water districts are considering asking for statewide legislation that would require all water authorities, urban and agricultural, to price their water at no less than each district's marginal cost.

QUESTIONS

1. What determines the farm sector's demand for water? To what extent is this demand ultimately derived from the willingness of consumers to pay for the extra farm output that the water helps to produce?
2. Based on the facts provided in this case, was there any correlation between the average price of water in a particular California water basin and the percent of votes in that water basin against marginal cost pricing? Would you have expected any such correlation? Why or why not?
3. Exhibit 8 shows the amount of variation among regions in the ratio of water to nonwater inputs. Why is it important to know how much variation of this sort exists? Does this influence how much effect marginal cost pricing of water would

have on a crop's total cost in California?

4. Exhibit 9 shows California's percentage of U.S. production of various crops. Why is this information important in predicting the effect of marginal cost pricing of water on California's output of these crops?
5. Is it likely that California lettuce growers would be hit as hard by marginal cost pricing of water as would California rice growers? Why or why not?
6. Is it possible that marginal cost pricing of water would reduce the demand for croplands? Why or why not?
7. How will the 1992 federal law regarding the Central Valley Project (described in problem 20.10 in the text) affect the situation described in this case? Will it be possible for farmers to sell their water? If so, what will be the effects?